BIOLOGY OF RICE-FIELDS BLUE-GREEN ALGAE

S.C. SANTRA

2027

Daya Publishing House®

A Division of

Astral International Pvt. Ltd.

New Delhi – 110 002

First Published, 1993
Reprinted, 2027

ISBN: 978-81-7035-115-3 (HB)

Published by : **Daya Publishing House®**
A Division of
Astral International Pvt. Ltd.
– ISO 9001:2015 Certified Company –
4736/23, Ansari Road, Darya Ganj
New Delhi-110 002
Ph. 011-43549197, 23278134
e-mail: info@astralint.com
Website: www.astralint.com

Digitally Printed at : **Replika Press Pvt. Ltd.**

PREFACE

Researches on blue green algae in India has a long heritage as because of the fact, there were a large number blue green algal taxa growing abundantly in this country particularly in humid tropical areas. Water logged rice field is one of the major habitat for growth of a wide variety of blue green forms and their significant importance in rice ecosystem can not be ignored.

There were a number of reports published so far on taxonomy of rice field blue greens, but no attempt has so far been made on the detailed biology and taxonomical account of rice field blue green algal system.

Considering these facts, this book was prepared on the basis of detailed investigation made during 1985-1990 on rice field blue green algal collections from West Bengal in particular.

In preparing this book for publication I wish to thank M/s Daya Publishing House, New Delhi for expressing their eagerness to publish the book, and my beloved student Dr. U.C. Pal for providing constant support during the course of investigation as well as preparation of this manuscript. My family members also helped a lot by putting moral support to undertake this task, otherwise the book would not be completed.

S.C. Santra

CONTENTS

	Preface	iii
	Introduction	vii
Chapter 1.	Biology of Blue-Green Algae (Cyanobacteria)	1
Chapter 2.	Isolation, Purification and Cultivation of Blue-Green Algae	21
Chapter 3.	Taxonomic Enumeration of Rice-Fields Blue-Green Algae	29
Chapter 4.	Importance of Blue-Green Algae	147
	Bibliography	157
	Glossary	165
	Index	169
	Illustration of Plates	175

INTRODUCTION

Cyanophycean algae or blue-green algae are considered to be one of the remarkable group of photosynthetic simple plant forms, which represent a link between bacteria and green plants. The cellular organisation of this form of algae, is prokaryotic which is characterised by the lack of membrane-bound organelles including nucleus, chloroplast or mitochondria. Thus they are very much identical to photosynthetic bacterial forms, thus it is very often called blue-green algae as Cyanobacteria.

It is further interesting to note the fact that blue-green algae apparently have a long history on earth. Their fossil records were traced back from early Precambrian period, i.e. over three billion years ago. Most probably, they were responsible for a major evolutionary transformation leading to the development of aerobic metabolism and to the subsequent rise of higher plant and animal forms.

Blue-green are common in all kinds of natural habitats. Many species are cosmopolitan, distributed throughout the world. Certain blue-greens are more or less permanent and often dominant components of the plankton of tropical, temperate and frigid lakes and oceans; others are characteristic of streams, rivers, hot springs or rocky seashore. Some one well adapted to terrestrial habitats. They are also known to be amongst the earliest colonizers of arid land.

Finally, the study of Cyanobacteria became 'fashionable' in the last two decades and there are many academic and practical reasons for the recent upsurge of interest in them. First is the recognition of their important role in the natural environment. Another is the growing appreciation of the part of blue-green algae might play in Man's economy, particularly in solving the World's desperately urgent food and energy problems. A third and not least important reason for the rapidly growing popularity of blue-green is the realization that they provide a relatively simple and useful model system for the study of fundamental cellular process such as macromolecular synthesis, regulation of gene expression, cell differentiation and ordered developmental pattern formation (Carr and Whitton, 1973; 1982; Fogg *et .al*, 1973).

1

BIOLOGY OF BLUE-GREEN ALGAE (CYANOBACTERIA)

Habitat and Distribution

Blue greens have wider distribution (Table 1.1): they are more or less permanent and often dominant components of aquatic bodies—freshwater and also in marine environment either as planktons or benthic or floating forms. A number of blue green algae were also recognised to be grown in thermal water. They can often be seen on the surface of moist soil and growing beneath the soil surface.

Table 1.1: Natural Habits of Blue-Green Algae

Types of Habitats	*Algae forms*
I. Freshwater	
(a) Planktonic forms	*Anabaena, Gloeotrichia, Oscillatoria, Gloeocapsa, Aphanizomenon, Microcystis,* etc.
(b) Attached and benthic forms	*Phormidium Scytonema, Calothrix, Gloeotrichia, Lyngbya, Nostoc,* etc.
II. Thermal water	*Oscillatoria, Pleurocapsa, Calothrix, Synechococcus,* etc.
III. Marine environment	
(a) Planktonic forms	*Oscillatoria, Trichodesmium, Synechococcus,* etc.
(b) Benthic forms	*Phormidium, Calothrix, Rivularia, Lyngbya,* etc.
IV. Terrestrial forms	*Schizothrix, Microcoleus, Oscillatoria, Scytonema, Nostoc,* etc.
V. Symbiotic associations:	Association between blue-green algae
(a) Association between fungi and blue-green algae in lichens	like *Nostoc, Calothrix, Stigonema, Scytonema* with various fungus to form lichens.
(b) Association between liverworts and blue-greens	Associations between *Nostoc* and the bryophytes *Blasia, Anthoceros* and *Cavicularia.*
(c) Association between ferns and blue-greens	Associations of *Anabaena* with aquatic fern like *Azolla.*
(d) Blue-greens associated with higher plants	Association of *Nostoc* with roots of *Cycas* and *Gunnera.*
(e) Association of blue greens with invertebrates	Association of *Aphanocapsa* with spores.

In addition, a number of blue green algae engage themselves in symbiotic associations with a wide range of organisms, viz. lichens, liverworts, aquatic fern and also in roots of some phanerogamic plants. Blue-greens also form associations with several specialised intervertebrates.

Thallus Morphology

Blue-green algae display a great diversity of forms which differ in their morphology, structure and function, and in their mode of response to environmental stimuli. The major thallus forms are of five major groups viz. unicellular, colonial, simple filamentous, branched filamentous, multiseriate parenchymatous forms (Table 1.2).

Table 1.2: The Major Thallus Forms of Blue-Green Algae

Groups	*Basic morphology*	*Modes of reproduction*	*Representative order & genera*
I	Unicellular	Binary fission	Chroococales : *Gloeotheca, Synechococcus, Gloeocapsa, Microcystis, Merismopedia,* etc.
II	Colonial	Budding Multiple fission	Chamaesiphonales: *Chamaesiphon, Dermocarpa.* Pleurocapsales: *Xenococcus, Myxosarcina, Pleurocapsa.*
III	Filamentous unbranched non-differentiated	Trichome fragmentation hormogonia	Nostocales: *Oscillatoria, Spirulina, Lyngbya, Phormidium.*
	Filamentous unbranched heterocystous	Trichome fragmentation, hormogonia, akinetes	Nostocales: *Anabaena, Nostoc, Nodularia, Cylindrospermum, Calothrix, Rivularia, Gloeotrichia, Scytonema.*
IV	Filamentous branched heterocystous uniseriate	Trichome fragmentation, hormogonia akinetes	Stigonematales: *Nostocopsis*
V	Filamentous branched heterocystous multiseriate parenchymatous	Trichome fragmentation, hormogonia, akinetes	Stigonematales: *Westiella, Stigonema, Mastigocladus, Hapalosiphon, Chlorogloeopsis.*

The simplest blue-greens appear to be the unicellular forms, most of which were placed in the order Chroococcales. Their cells are spherical, ovoid or cylindrical

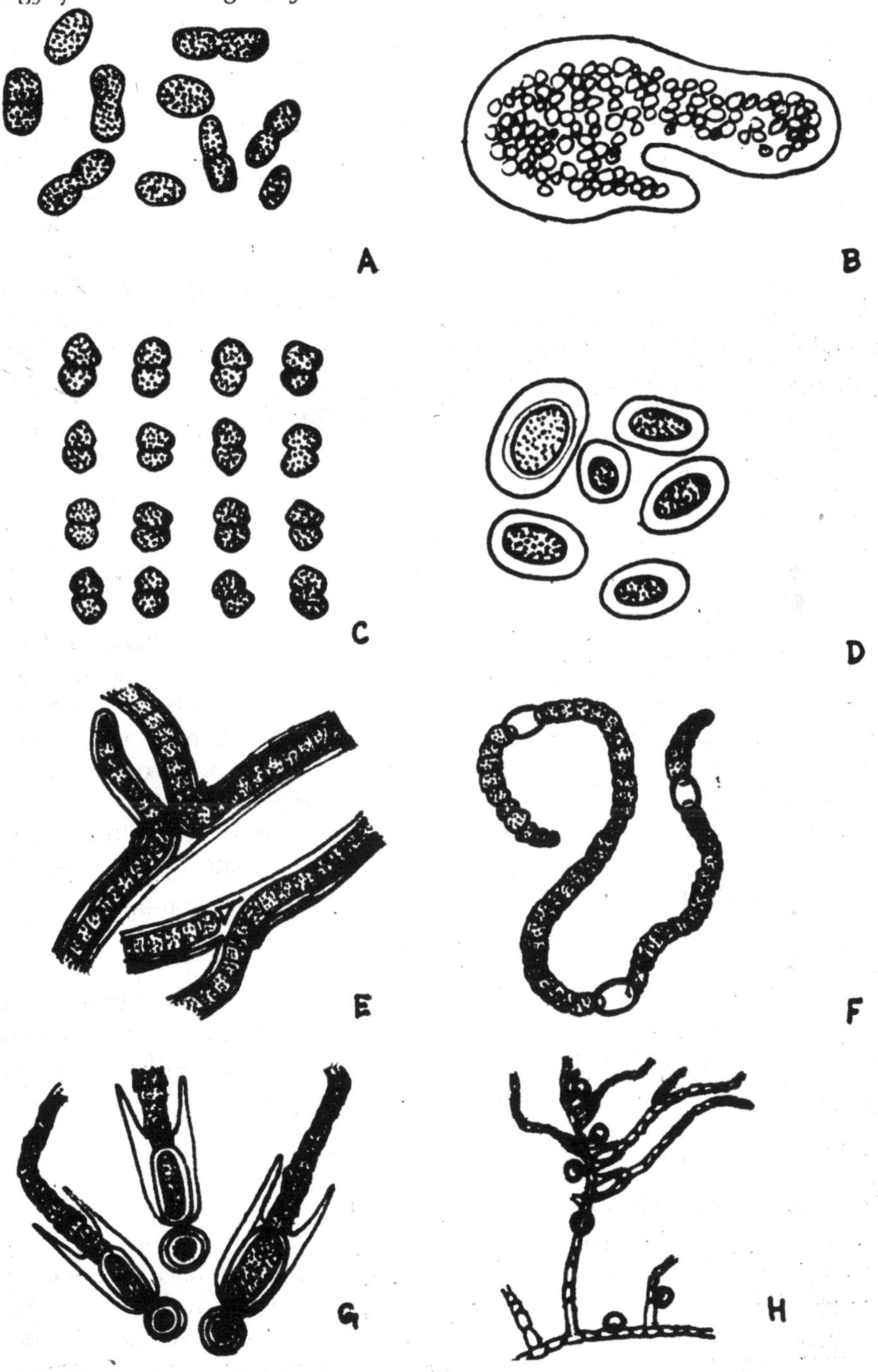

Fig. 1.1: Thallus forms (light micrograph):
(a) Synechococcus, (b) Microcystis, (c) Merismopedia, (d) Gloeothece, (e) Scytonema, (f) Anabaena, (g) Gloeotrichia, (h) Nostocopsis.

and they all reproduce by binary fission. But in a number blue-greens cells are aggregated to form colonial thallus forms, which very often bounded by a sheath.

A great number of blue-green taxa display a filamentous morphology formed by the daughter cells remaining in close contact following repeated cell divisions which occur in a single plane at right angles to the main axis of the filament. This process give rise to a simple multicellular structure which consists of a chain of cell and called a trichome. The trichome may be straight, or is coiled in regular spiral, or falsely branched or even truely branched. In the majority of the filamentous forms, the trichome is usually uniseriate and covered by distinct sheath. But in the members of order Stigonematales the main branch of trichome becoming multiseriate while side branches were uniseriate and even cells in the main branch and in the side branches may differ in size and shape.

Cellular Ultrastructures

The cells of most blue-greens measures from 2-5 μm (1μm = 0.001 mm) in diameter. The detailed structure of such minute cells can be revealed only by the use of an electron microscope, which provides a magnification of between 10000 and 250000 times, sufficient to resolve the basic pattern of cellular construction. On the whole, the cells of blue-greens show a typical prokaryotic organisation, profoundly different from the more complex eukaryotic architecture of plant and animal cells (Lang, 1968, 1972). In blue-green algal cells, the photosynthetic, respiratory and genetic apparatuses are present but are not delimited from each other by any bounding membrane of their own. No sharp boundaries divide the cell into special regions. Rather, there are several cellular components with characteristic fine structure. These are distributed through the cell in patterns varying from species to species and also in different developmental stages in the same species (Fig.1.2).

In general, the central (or nucleoplasmic) region is lightly granulated due to the presence of ribosomes, and contains the extremely folded thin thread of the prokaryotic type of 'circular' chromosome. The peripheral (or chromatoplasmic) region is traversed by thin sheets of the photosynthetic membrane system, incorporating the photosynthetic pigments. The cell is enclosed by a membrane, the plasmalemma, which is fortified by a multilayered cell wall. External to the wall, the cell may be surrounded by a gelatinous sheath or a more firm envelope.

(A) Surface Structures: Cell Wall and Sheath

The protoplast of blue-greens is surrounded by a cell wall, the construction and chemical composition of which is similar to the wall of Gram negative bacteria. In general, cell walls are composed of four layers, designated from inside out as the LI, LII, LIII and LIV with the LI being next to the plasmalemma. The LI and LIII layers are electron transparent and the LII and LIV layers are electron-opaque.

The wall is composed of two principal layers, the inner murein or petidoglucan and the outer lipoprotein layers. The murein layer which forms a polymeric fibrous mesh is primarily responsible for the mechanical strength of the wall; it confers the shape of the cell and protects it against osmotic damage. The lipoprotein layer

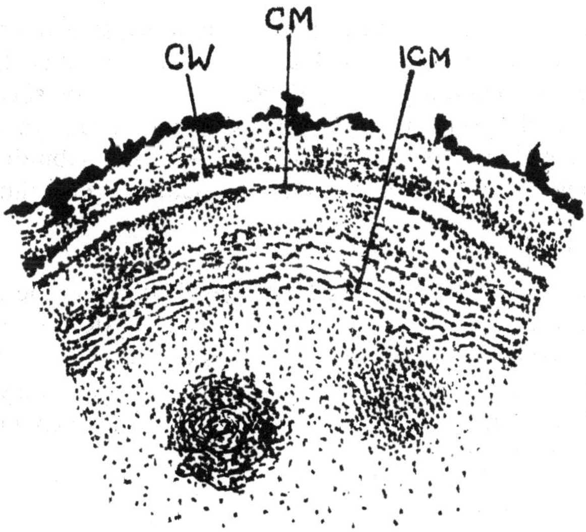

Fig.1.2: Ultrastructure of Blue green algal cell:
Section of *Synechocystis:* CW - cell wall; CM - cell membrane; ICM - Intracytoplasmic membrane.

probably controls the transport of solutes as in bacteria. The space between the two layers, termed periplasmic space, appears to have a similar content of lipopolysaccharides and degradative enzymes as in Gram negative bacteria.

The cell wall is often covered by a voluminous fluffy or dense, sometimes laminated sheath or capsule. It is essentially a slime layer composed of polysaccharides and incorporates fine fibrillar elements. Sheaths are often coloured, with red sheaths found in algae from highly acid soils and blue sheaths characteristic of algae in basic soils. Yellow and brown sheaths are common in specimens from habitats of high salt content, particularly after the algae dry out. The function of this envelope is poorly understood but it has been suggested that it may protect the cells from desiccation, or against invasion of pathogens or against phagocytosis by grazing animals. It may also render toxic or harmful substances ineffective and promote the attachment of the organisms to solid substrates. Large quantities of metabolic products may be investigated in the formation of extensive sheaths. The synthesis of sheath polysaccharides is under genetic control.

Blue-green algae lack flagella and until recently it is also believed to lack pili. However many cyanobacteria are now known to bear certain surface structures outside their outer cell membrane. Vaara (1982) has worked on several species of Chroococcacean cyanobacteria and show piliation. In general, cyanobacterial pili are long, stiff threads of even width and seem to possess a subunit structure, the subunits being made of pilin (molecular weight is about 21,000 daltons).

(B) Plasma Membrane

It covers the surface of the cytoplasm and lies just below the cell wall. It is typically 80 Å thick, consisting of two dark lines separated by a light area in sectional images. It forms several projections or invaginations into the cytoplasm. Being a unit membrane structure, it consists of a lipid bimolecular layer containing abundant proteins. It is semipermeable and assumed that the enzymes involved in respiration or in cell wall synthesis are located on it.

(C) Protoplasmic Structures

It is broadly divided into two zones—central hyaline granulated nucleoplasmic region and peripheral chromoplasmic region with photosynthetic membrane system. Many of the protoplasmic structures found in the bacteria occur in the cyanophyceae.

(a) *Photosynthetic membrane system*: The peripheral region of the cytoplasm contains the photosynthetic apparatus of blue-greens. In unicellular forms this commonly consists of a few membrane layers which extend in concentric sheets beneath the cell membrane and surround the central nucleoplasmic region. The more elaborate membrane structure seen in the cells of filamentous species is apparently formed by means of expansion and invagination of the membranes. The basic structure of the membranes resembles a flattened sac and is known as the thylakoid. The appressed membranes of the thylakoid may occasionally or partially separate displaying the inner cavity of the thylakoid sac. Each thylakoid is around 20 nm thick. It consists of two closely appressed unit membranes separated from one another by a space of 3-6 nm.

The lipid bilayer of the thylakoid membrane incorporates the lipophilic photosynthetic pigments, chlorophyll a and the various carotenoids. It also incorporates the components of electron transport chain viz. cytochromes, plastocyanin and ferredoxin. The major photosynthetic pigments along with absorption bands (nm) of blue green algae are presented in Table 1.3. The relative quantities of these pigments may vary according to species and the spectral composition of light, and will determine the coloured appearance of these organisms.

Table 1.3

Photosynthetic Pigments (with major absorption bands) of Blue-Green Algae

Pigment Groups	*Major absorption band (nm)*	
CHLOROPHYLLS	Chlorophylla	435, (670), 680, (700)
CAROTENOIDS	—carotene	(431), 450-454, 478-480
	Echinenone	455-459, (475)
	Zeaxanthin	(430), 453, 479
	Canthaxanthin	466
	Myxoxanthophyll	504-508, 474-476, 450-452
PHYCOBILIPROTEINS	Allophycocyanin	650
	Phycocyanin	610-625
	Phycoerythrin	555-565

As it is wellknown that environmental conditions particularly the intensity and quality of light and mineral nutrients are known to affect pigment composition. In blue-greens algae, phycobili-protein synthesis is particularly susceptible to environmental influences. Change in pigmentation is largely affected by the change in the phycocyanin: phycoerythrin ratio. Certain strains will appear distinctly blueish green when grown in red light because phycocyanin, which absorbs mainly red light, it present in the greatest quantities; whilst if they are grown under a green light regime, they will appear rather reddish due to the predominance of the green light absorbing phycoerythrin in their cells. Thus they produce mainly the pigment which absorbs efficiently the light present. The shift between the red light and green light effects was found to occur at about 590 nm wavelength. This phenomenon has been termed as *"Complementary Chromatic adaptation"*.

Phycoerythrin synthesis occurs at maximum rate in green light (about 540 nm) and is reduced on completely suppressed in red light (about 650 nm). Rifamycin, an inhibitor of RNA polymease, prevents subsequent phycoerythrin synthesis. This indicates that chromatic adaptation is genetically controlled at the level of

DNA transcription. The control is mediated by photoreceptive pigments called 'adaptochromes', which appear to be closely associated with the phycobilisomes; they sample the quality of light and induce the changes in pigment synthesis. Adaptochromes are probably present in two forms, which interchanges reversely, one absorbing green light and the other red light; thus they display green versus red antagonism, in contrast to the red/far-red reversible transformation characteristic of higher plant phytochromes. Phycocyanin synthesis on the other hand in some species of blue green algae is independent of the spectral composition of light, in others it may be under chromatic control, occurring at maximum rate in red light and at its lowest rate in green light. Allophycocyanin synthesis appears to be unaffected by light quality. Three groups of blue-greens can be thus recognised according to their response to the spectral composition of light. They are as follows:-

- In the first group, phycoerythrin is always present and its synthesis is increased in green light; phycocyanin levels are unchanged.
- In the second group, phycoerythrin synthesis is induced in green light and suppressed in red light; phycocyanin is always present and its synthesis is enhanced in red light.
- In the third group, neither phycocyanin synthesis nor phycoerythrin production are affected by the spectral composition of light.

It is further interesting to note the fact that the photosynthetic pigments of blue green algae are located in granular supramolecular complexes, called *phycobilisomes.* They contain the chromatophore-bearing water-soluble phycobiliprotiens, i.e., the blue phycocyanin (λ_{max} 620 nm) and allophycocyanin (λ_{max} 650-670 nm) and the red phycoerythrins (λ_{max} 550-570 nm). Phycobiliproteins represent the primary light—harvesting ('antenna') pigment for photosystem II in blue greens. Light energy trapped by these pigments is transferred to chlorophyll with high efficiency. Each phycobilisome consists of triangular core surrounded by six peripheral rods, composed of disc-shaped subunits. The triangular core, which is in direct contact with the thylakoid membrane, consists of allophycocyanin while the rod contain phycocyanin and phycoerythrin. Phycobilisomes are arranged in parallel rows on the external surface of the thylakoid membrane. The close proximity to the thylakoid membrane is essential for maximum energy transfer by resonance in the primary act of photosynthesis.

(b) *Central hyaline nucleoplasmic region (Nuclear apparatus):* As first demonstration by cytochemical studies and confirmed by electron microscopy (Hopwood and Glauert, 1960; Ris and Singh, 1961; Pankratz and Brown, 1963; Leak, 1967), the nuclear regions of cyanobacteria are comparable with those of other prokaryotes in lacking both a nuclear membrane and compact chromosomes of the eukaryotic type. The apparent absence of histones was consistent with the prokaryotic nature of Cyanobacteria. The assumption that the fibrils observed in electron micrographs of the nuclear regions were composed of DNA was confirmed by their sensitivity to DNAase and incorporation of ^{3}H thymidine (Leak, 1965). Studies performed during the past 30 years revealed the fact that there are presence of the four normal nucleotide bases and the characteristic hyperchromicity during thermal denaturation. This indicate the double stranded nature of the Cyanobacterial genome. The overall range of DNA base composition, 35-71 mol % GC, corresponds to the diversity found in other bacteria. This diversity, however, is confined to two major groups: Unicellular Cyanobacteria dividing by binary fission (35-71 mol % GC), and filamentous, a heterocystous cyanobacteria (40-67 mol % GC). All other strains, including the heterogeneous pleurocapsalean and heterocystous forms, contain DNA with a narrow (38-47 mol % GC) range of base composition.

The cyanobacteria in their entirety differ from all other bacteria in exhibiting an extremely wide range of genome size (genetic complexity). All other bacteria examined contain genomes of relatively uniform size, $1.0x10^9$ to $3.6x10^9$ daltons mol. wt., but cyanobacteria revealed a much wider diversity in genetic information content ($1.6x10^9$ to $8.6x10^9$ daltons genome mol. wt.) which in general was correlated to the morphological and/or physiological complexity of the organisms.

A number of strains contained DNA which renatured rapidly. Such DNA can be interpreted either as reiterated chromosomal DNA or small extrachromosomal elements (phages or plasmids) present in high copy number (Fig. 1.3). Subsequently, plasmids were characterized in a large number of Cyanobacteria. Six different sizes of covalently closed circular (CCC) DNA were observed as plasmid. The base composition of plasmid DNA was significantly different from that of chromosomal DNA.

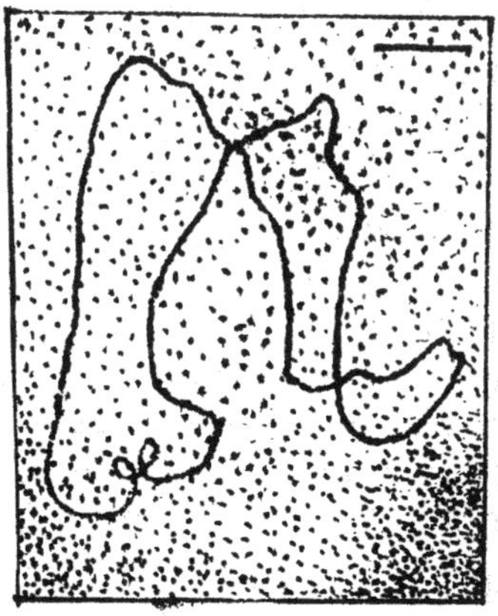

Fig. 1.3: Electron micrograph of plasmids isolated from Nostoc.

(c) *Ribosomes*: Ribosomes in blue greens are seen diffusely distributed throughout the cytoplasm and particularly concentrated in the nucleoplasmic region. Their size, about 10 to 15 nm in diameter; and their sedimentation properties, 70S (Svedberg Units), are typically prokaryotic. The ribosomes dissociate into large 50S and small 30S subunits, and contain 5S, 16S and 23S ribosomal RNAs with molecular weights of about 0.04, 0.55 and 1.05 x 10^6 daltons, respectively.

(d) *Cytoplasmic inclusions and reserve products:* The light microscopy is inadequate for resolving the subcellular structure of blue greens, it nevertheless shows clearly that their cytoplasm is heterogeneous and usually incorporates granular structures of considerable density. Electron microscopic examination implies that these optical features arise from the presence of various cytoplasmic inclusions. In general an EM studies, the cytoplasm contains diverse kinds of granular bodies of various size and shapes. These are as follows:-

(i) Cyanophycin granules,

(ii) Carboxysomes,

(iii) Polyphosphate bodies,

(iv) Lipid globules,

(v) Glycogen granules, and

(vi) Poly β-hydroxybutyrate particles, etc.

Cyanophycin (or structured) granules are relatively large in size (up to 500 nm in diameter) and are made of high molecular weight (25-100x10^3). These granules are deposited mainly at the cell periphery. They consist of a unique polypeptide composed only of two amino acids, arginine and aspertic acid. The arginyl residues are attached at each free carboxyl group of the polyaspertate core. They are produced, unlike other polypeptides, by a ribosome-independent mechanism, and function in nitrogen turnover. They stain with neutral red or carmine but not with methylene blue or toluidine blue.

Polyphosphate bodies are of course fairly large structures and appear highly electron-dense even in thin section. These bodies stain reddish blue with toluidine blue due to their metachromasy. These granules are earlier termed as *metachromatic* or *volutin* granules. These granules are acting as a phosphate store.

Carboxysomes or *polyhedral bodies* are semicrystalline structures about 200-300 nm in diameter. They contain a reserve form of the primary photosynthetic enzyme, ribulose 1-5 biphosphate carboxylase (RuBPCase), which in its active form, catalyzes photosynthetic CO_2 fixation into ribulose- 1-5 biphosphate (RuBP). They occur usually in the central or core region of the cell.

Lipid globules, which are spherical, variable in size (30 to 90 nm in diameter), most abundant near the cell surface, and possibly represent lipid stores for use in membrane synthesis.

Glycogen granules are minute (about 30 x 65 nm in size), ovoid or rod-shaped structures deposited primarily in the cytoplasm between the thylakoids and serve as a carbon and energy source. Some blue greens such as *Chlorogloea fritschii* and *Microcystis* species contain *poly-β-hydroxybutyrate granules* which may be distinguished from the closely resembling polyphosphate granules by the presence of a 3 nm limiting monolayer.

Several other inclusions like microtubules, microfilaments, wall bodies, crystalline and trilamellar bodies, and also granular and fibrillar crystals, have been observed in thin sections from various blue greens but so far nothing is known about their function.

According to Jensen (1984), all the above mentioned cytoplasmic inclusions of cyanobacterial cell may be broadly classified into two categories: those of regular occurrence and those which occur irregularly. The first category includes thylakoids with associated phycobilisomes, DNA, ribosomes, lipid droplets, glycogen, cyanophycin granules, polyhedral bodies and polyphosphate bodies. These regular inclusions occur in most or all cyanobacteria. A few species contain, besides the above regular inclusions, the so-called irregular inclusions viz. membrane monolayer inclusions (e.g. gas vacuoles and poly-β-hydroxybutyrate granules), membrane limited inclusions (e.g. prolamellar like bodies found in some species of *Anabaena* and *Nostoc*) and tubular and filamentous type.

Many planktonic bacteria characteristically possess *gas vacuoles* which appear as darkish objects under low power of the microscope and as reddish granules in high power. The most important property of gas vacuoles is that they vanish completely when subjected to pressure. Since they contain certain gases, the presence of gas vacuoles in a cell makes it buoyant, i.e. float on the surface water. Likewise, disappearance of the vacuoles makes the organism sink down.

A single gas vesiele is a cylindrical (70 nm in diameter and about 350 nm long) closed at both ends by conical caps. The vesicle is formed by an extremely thin (2 nm) and rigid proteinaceous membrane enclosing a gas-filled space. Many hundreds of vesicles are assembled in tight groups or bundles in the cytoplasm seen as bright areas of the cell called *gas vacuoles*.

The reserve products which produce cytoplasmic inclusions are accumulated under conditions of excess supply of particular nutrients (orthophosphate in case of polyphosphate granules or combined nitrogen for cyanophycin granules). Identically, reserve products are mobilized during periods of nutrient shortage or under conditions when photosynthesis is not possible. Polyphosphate bodies are depleted in the absence of phosphate in the environment, cyanophycin granules are degraded under nitrogen-limited growth conditions, and glycogen granules are utilized as a source of carbon in biosynthetic processes or for energy production in the dark. Glycogen accumulated during nitrogen depletion is degraded upon fresh supply of a nitrogen source thus providing carbon skeletons to incorporate the nitrogen assimilated. It is further noted that the accumulation and degradation of reserve material is catalyzed by specific enzymes. Cyanophycin polypeptide is largely absent in exponentially growing cell material, it accumulates in stationery phase cultures, and is rapidly degraded during re-newed growth when cells are transferred into fresh nutrient medium.

Reproduction

Most unicellular blue greens reproduce by binary fission. An exception is seen in the genus *Chamaesiphon* where a new cell is produced in the process of an unequal fission or budding. Spherical cells are budded off at one end (the apical pole) of the ovoid or elongated sessile vegetative cell, which displays a distinct, basal to apical, polarity.

In the order Pleurocapsales a characteristic mode of reproduction is by multiple fission. This occurs through a series of rapid divisions in three planes of the enlarged mother cell, and is not accompanied by cell growth. The small spherical baeocytes produced are released by the rupture of the mother cell wall. There are various modes of reproduction in the filamentous forms and some involve the formation and germination of specific reproductive structures. The simplest and in some genera the commonest way of reproduction is by random trichome fragmentation. In certain filaments species hormogone development takes place and each hormogone detached from the filament and produce new filament.

In addition there are various modes of asexual spore body formation takes place. These are named as akinetes, heterocysts, exspores and endospores, etc. The major modes of reproductions are discussed below:

(a) *Cell division (or binary fission):* The most simplest mode of reproduction is by cell division which is followed by growth and expansion. Cell division takes place in different directions—

(i) In a single plane perpendicular to the long axis of the ovoid or rod-shaped cell (i.e. transverse binary fission). This kind of cell division found in *Synechococcus (Anacystis, Agmenellum, Aphanotheca, Coccochloris), Gloeotheca,* and *Synechocystis (Aphanocapsa, Microcystis).*

(ii) The cubical colonies of *Eucapsa* arise through cells dividing successively in three planes at right angles to each other, and the daughter cells expanding in a direction parallel to the previous plane of cell division.

(b) *Trichome fragmentation:* This is one of the simplest mode of reproduction found in filamentous blue greens, where trichome fragments may vary in size but even a single detached cell may be capable under favourable conditions of reproducing the characteristic filamentous morphology of a particular species.

(c) *Hormogonia* (or *hormogones*): Filament fragmentation is often a more ordered process associated with the production of distinct reproductive trichome segments called *hormogones* (or *hormogonia*). These are short 5 to 15 celled segments which separate by the rounding off of their end cells within the trichome envelope. They exhibit active gliding motion upon their liberation, and develop into a new trichome. Hormogone development in certain species is initiated by the formation along the trichome of *separation discs* (*necridia*). These are cells which undergo lysis and dehydration, and serve as preformed breaking points for hormogone detachment. While in some genera, like *Oscillatoria,* several hormogonia may be produced along a single trichome, in others hormogonia develop in more defined positions, for example, below the apical 'hair' in *Rivularia* or at the end of the lateral branches in *Fischerella.*

(d) *Akinetes*: Many of the heterocystous blue greens are also capable of producing perennating structures, *akinetes* or *spores*, under adverse environmental conditions which develop through the transformation of vegetative cells. During the process of akinete differentiation the cell increase several fold in size, accumulates large reserves of cyanophycin polypeptide and glycogen and deposits a thick and elaborate extracellular capsule which completely surrounds the mature akinete. The morphological-structural changes are accompanied by profound changes in the biochemistry and physiology of the cell. Polyphosphate disappears, pigment composition changes, and RNA content is greatly increased. In addition, photosynthetic and biosynthetic activities are reduced and may cease altogether upon completion of akinete differentiation but respiratory activity remains relatively high. The mature akinetes have a normal cell wall surrounded by a wide three-layered coat with a protoplasm full of food reserves. During favourable environmental season, the process of akinete germination begins with gradual disappearance of reserve products, and this is accompanied by the synthesis of new DNA, RNA, pigment and protein and the formation of new thylakoids. These are followed by one or two transverse divisions, and finally by the local dissolution of the akinete envelope, enabling the emergence and release of the short germling.

Akinete formation and germination may have an important role in the survival of planktonic blue-greens which frequently undergo photo-oxidation and rapid breakdown in surface blooms. Akinetes were reported to survive several years of deposition in the anaerobic mud sediments and to be capable of germination after many years of storage in herbarium specimens.

(e) *Heterocysts:* Heterocysts are larger than vegetative cells and appear empty in the light microscope. Heterocysts are formed from vegetative cells by the dissolution of storage granules, the deposition of a multilayered envelope outside the cell wall, the break-down of the photosynthetic thylakoids and the formation of new membranous structures. Heterocysts are produced under nitrogen-depleted conditions by the gradual transformation of certain vegetative cells (Fig. 1.4).

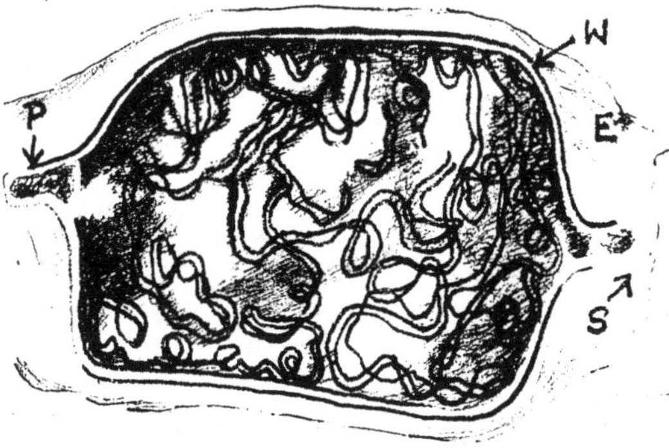

Fig. 1.4: Electron micrograph of a heterocyst of *Anabaena*: E - envelope; W - cell wall; S - septum; P - pore channel; T - thylakoid.

Heterocysts contain one or two polar granules made of cyanophycin. They contain much less of phycocyanin than do vegetative cells. They show higher respiratory activity than vegetative cells but have a reduced ability to fix CO_2.

Heterocyst usually functions as anaerobic factories for nitrogen fixation under externally aerobic conditions. In addition, heterocyst may act a resting spore bodies, which on germination may give rise to new filament. It is further noted that intercalary heterocysts are responsible for breakage of older filament into several newer filament and thus multiplication takes place. The developmental sequences of heterocystous stage and vegetative stage was studied in a number of genera under different environmental conditions. During the well documented development of *Nostoc muscorum* two separate cycles, the heterocystous and the sporogenous cycle can be distinguished. When motile hormogonia produced in the heterocystous cycle come to rest, the terminal cell of the juvenile filament develops into a heterocyst, whilst the remaining cells continue to grow and divide, each giving rise to cluster of four cells surrounded by the parent envelope: this is referred to as the "aseriate" stage of development. When the organism is grown on organic substrates in the dark, the process is arrested at this stage. Further development requires illumination, but it may continue in the dark after only a brief exposure to light. Light activation induces the liberation of the cell clusters from the mother cell envelope, which emerge in the form of short motile trichomes. They grow into hormogonia and later form intercalary heterocysts. Their release is the beginning of a new heterocystous cycle. In an alternative sequence when the heterocystous trichome undergo excessive sporulation, akinetes are formed. The akinetes eventually germinate and the emerging germling continue their development through the heterocystous cycle.

(f) *Endospores:* These true asexual spores formed within two cells by internal division of the protoplast, either forming a row of spores or an irregular mass.

(g) *Baeocytes:* In the order Pleurocapsales a characteristic mode of reproduction is by multiple fission. This occurs through a series of rapid divisions in three planes of the enlarged mother cell, and is not accompanied by cell growth. The small spherical baeocytes produced are released by the rupture of the mother cell wall.

Sexuality and Cell Cycle

Though the members of blue-green algae lacks sex organs and gamete production, yet there are a number ways by which genetic recombination can takes place, like that of bacteria. Much of the genetic work published to date has been concerned with the induction of mutants, recombination studies and genetic mapping.

In 1962, Kumar attempted to demonstrate the genetic recombination in *Anacystis nidulans.* He used two mutant strains, one resistant to penicillin, the other to streptomycin and cultured the two strains together in the absence of antibiotics. Then a double resistant, recombinant strain was isolated by subculture into medium containing both antibiotics. Singh and Sinha (1965) also attempted to demonstrate genetic recombination in the filamentous form *Cylindrospermum majus*. They used non-sporulating Streptomycin resistant and spore-forming penicillin resistant strains and isolated double resistant spore forming strains from a medium containing both antibiotics. These observations suggest the fact that genetic recombination takes place between two strains during incubation. The mechanism however seems to be transformation or conjugation as noticed in bacteria.

The mutational studies of blue green algae further suggest that the fact that like bacteria, mutation can be developed in blue-greens spontaneously or induced by mutagenetic agents (NTG, UV-rays etc.). Both genetic recombination and mutational changes leads to the genetic variations in blue greens.

In 1970, Shestako and Khyen reported the genetic transformation in *Anacystis nidulans* in cell free system. Subsequently after the isolation of Cyanophage virus, the possibility of recombination by transduction phenomenon was suggested by a number of workers.

Using batch-culture technique, cell cycle of several blue greens were studied during past three decades. The life history of *Chlorogloea fritschii* is most simpler and has four stages I, II, III & IV (stage I - hormogonium stage, stage II - long filamentous stage, stage III - colonial stage and stage IV - endospore stage).

In *Nostoc muscorum*, there are two main cycles of development, the so called heterocystous and sporogenous cycle. These cyclical changes dependent on light. In certain species of *Nostoc*, the motile (reproductive) and non-motile (vegetative) phases occur regularly, clearly defining the life cycle.

Many blue-greens have plasmids (extrachromosomal covalently closed circular DNA molecules) which are fragments of DNA of viral or bacterial origin that lie in the host cytoplasm and replicate independently. Plasmids appear to be quite common companions of the host blue green genome more than half of the strains tested contained one or more plasmids. The three plasmids (pDu_1, pDu_2 & pDu_3) of *Nostoc* strain PCC 7524 had molecular weights of 4, 8 and 28 Md. respectively. The character of plasmids in the phenotypic expression of inherited properties, in the growth and natural distribution of species, or in the evolution of blue greens is as yet unknown.

Observations of the growth of blue greens in nature and in culture suggest that reproductive structures (viz. hormogonia, akinetes or baeocystes etc.) are produced at some stage in the growth. Distinct alteration of morphological forms may occur during the developmental cycle of most colonial and filamentous genera, though detailed investigation of developmental sequences have been undertaken with relatively few species.

In the cell cycle of *Nostoc muscorum* there are two distinct cycle: the heterocystous and the sporogenous cycle. Both phases alternate and influenced by light. Motile hormogonia produced in the heterocystous cycle when come to rest, the terminal cell of the juvenile filament develops into a heterocyst, whilst the remaining cells continue to grow and divide, each giving rise to cluster of four cells surrounded by the parent envelope: this is referred to as the "aseriate" stage of development. Further development requires light but it may continue in the dark after only a brief exposure to light. Light activation induces the liberation of the cell clusters from the mother cell envelope, which emerge in the form of short motile trichomes. They grow into hormogonia and later form intercalary heterocysts. Their release is the beginning of a new heterocystous cycle. The developmental process in *N. muscorum* may take an alternative sequence when the heterocystous trichomes undergo excessive sporulation. This is called the sporogenous cycle. The akinetes eventually germinate and the emerging germlings continue their development through the heterocystous cycle. The above stated light induced (i.e. photoinduction) filamentous development in the Nostocaceae depends on particularly red light (650 nm).

Cyanobacterial Genes

Although the available knowledge of cyanobacterial gene organisation is not as extensive as that for bacteria genes, there is reason to believe that they are very similar and probably regulated identically. The genes of the nitrogenase complex of Cyanobacterium, *Anabaena* are arranged in the nif HDK operon which contains three open reading frames like that of *Klebsiella* (Fig. 1.5).

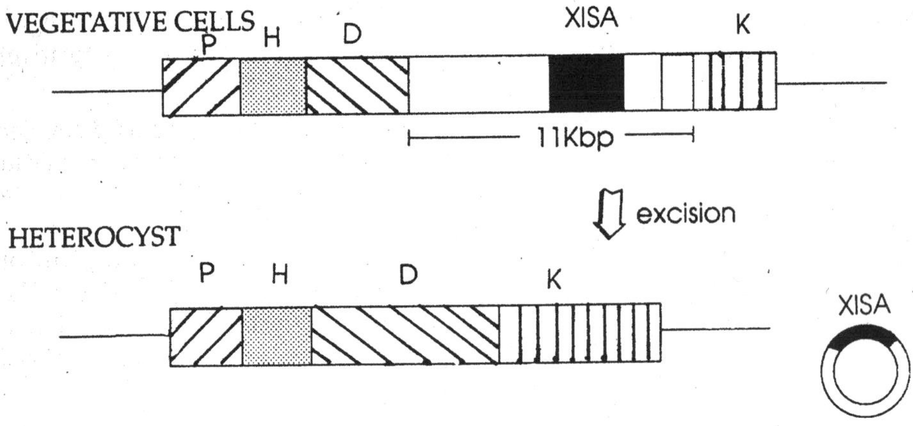

Fig. 1.5:Gene rearrangement in the nif operon of *Anabaena*. In vegetative cells which do not fix nitrogen, nif D is interrupted by an 11 kbp insert, which is excised during cellular differentiation to heterocysts. Excision is performed by the gene product of xisA, itself located within the insert. Excision results in a functional operon. P - promoter; H, D and K are open reading frames.

During heterocyst differentiation an interesting gene rearrangement occurs which results in a novel means of gene regulation. In vegetative cells the nif D ORF is interrupted by 11 kilobase pairs (kbp) of DNA, effectively preventing any possibility of transcription. Removal of the insertion creates the mature functional operon in the heterocyst.

Evolution & Phylogeny

The position of blue-green algae in the hierarchy of living organisms is a very problematic fact. It is clear that they resemble the bacteria more than any other plant group.

(a) *Relationships to Bacteria*

The chief points of similarity between the two groups are as follows:

(1) *Cellular organisation* together with the bacteria, the blue-green algae stand in sharp contrast to the rest of living organisms in having a prokaryotic cellular organisation. As well as in the absence of membrane bound organelles and other general prokaryotic features there is resemblance in the details of the structure of gas vesicles, which are only found in certain bacteria and blue-green algae and no where else.

(2) *Cell wall structure and chemistry:* The fine structure as seen in electron micrographs of sections of the walls of blue-green algae bears a general resemblance to that of bacteria, particularly that of the gram-negative kinds in that both have a mucopeptide as a major cell wall component. This contains α-ε-diaminopimelic acid and muramic acid and is susceptible to lysozyme treatment.

(3) *Cell division:* This process, which is accomplished by diaphragm-like in growths form the periphery towards the centre, is superficially at least, similar in the two classes and distinct from other modes of division found in the higher plants and animals.

(4) *Ribosomes:* There is good evidence that there are two forms of these bodies, characterized by different sedimentation properties in the ultracentrifuge and by the RNA sps. which they contain, and that one type is characteristic of bacteria, blue-green algae, higher plant chloroplasts and mitochondria, whereas the other is found only in the cytoplasm of higher plants and in animals.

(5) *Poly-β-hydroxybutyric acid*: This reserve product occurs in bacteria and in blue-green algae, but is absent from eukaryotic organisms.

(6) *Nitrogen fixation*: The capacity to assimilate elemental nitrogen is found only in certain free living bacteria and blue-green algae, and in symbiotic systems containing bacteria and blue-green algae and has not been demonstrated convincingly in any eukaryotic organism without symbiotic microorganisms.

(7) *Survival at high temperature.:* In general bacteria and blue-green algae are capable of survival and growth at temperatures about 20°C higher than the highest tolerated by eukaryotes.

(8) *Susceptibility to antibiotics:* Many antibiotics, such as penicillin and streptomycin, inhibit the growth of some bacteria and blue-green algae more severely than that of other micro-organisms.

(9) *Genetic recombination*: In both bacteria and blue-green algae this is accomplished by parasexual processes and not by sexual reproduction involving syngamy and meiosis as in eukaryotes.

The prokaryota therefore are a well-defined group and there is some justification for considering the blue-green algae as a specialized group of bacteria.

On the other hand, it is equally clear that there are important differences between the two groups, including the following:

(1) The blue-green algae show much greater structural diversity than the bacteria. Multicellular organisms with protoplasmic connexions between the cells are found among blue-green algae but not amongst the bacteria with the possible exception of the flexi-bacteria, which are of uncertain taxonomic position. The branched multiseriate filamentous thallus of algae such as sps. of *Stigonema* is much more elaborate than anything found in the bacteria. The bacteria do not as a rule form distinct colonies under natural conditions whereas most blue-green algae are aggregated into colonies of more or less characteristic form and structure. Only in a few instances, for example, with *Merismopedia* in the blue-green algae and *Thiopedia* amongst the bacteria, there is some parallelism.

(2) Differentiation of the cell is dissimilar: blue-green algae do not produce bacterial type endospores and there are no structures resembling the heterocysts or akinetes of the blue-green algae to be found in bacteria.

(3) Blue-green algae never produce flagella of any sort whereas the bacteria, although they do not produce the 9+2 stranded type, characteristic of the Eukaryota, include many forms with single-stranded flagella. It is however, possible that bacterial flagella may be homologous with the contractile fibrils which are postulated as occurring at the surface of some gliding blue-green algae.

(4) Although there is probably a close similarity between the basic photochemical mechanism of the photosynthetic bacteria and of blue-green algae, the pigments which mediate them are chemically distinct and whereas blue-green algae are capable of utilizing water as the ultimate hydrogen donor with consequent evolution of O_2, no bacteria are capable of doing this.

(5) The fatty acid compositions of blue-green algae are different from those of bacteria: polyunsaturated fatty acids are absent from the photosynthetic bacteria and the mono-unsaturated acid present is mostly the Δ-11 C_{18} vaccenic acid, which is also found in non-photosynthetic bacteria. Polyunsaturated fatty acids are present in the morphologically more complex forms and the monoenoic acid found is the Δ-9 compound oleic acid.

(b) *Relationship with Red Algae*

There are furthermore, some resemblances which seem to be of significance between the cyanophyceae and a eukaryote group, the Rhodophyceae. The most obvious are in the photosynthetic apparatus. The principle pigment, Chlorophylla, is the same in the both groups and β-carotene and phycobiliproteins are common accessory pigments. The phycobiliproteins, while not absolutely identical in these two algal groups, are closely related, according to immunological studies and are arranged in similar structures, the phycobilisomes. The phycobiliproteins of the Cryptophyceae, the only other group where they occur, are immunologically distinct and do not form phycobilisome aggregates. There are, of course, similar proteinaceous pigments, phytochromes; which occur widely in higher plants and in some algae, having a similar, though distinct, bilin prosthetic group. The thylakoids also show a resemblance in occurring singly rather than being closely stacked, as in higher plants and other algal groups. Furthermore, it is striking that of all the eukaryote groups the Rhodophyceae is the only one having no flagellated forms and showing no indication that such organelles were ever possessed in the ancestry of the group. There are further resemblances in chemical features which may or may not be significant. The major reserve polysaccharide of blue-green algae has a highly branched molecule consisting mainly of α-1-4 linked glucose units and resembling the amylopectin of higher plants. Rhodophycean starch is very similar. The trisaccharide, trehalose, which is otherwise uncommon in the plant kingdom, occurs in blue-green algae and in certain red algae.

Although such similarities exist it is important to stress that these should not be interpreted too strongly in terms of phylogeny. Indeed, it would be naive to suppose that the blue-green algae evolved directly from organisms resembling a present-day bacterium or that present-day red algae are derived directly from an ancestral blue-green alga. Klein (1970) for example, who pointed out that the unicellular eukaryotic hot spring alga, *Cyanidium caldarium*, has many characteristics intermediate between blue-green and red algae, was properly cautious in not suggesting a direct phylogenetic link.

Phylogeny and Fossil Record

A great deal of attention has been paid since the early 1950s to the study of fossil blue-green algae and over three-quarters of the known records have been obtained since then. These fossils have been found predominantly in Precambrian rocks, which have been studied extensively. A well preserved fossil blue-green alga is the heterocystous alga *Kidstoniella fritschii* which was present in Middle Devonian material from the Rhynie Chert, Scotland. In the Middle Precambrian well developed *Nostocaceae*-like and *Oscillatoria* like forms can be recognized easily. It seems, according to Schopf (1974), that Nostocacean-like blue-green algae may have developed before *Oscillatoria* like types but that by the Late Precambrian, the latter were the more prominent. Similarly, certain other algal form also obtained from various rocks.

Geochemical and Biochemical Evolution

Associated with morphological fossils just described are chemical fossils such as the hydrocarbons prytane, pristane and L-amino acids. These are generally believed to be of biological origin rather than the result of secondary contamination of the rocks and the range found in Precambrian sediments is similar to that found in contemporary blue-green algae. Thus, there is geochemical evidence which supports the view that photosynthetic micro-organisms had certainly evolved by the Middle Precambrian.

Phylogeny and Living Cyanophyceae

Phylogeny within the Cyanophyceae is problematical. Among the early works, Desikachary (1959) has discussed in more detail the phylogenetic views of many of the blue-green algae.

In any attempt to delineate evolutionary trends in the blue-green algae there are at least 3 possible approaches—(1) one can consider the fossil evidence; (2) one can look for morphological intermediates; (3) biochemical differences and similarities between living forms can be compared.

Among these three approaches, information about the 3rd way is very scanty. Recently Edelman et al. (1967), and Stanier et al. (1971) categorised the members on the basis of DNA-base composition.

Symbiotic Origin of Chloroplasts

There are many resemblances between blue-green algae and chloroplasts. Both are bounded by semipermeable membrane about 7 nm in thickness and contain thylakoids with membranes of 7-8 nm thickness. Both contain DNA which is not organized in chromosomes but as 2.5 nm diameter fibrils not associated with histone. The total amount of DNA present in *Anacystis nidulans* and *Anabaena variabilis* cells is of the same order as that in chloroplasts from *Euglena* and from two higher plant species, although much higher than in *Acetabularia* chloroplasts on the detailed biochemical level, apart from the similarity in the photosynthetic pigments and products, there are also other striking correspondences. The major ribonuclear component in plant cells is of 80S size, as in other Eukaryota but chloroplasts contain only 70S ribosomes—which are the size of those found in prokaryota. Furthermore, the 80S ribosomes of Eukaryota contain one mole of 25-28 S RNA hydrogen bonded to one mole each of 5.8S, 18S and 5.5S RNA, whereas the 70S ribosome of bacteria, blue-green algae and chloroplasts contain only 3 RNA moles; one each of 23S, 16S and 5sRNA. Chloramphenicol, which inhibits protein synthesis by prokaryotic cells has been found to bind to ribosomes from chloroplasts and to those from *Oscillatoria* but not to eukaryotic ribosomes. Moreover, the specificity of chloramphenicol isomers is the same for prokaryotic and chloroplast ribosomes.

There are, however, some facts which are not easy to reconcile with the endosymbiotic view of the origin of the chloroplast. Although chloroplasts have all the genetic apparatus for protein synthesis it appears that they are not entirely autonomous in this respect, as perhaps might be expected if they were derived from once free-living cells, but that they depend to some extent on mRNA of nuclear origin. Using the inhibitor lincomycin, which is highly specific for protein synthesis of chloroplast and prokaryotic ribosomes, Ellis and Hartley (1971) have shown that, of all the chloroplast enzymes they investigated, only RuBP-carboxylase synthesis was carried out by chloroplast ribosomes. Others even though located exclusively in chloroplasts, depend on cytoplasmic ribosomes for synthesis. However, in the course of evolution of symbiotic cells into chloroplasts, it may well be that their autonomy would be restricted and put of the genetic control of their development taken over by the nucleus of the most cell. A more weighty objection to the theory is that the DNA base ratios of chloroplasts show a wide variation in 25 to 60 mol % of guanine and cytosine, which seems to imply the unlikely possibility that chloroplasts of different species have originated from different prokaryots (Carr and Craig, 1970).

2

ISOLATION, PURIFICATION AND CULTIVATION OF BLUE-GREEN ALGAE

Blue-greens are common in all kinds of natural habitats. Many species are cosmopolitan, distributed throughout the world. Certain blue greens are more or less permanent and often dominant components of the plankton of tropical, temperate and frigid lakes and oceans, others are characteristic of streams, rivers, hot springs or rocky seashores. Some are well adapted to terrestrial habitats, and are particularly abundant in tropical soils and in water-logged fields.

From all sorts of habitats, blue-greens could be isolated through aseptic technique and thereafter cultured in artificial medium.

Culture Media

The composition of some media for blue-green algae culture are given below. In all these cases, the phosphate solution should be sterilised separately and added to the other ingredients of the medium after cooling to avoid precipitation.

1. *Modified Benecke's Medium* (De, 1939)

Salts	*Amount* (g/l)
KNO_3	0.2
$MgSO_4, 7H_2O$	0.2
K_2HPO_4	0.2
$CaCl_2$	0.1
$FeCl_3$ (1%)	2 drops.

2. *Chu No. 10* (Chu, 1942)

Salts	*Amount* (g/l)
$Ca(NO_3)_2$	0.04
K_2HPO_4	0.01
$MgSO_4, 7H_2O$	0.025
Na_2CO_3	0.02
Na_2SiO_3	0.025
$FeCl_3$	0.8 mg/1
pH	6.5-7.0

3. *Fogg's nitrogen free medium* (Fogg, 1949)

Salts	*Amount* (g/l)
KH_2PO_4	0.2
$MgSO_4, 7H_2O$	0.2
$CaCl_2, 2H_2O$	0.1
Fe-EDTA	1.0 ml *
A_5-Solution	1.0 ml **

* *Fe-EDTA solution*: Take 26.1g of EDTA in 268 ml of 1N KOH soln. and add to it 24.9 gms of $FeSO_4$, $7H_2O$; dissolve the contents by stirring. Aerate the solution overnight (16-18 hr). Store the Fe-EDTA (dark straw coloured) in amber coloured reagent bottle.

** *A_5 solution (Stock)*:

Salts	*Amount* (g/l)
$MnCl_2, 4H_2O$	1.81
MoO_3	0.0177
$ZnSO_4, 4H_2O$	0.222
$CuSO_4, 5H_2O$	0.079
H_3BO_3	2.86

4. *Gerloff's medium* (Gerloff et al., 1950)

Salts	*Amount* (g/l)
$NaNO_3$	0.0413
Na_2HPO_4	0.0082
KCl	0.0086
$MgCl_2, 6H_2O$	0.02
Na_2SO_4	0.015
$CaCl_2, 2H_2O$	0.359
Ferric citrate	0.003
Citric acid	0.003
Na_2CO_3	0.02
$Na_2SiO_3, 9H_2O$	0.025
A_5 solution	1.0 ml **
pH	7.0

5. *Allen and Arnon's medium* (Allen & Arnon, 1955)

Salts	*Amount* (g/l)
K_2HPO_4	0.358
$MgSO_4, 7H_2O$	0.25
NaCl	0.232
$CaCl_2, 6H_2O$	0.11
Micro-elements (A4 + H7)	1 ml •
KNO_3	0.02 M

• [Microelements - A_4 + H7 contains ppm metals—

$MnSO_4,4H_2O$ (0.50); MoO_3 (0.10); $ZnSO_4,4H_2O$ (0.05); $CuSO_4,5H_2O$ (0.02); H_3BO_3 (0.50); NH_4NO_3 (0.01); $Co(NO_3)_2$, $6H_2O$ (0.01); $NiSO_4,6H_2O$ (0.01); $Cr_2(SO_3K_2SO_4$, $24H_2O$ (0.01); $Na_2SO_4,2H_2O$ (0.01), $TiO(C_2O_4)x\ H_2O$ (0.01), Fe-EDTA (4.00)].

6. *Medium for Spirulina* (Zarrouk, 1966)

Salts	*Amount* (g/l)
$NaHCO_3$	18.0
K_2HO_4	0.5
$NaNO_3$	2.5
K_2SO_4	1.0
NaCl	1.0
$MgSO_4,7H_2O$	0.2
$CaCl_2$	0.04
$FeSO_4$	0.01
EDTA	0.08
A_5-solution**	1.0 ml

7. Kratz and Myer's media (Kratz and Myer, 1955):

It is specific for *Anacystis nidulans, Anabaena variabilis* and *Nostoc muscorum* (pH 7.8)

Salts	*Amount* (g/l)
KNO_3	1.0
$Ca(NO_3)_24H_2O$	0.025
K_2HPO_4	1.0
$MgSO_4,7H_2O$	0.25
Sodium citrate	0.165
$Fe_2(SO_4)_3,6H_2O$	0.004
**A_5	1 ml

8. *Synthetic medium for non-N_2 fixing blue-green algae* (Watanabe, 1960)

Salts	*Amount* (g/l)
KNO_3	1.0
K_2HPO_4	1.0
$MgSO_4,7H_2O$	0.25
$Ca(NO_3)_2,4H_2O$	0.025
Sodium citrate	0.165
$FeSO_4,7H_2O$	0.02
**A_5 solution	1.0 ml

9. *Bold Basal medium* (Nichols and Bold, 1965)

(a) Solution I: Macronutrients

Salts	*Amount* (g/400 ml)
$NaNO_3$	10
$CaCl_2,2H_2O$	1

Salts	*Amount* (g/400 ml)
$MgSO_4,7H_2O$	3
K_2HPO_4	3
KH_2PO_4	7
NaCl	1

(b) Solution II: EDTA solution

Salts	*Amount* (g/l)
EDTA	50
KOH	31

(c) Solution III: Iron solution

Salts	*Amount* (g/l)
$FeSO_4,7H_2O$	4.98
H_2SO_4	1.0 ml

(d) Solution IV: Boron solution

H_3BO_3	11.42 g/l

(e) Solution V: Micronutrients

Salts	*Amount* (g/l)
$ZnSO_4,7H_2O$	8.82
$MnCl_2,4H_2O$	1.44
MoO_3	0.71
$CuSO_4,5H_2O$	1.57
$Co(No_3)_2,6H_2O$	0.49

Five above stated individual stalk solutions is now mixed in the proportion as stated below:

Solution	*Amount* (ml)
I	10
II	1
III	1
IV	1
V	1

Then final volume was made upto 1 lit and pH was adjusted at 6.6

10. *Medium for thermophilic (45-70°C) blue-green algae* (Castenholz, 1970)

Salts	*Amount* (g/l)
Nitrilotriacetic acid	0.1
$CaSO_4,2H_2O$	0.06
$MgSO_4,7H_2O$	0.10
NaCl	0.008
KNO_3	0.103
$NaNO_3$	0.689
$NaHPO_4$	0.111
$FeCl_3$ (prepared as 0.2905 g/lit.)	1 ml

** Micronutrient solution—0.5 ml, pH 8.2

Micronutrient solution:

Salts	*Amount* (g/l)
$MnSO_4,H_2O$	2.28
$ZnSO_4,7H_2O$	0.50
H_3Bo_3	0.50
$CuSO_4,5H_2O$	0.025
$Na_2MoO_4,2H_2O$	0.025
$CoCl_2,6H_2O$	0.045
H_2SO_4 (conc.)	0.5 ml

Procedure for Isolation

1. Sterile media were prepared for isolation of blue-greens.

2. 5 ml of each medium is dispensed in test tubes and plugged and sterilised.

3. Liquid inocula should be homogenized by agitation for 30 min before plating. Soil specimens should be mixed with medium and also agitated for 30 min. Serial ten-fold dilutions are prepared from each water or soil suspension sample and used to inoculate test tubes containing selective sterile medium.

4. Inoculated tubes should be then incubated at 35°C without aeration, for 2-3 weeks. The light intensity should be maintained at 3240-6480 lux (300-600 flc) with fluorescent lights.

For purification of algal culture usually two methods are employed.

(a) In case of unicellular blue-green algae—

(i) The algal samples are transferred from primary enrichment tubes to fresh liquid medium and incubated.

(ii) Then from the secondary tubes, the algae is streaked on mineral agar plates and incubated.

(iii) After incubation, the colonies are isolated with a sterile loop into new media and incubated.

(iv) In this way through successive transfers from liquid to solid media, unialgal/or axenic cultures are yielded.

(b) In case of filamentous blue green algae—

(i) Normally the unialgal culture of filamentous forms can be obtained simply by repeated liquid transfer of small amounts of material.

(ii) Certain filamentous forms can be homogenized and a suspension of short filaments is made. Then by using such suspension isolated colonies can be developed on selected agar medium surface.

Surface contamination of algal sample can also be removed either by antibiotics treatment (20-500 μg/lit) or by UV radiation treatment (λ 2750 A°) for 20-30 min.

All strains should be tested for its purity. It should be grown in nutrient broth or peptone-glucose medium at different temperature and observations be made after 24 hrs.

Culture Methodology

A 'culture' is defined as the active growth or cultivation of microorganisms in or on nutrient media.

(A) Pure Culture

It consists of a population of only one species of micro-organism, all derived from a single parent microorganism. The isolation of one kind of microorganisms from a mixture of many different kinds is called the pure culture technique.

(B) Enrichment Culture

It is one of the special methods used for the isolation of a specific type of microorganism from a mixed population. This method is often used if the microorganism to be isolated is present in relatively small numbers or its growth is slow with respect to other species present in the mixed sample. The enrichment culture provides a specially designed cultural environment by incorporating a specific nutrient in the medium and by modifying the physical conditions of incubation. This will favour the growth of the particular type of microorganism sought.

For enrichment culture of algae, in the culture media lake or sea water, soil extract, or other special substances can be added.

(C) Continuous Culture

For continuous culture of different species of algae a special type of glass apparatus is often used. In this apparatus air circulation, temperature regulation and illumination for algal growth was properly maintained. At a regular interval fresh culture media is added and algal cells are harvested.

(D) Synchronous Culture

Synchrony in cell division is often not seen in normal culture process. By shifting the incubation temperature or by light-dark treatment of culture tubes, synchronous culture of different algal species can be possible. In fact there is no single method which can be applied universally to all groups of algae.

(E) Mass Culture Techniques for Field Use

Production of blue-green algae in large quantities for its application in the rice fields is a multistep process involving both laboratory and outdoor cultivation.

(a) *Laboratory culture*: A stock culture nitrogen fixing blue-green algae is maintained in Fogg's medium (1949). Then mass culture is done in larger vials with aerations.

(b) *Outdoor culture*: A shallow tray of galvanized iron sheet or cement tank (measuring 2m x 23 cm) is prepared and about 8-10 kg. of sterile soil mixed with 200g superphosphate is poured on each tray. Then trays are filled with water upto 15 cm. The soil reaction must be maintain neutral if required lime should be added. Simultaneously to prevent insect pest attack pesticides like Carbofuran (3% granules) at the rate of 25g per tray or BHC or Malathion (2-5 ml per tray) is added. Then algal culture (5-10g dry weight basis) is sprinkled on the standing water in each tray and exposed the trays in sunshine. After 7-10 days, a thick mat of algal culture will be obtained. During the algal growth water is to be added regularly in each tray.

When the algal growth is sufficiently thick, the water of the tray is allowed to evaporate and allowed to dry the algal mat. Dry algal flakes are then collected from the soil surface of the tray and store in bags (in measure quantity) for use in the fields.

(c) *Field production*: At first an area of 40 m^2 is demarked by strong bunds of 15 cm and puddled properly to facilitate water logging. Then the field so demarcated is flooded with water of 2.5 cm depth and superphosphate (@ 12 kg/40m^2) is applied uniformly. Thereafter composite algal culture of blue-green stock mixture is sprayed in the field @ 5 kg/40m^2. For prophylactic control of predators like snails, pesticides like carbofuran (3%) is also added @ 250g/40 m^2. Then algal mat will formed within two weeks. The mat is then removed and dried in air for field application. The same area may once again used for further algal culture in the way mentioned earlier.

During summer months, the average yield of blue-green algae per harvest ranges from 1 to 3 kg/m^2.

3

TAXONOMIC ENUMERATION OF RICE FIELDS BLUE-GREEN ALGAE

Introduction

Soil is one of the most potential habitat for algal growth particularly in moist or water-logged situation. The algal flora of the soil includes members of the Cyanophyceae, Chlorophyceae, Bacillariophyceae, Euglenophyceae, Chrysophyceae and Rhodophyceae. Among these blue green algal forms play a significant contribution in maintaining soil fertility and so also soil reclaimation. Like other countries, soil algae particularly the rice field blue green algae were studied in India for past half centuries by various workers (Aiyer, 1965; Aiyer et al., 1972; Anand, 1980; Anand and Revathi, 1987; Anand et al., 1987; Anand, 1989; Banerjee, 1935; Biswas, 1926; Bongale, 1981, 1984; Bongale and Bharathi, 1980; Chatterjee and Chatterjee, 1983; De, 1939; De and Mandal, 1956; Desikachary, 1959; Dutta and Venkataraman, 1958; Gonzalves and Gangla, 1949; Goyal et al., 1984; Grover and Pandhol, 1975; Gupta, 1957; Gupta, 1964; Gupta, 1972; Gupta and Kumar, 1968; Jha et al., 1986; Kamat 1963; Khan, 1985; Khan and Mathur, 1976; Khurana and Venkataraman, 1968; Kolte and Goyal, 1985; Kumar, 1970; Kusum, 1965; Laloraya and Mitra, 1973; Maity and Santra, 1985; Maity, Bandopadhyaya and Santra, 1987; Marathe, 1963, 1964, 1967; Mishra and Purohit, 1979; Mukhopadhyaya and Chatterjee, 1981; Pal and Yadav, 1974; Pal and Santra, 1982; Pal and Santra, 1985; Pandey, 1965a, b; Pandey and Mitra, 1965; Parukratz, 1940; Patil and Satav, 1985; Prasad et al., 1976, 1977, 1978, 1979; Prasad and Mehrotra, 1980; Rao, 1938; Ramakant and Pandey, 1984; Reddy et al., 1985; Santra, 1984, 1987, 1990; Santra et al., 1988; Sardeshpande and Goyal, 1981; Sarma and Swarn Kanta, 1978; Sharma et al., 1979; Sengar and Pandey, 1985; Shivaprakasha et al., 1983; Shukla, 1971; Singh et al., 1970; Singh, 1939, 1942, 1950; Sinha and Mukherjee, 1975a,b, 1984, Subba Raju, 1972; Tiwari, 1972, 1975; Tiwari and Pandey, 1976; Venkatarkishnan and Narayanan, 1982). In tropical countries, prokaryotic organisms like free living and symbiotic blue-green algae has enormous role to play in biological nitrogen fixation. Water logged rice field is the ideal habitat for most of the blue-green algal forms those are capable of nitrogen fixation. As such many tropical rice fields receive no chemical fertiliser nor organic manure, yet they remain productive. Among the soil algae blue-green algae however is the most dominant and these forms played a significant role in soil fertility, soil reclamation and soil conservation.

Summarising 30 years of studies on algal distribution in rice field of tropical and temperate countries, it was concluded that blue-green algae constitute about 15% of the total algal flora in the tropics, and about 2% in the temperate climate (Watanabe and Yamamoto, 1970). The most common tropical rice field nitrogen fixing blue-green algal inhabitants were species of *Anabaena, Nostoc, Aulosira,*

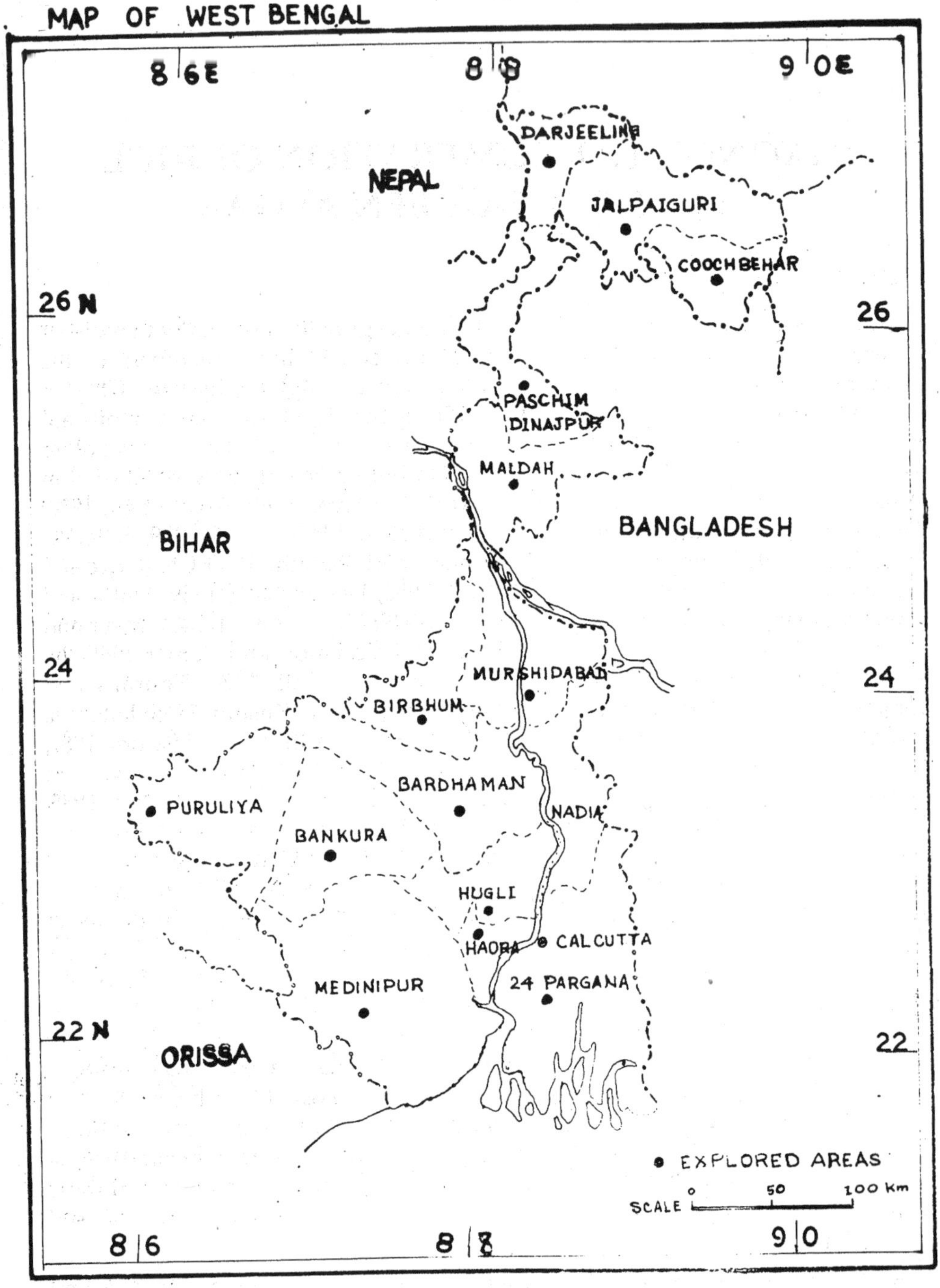
MAP OF WEST BENGAL
86E
88
90E
26 N
26
24
24
22 N
22
86
88
90
DARJEELING
NEPAL
JALPAIGURI
COOCHBEHAR
PASCHIM DINAJPUR
MALDAH
BIHAR
BANGLADESH
MURSHIDABAD
BIRBHUM
BARDHAMAN
PURULIYA
BANKURA
NADIA
HUGLI
HAORA
CALCUTTA
24 PARGANA
MEDINIPUR
ORISSA
EXPLORED AREAS
SCALE
0
50
100 Km
BASED UPON SERVEY OF INDIA MAP

Map 3.1

Calothrix, Cylindrospermum, Gloeocapsa, Rivularia, Scytonema and *Tolypothrix* (Singh, 1961; Venkataraman, 1975; Santra, 1987).

The distributional pattern of rice field blue-green algae in various States have been carried out sporadically. In West Bengal, rice field algae were studied in the past by Biswas (1926), De (1939), Saha and Mandal (1980), Sinha and Mukherjee (1975, 1984), Mukhopadhyay and Chatterjee (1981), Pal and Santra (1982), Pal and Santra (1985), Maity and Santra (1985) and Santra (1987, 1991). Identically considering the importance of blue-green algae in nitrogen economy of rice field soil, a number of trials were made in several tropical countries like India. In this chapter a comprehensive survey report on the taxonomy of the rice field blue-green algae is attempted.

West Bengal is pre-eminently a land of river basin and also have a large number of freshwater bodies: ponds, streams, lakes and water logged crop fields. These freshwater habitats are very rich in algal flora in terms of population density and diversity. During past few decades the algal flora of West Bengal were studied by various workers and more than 450 species were so far reported. The algal flora constitute higher amount of blue-green algae followed by green forms, particularly in moist soils, water logged crop fields, streams, ponds and effluent channels.

A sketch of geographical boundary of West Bengal showing various agro-climatic zonations are given in the Map 3.1. The algal samples were collected from rice field of different districts of West Bengal at different seasons for a period of 5 years. The samples were duly preserved in FAA and examined under microscope for identification and description. In addition, for isolation of some blue-green algae from rice field soil samples, aseptic pure culture techniques were employed using modified Chu-10 and Fogg's medium. At present, a total of 240 Taxa of rice field blue-green algae were identified and described. Most of the species were heterocystous and nitrogen fixer. A detailed check list is mentioned in Table 3.1.

Table 3.1: A check list of rice field blue-green algae of West Bengal

(A) Order CHROOCOCCALES:

1. *Microcystis elongata* Desikachary, 2. *M. aeruginosa* Kütz. 3. *M. flosaquae* (Wittr.) Kirchner, 4. *M.bengalensis* Banerji, 5.*Chroococcus minutus* (Kütz) Näg 6.*C. pallidus* Näg 7. *C. cohaerens* (Bréb.) Nag 8. *Gloeocapsa polydermatica* Kütz. 9. *G. calcarea* Tilden 10. *G. decorticans* (A. Br.) Richter 11. *G. rupestris* Kütz. 12. *G. montana* Kutz. 13. *G. luteo-fusca* Martens 14. *G. crepidinum* Thuret 15. *Gloeothece samoensis* var. *major* Wille 16. *G. membranacea* (Rabenh.) Born. 17. *G. rupestris* (Lyngb.) Born. 18. *G. rupestris* var. *maxima* West, W. 19. *Aphanocapsa roeseana* De Bary 20. *A. roeseana* forma *major* Laloraya et Mitra 21. *A. banaresensis* Bharadwaja 22. *A. biformis* A. Br. 23. *A. littoralis* Hansg. 24. *A. pulchra* (Kütz.) Rabenh. 25. *A. thermalis* (Kütz) Brugg. 26. *A. grevillei* (Hass.) Rabenh. 27. *Aphanothece bullosa (Menegh.) Rabenh.* 28. *A. microscopica* Näg. 29. *A. pallida* (Kütz) Rabenh. 30. *A. stagnina* (Spreng.) A. Br. 31. *A. naegelii* Wartm. 32. *Merismopedia convoluta* Breb. 33. *M. tenuissima* Lemm. 34. *Johannesbaptistia pellucida* (Dickie) Taylor et Dr. 35.*Chlorogloea fritschii* Mitra

(B) Order CHAMAESIPHONALES:

36. *Stichosiphon sansibaricus* (Hieron.) Drouet et Daily

(C) Order PLEUROCAPSALES:

37. *Myxosarcina spectabilis* Geitler

(D) Order NOSTOCALES:

38. *Arthrospira jenneri* Stizenb. ex Gomont 39. *A. massartii* Kuffareth 40. *A. massartii* var. *indica* Desikachary 41. *A. gomontiana* Setchell 42. *A. platensis* (Nordst.) Gomont var. *tenuis* (Rao) Desikachary 43. *Spirulina major* Kütz. ex Gomont 44. *S. subsalsa* Oerstd. ex Gomont 45. *S. princeps* West, Wet. G.S. 46. *S. meneghiniana* Zanard. ex Gomont 47. *Oscillatoria annae* van Goor 48. *O. subbrevis* Schmidle 49. *O. subbrevis* forma *major* West, G.S. 50.*O. acuta* Bruhl et Biswas, orth. mut. Geitler 51. *O. brevis* (Kutz.) Gomont var. *non-granulata* Kamat 52. *O. earlie* Gardner 53. *O. acuminata* Gomont 54. *O. schultzii* Lemm. 55. *O. okeni* Ag. ex Gomont 56. *O. amphibia* Ag. ex Gomont 57. *O. obscura* Bruhl et Biswas 58. *O. princeps* Vaucher ex Gomont 59. *O. curviceps* Ag. ex Gomont 60. *O. pseudogeminata* Schmid, G. 61. *O. pseudogeminata* var. *unigranulata* Biswas 62. *O. amphigranulata* van Goor 63. *O. proteus* Skuja 64. *O. quadripunctulata* Bruhl et Biswas 65. *O. anguina* (Bory) Gomont 66. *O. anguina* (Bory) Gomont forma 67. *O. proboscidea* Gomont 68. *O. chalybea* (Mertens) Gomont 69. *O. amoena* (Kutz.) Gomont 70. *O. formosa* Bory ex Gomont 71. *O. splendida* Grev. ex Gomont 72. *O. animalis* Ag. ex Gomont 73. *O. tenuis* Ag ex Gom. 74. *O. raoi* De Toni 75. *O. limosa* Ag. ex Gomont 76. *Trichodesmium thiebautii* Gomont 77. *Crinalium magnum* Fritsch et John 78. *Porphyrosiphon notarisii* (Menegh.) Kütz. ex Gom. 79. *Phormidium corium* (Ag.) Gomont 80. *P. corium* var. *capitatum* Gardner 81. *P. corium* forma *majoro* Santra et Pal. 82. *P. ambiguum* Gomont 83. *P. ambiguum* var. *major* Lemm. 84. *P. fragile* (Meneghini) Gomont 85. *P. hohneri* Schmidle 86. *P. jadinianum* Gomont 87. *P. molle* (Kütz.) Gomont 88. *P. rotunda* Santra et Pal 89.*P. laminosum* Gomont 90. *P. tenue* (Menegh) Gomont 91. *P. purpurascens* (Kutz.) Gomont 92. *P. valderianum* (Delp.) Gomont 93. *P. stagnina* Rao, C.B. 94. *P. papyraceum* (Ag.) Gomont 95. *P. subfuscum* Kütz. ex Gomont 96. *P. angustissimum* West, W. et. G.S. 97. *P. foveolarum* (Mont.) Gomont 98. *P. anomala* Rao, C.B. 99. *P. favosum* (Bory) Gomont 100. *P. retzii* (Ag.) Gomont 101. *P. lucidum* Kütz. ex Gomont 102. *P. uncinatum* (Ag.) Gom. 103. *Lyngbya laxespiralis* Skuja 104. *L. spiralis* Geitler 105. *L. lagerheimii* (Möb.) Gomont 106. *L. lagerheimii* forma *lacerata* Prasad et al. 107. *L. birgei* Smith, G.M. 108. *L. ceylanica* Wille 109. *L. ceylanica* var. *bankurensis* Sinha and Mukherjee 110. *L. confervoides* C. Ag. ex Gomont 111. *L. majuscula* Harvey ex Gomont 112. *L. martensiana* Menegh. ex Gomont 113. *L. hieronymusii* Lemm. 114. *L. allorgei* Fremy 115. *L. cryptovaginata* Schkorbatow 116. *L. stagnina* Kütz. 117. *L. major* Menegh. ex Gomont 118. *L. perelegans* Lemm. 119. *L. aerugineo-coerulea* (Kutz.) Gomont 120. *Schizothrix telephoroides* (Mont.) Gomont 121. *S. friesii* (Ag.) Gomont 122. *S. muelleri* Näg. ex Gomont 123. *S. penicillata* (Kütz.) Gomont 124. *S. lacustris* A. Br. ex Gomont 125. *S. arenaria* (Berk.) Gomont 126. *S. lamyi* Gomont 127. *Symploca muralis* Kütz. ex Gomont 128. *Microcoleus chthonoplastes* Thuret ex Gomont 129. *M. subtorulosus*

(Breb.) Gomont 130. *M. vaginatus* (Vaucher) Gomont 131. *M. lacustris* (Rabenh) Farlow 132. *M. paludosus* (Kutz.) Gomont 133. *Hydrocoleum lyngbyaceum* Kütz. ex Gomont 134. *Cylindrospermum muscicola* Kutz. ex Born. et Flah. 135. *C. muscicola* var. *longispora* Dixit 136. *C. stagnale* (Kütz.) Born. et Flah. 137. *C. licheniforme* Kütz. ex Born. et Flah. 138. *C. majus* Kütz. ex Born. et Flah. 139. *C. indicum* Rao, C.B., orth mut. De Toni 140. *C. alatosporum* Fritsch 141. *C. michailouskoense* Elenkin 142. *C. sphaerica* Prasad 143. *Wollea bharadwajae* Singh 144. *Nostoc humifusum* Carm. ex Born. et Flah. 145. *N. calcicola* Breb. ex Born. et Flah. 146. *N. linckia* (Roth) Born ex Born. et Flah. var. *arvense* Rao, C.B. 147. *N. paludosum* Kütz. ex Born. et Flah. 148. *N. piscinale* Kütz. ex Born. et Flah. 149. *N. punctiforme* (Kütz.) Hariot 150. *N. hatei* Dixit 151. *N. muscorum* Ag. ex Born. et Flah. 152. *N. ellipsosporum* (Desm.) Rabenh. ex Born. et Flah. 153. *N. comminutum* Kütz. 154. *N. depressum* Wood 155. *N. verrucosum* Vaucher ex Born. et Flah. 156. *N. commune* Vaucher ex Born. et Flah. 157. *N. spongiaeforme* Agardh ex Born. et Flah. 158. *Anabaena oryzae* Fritsch 159. *A. khannae* Skuja 160. *A. naviculoides* Fritsch 161. *A. fertilissima* Rao, C.B. 162. *A. ambigua* Rao, C.B. 163. *A. utermohlii* Geitler 164. *A. oscillarioides* Bory ex Born. et Flah. 165. *A. iyengarii* Bharadwaja 166. *A. Iyengarii* var *tenuis* Rao, C.B. 167. *A. Iyengarii* var *attenuata* Rao, C.B. 168. *A. orientalis* Dixit 169. *A. torulosa* (Carm.) Lagerh. ex Born. et Flah. 170. *A. fuellebornii* Schmidle 171. *A. volzii* Lemm. 172. A. spiroides Klebahn 173. *A. gelatinicola* Ghose 174. *A vaginicola* Fritsch et Rich 175. *A. vaginicola* f. *fertilissima* Prasad 176. *A. variabilis* Kütz. ex Born. et Flah. 177. *A. variabilis* var. *ellipsospora* Fritsch 178. *A. circinalis* var. *crassa* Ghosh 179. *Pseudanabaena schmidlei* Jaag. 180. *Nodularia spumigena* Mertens ex Born. et Flah. 181. *Aulosira prolifica* Bharadwaja 182. *A. aenigmatica* Fremy 183. *A. pseudoramosa* Bharadwaja 184. *A. implexa* Bornet et Flah 185. *A. fritschii* Bharadwaja 186. *A. fertilissima* Ghose 187. *A fertilissima* var. *tenuis* Rao, C.B. 188. *Plectonema notatum* Schmidle 189. *P. tomasinianum* (Kütz.) Born. ex Gomont 190. *P. nostocorum* Born. ex Gomont 191. *Scytonema simplex* Bharadwaja 192. *S. coactile* Montagne ex Born. et Flah. 193. *S.* fritschii Ghose 194. *S. hofmanni* Ag. ex Born. et Flah. 195. *S. fremyii* Desikachary 196. *S. mirabile* (Dillw.) Born. 197. *S. schmidtii* Gomont 198. *S. caldarium* Setchell 199. *S. javanicum* (Kütz.) Born. ex Born. et Flah. 200. *S. leptobasis* Ghose 201. *S. burmanicum* Skuja 202. *S. guyanense* (Mont.) Born. et Flah. 203. *S. stuposum* (Kütz.) Bornet ex Born. et Flah. 204. *S. cincinnatum* Thuret ex Born. et Flah. 205. *S. myochrous* (Dillw.) Ag. ex Born. et Flah. 206. *S. tolypothrichoides* Kütz, ex Born. et Flah. 207. *Tolypothrix nodosa* Bharadwaja 208. *T. tenuis* (Kütz.) Schmidt. J. 209. *Microchaete uberrima* Carter 210. *M. tenera* Thuret ex Born. et Flah. 211. *Homoeothrix juliana* (Menegh.) Krichn. 212. *Calothrix fusca* (Kütz.) Born. et Flah. 213. *C. fusca* var. *crassa* Rao, C.S. 214. *C. parietina* Thuret ex Born. et Flah. 215. C. *ghosei* Bharadwaja 216. *C. brevissima* West G.S. 217. *C. brevissima* var. *moniliforme* Ghose orth. mut. Geitler 218. C. *braunii* (A. Br.) Bornet et Flah. 219. *C. scytonemicola* Tilden 220. *C. castellii* (Massal.) Born. et Flah. 221. *C. castellii* var. *somastipurense* Rao 222. *C. elenkinii* Kossinskaja 223. *C. membranacea* Shmidle 224. C. *marchica* Lemm. 225. *Gloeotrichia natans* Rabenhorst ex Born. et Flah. 226. *G. raciborskii* Woloszynska 227. *G. raciborskii* var. *kashiense* Rao, C.B. 228. *G. intermedia* (Lemm.) Geitler 229. *G. inermedia* var. *kanwaensis* Rao, C.B. 230. *G. pisum* Thuret ex Born. et Flah. 231. *G. pilgeri* Schmidle 232. *G. ghosei* Singh, R.N.

(E) Order STIGONEMATALES:

233. *Mastigocladus laminosus* Cohn 234. *Hapalosiphon welwitschii* West, W. & G.S. 235. *H. fontinalis* (Ag.) Bornet 236. *H. luteolus* West, W. & G.S. 237. *H. intricatus* West, W. & G.S. 238. *Westiellopsis prolifica* Janet 239. *Camptylonema indicum* Schmidle 240. *Stigonema ocellatum* (Dillw.) Thuret ex Born. et Flah.

Successional Pattern

The successional pattern of rice field blue green algae are dependent on a number of factors viz. climatic condition of the area, nature of soil substratum, stages of crop growth and methods of crop management practice. The maximization of algal biomass productivity varies widely in different climatic zoneș. In Japan, highest algal biomass in rice fields has been observed within 2 weeks to 4 weeks (Ichimura, 1954; Kurasawa, 1965) before transplantation. Identically, maximum algal biomass in rice fields has been recorded just before tillering in Ukraine (Prikhod Kova, 1968) between tillering and panicle initiation in Senegal (Roger and Raynaud, 1976); after panicle initiation in some parts of India (Gupta, 1966); after harvesting during the wet season and after heading during the dry season in the Phillippines (Watanabe et al. 1977, 1978). In the present study identical survey was conducted. The survey revealed that fact that climax stage of BGA colonisation was noticed during tillering phase to panicle initiation stage. The detailed accounts of dominant blue-green algal succession in rice fields (aman-cultivation) of West Bengal is mentioned in Table 3.2.

Table 3.2: Dominant Blue-Green Algal Succession in Rice Fields of West Bengal

Stage of rice cultivation	*Dominant species*
(A) Beginning of wet season (June-July)	1. *Anabaena khannae*, 2. *A torulosa*, 3. *A. volzii*, 4. *Aulosira prolifica*, 5. *Cylindrospermum indicum*, 6. *C. muscicola*, 7. *Lyngbya allorgei*, 8. *L. lagerheimii*, 9. *Nostoc paludosum*, 10. *N. humifusum*, 11. *N. calcicola*, 12. *N. comminutum*, 13. *Oscillatoria acuta*, 14. *O. brevis* var. *non-granulata*, 15. *O. schultzii*, 16. *Porphyrosiphon notarisii*, 17. *Schizothrix muelleri*, 18. *S. friesii*, 19. *Scytonema simplex*.
(B) Puddling stage (July-August)	1. *Anabaena khannae*, 2. *Cylindrospermum indicum*, 3. *Nostoc paludosum*, 4. *N. calcicola*, 5. *Oscillatoria acuta*, 6. *Schizothrix friesii*, 7. *Scytonema simplex*.
(C) Tillering and panicle initiation stage (September November)	1. *Anabaena khannae*, 2. *A. orientalis*, 3.*A. volzii*, 4. *Aphanocapsa roeseana*, 5. *Aulosira prolifica*, 6. *A. pseudoramosa*, 7. *A. fertilissima*,

(Contd...)

Stage of rice cultivation	*Dominant species*
	8. *Nostoc calcicola,* 9. *Cylindrospermum indicum,* 10. *Gloeotrichia natans,* 11. *G. raciborskii,* 12. *G. intermedia* var. *kanwaensis,* 13. *Lyngbya hieronymusii,* 14. *L. laxespiralis,* 15. *L. allorgei,* 16. *L. birgei,* 17. *L. martensiana,* 18. *Microcoleus paludosus,* 19. *M. subtorulosus,* 20. *Oscillatoria acuta,* 21. *O. earlei,* 22. *O. brevis,* 23. *O. okeni,*24. *Plectonema notatum,* 25. *Scytonema simplex,* 26. *S. coactile,* 27. *Spirulina major.*
(D) Post-harvest phase (December-January)	1. *Anabaena volzii,* 2. *Aphanocapsa roeseana,* 3. *A. banaresensis,* 4. *Aulosira prolifica,* 5. *Chroococcus minutus,* 6. *Hapalosiphon luteolus,* 7. *H. welwitschii,* 8. *Lyngbya allorgei,* 9. *L. hieronymusii,* 10. *Microcoleus paludosus,* 11. *Nostoc linckia* var. *arvense,* 12. *Plectonema notatum,* 13. *Phormidium molle,* 14. *P. ambiguum,* 15. *P. corium,* 16. *P. jadinianum,* 17. *P. bohneri,* 18. *Scytonema hofmanni,* 19. *S. fremyii,* 20. *S. mirabile,* 21. *Symploca muralis,* 22. *Schizothrix telephoroides,* 23. *S. friesii,* 24. *Stigonema ocellatum.*

It is further interesting to note the fact that some blue-green algae species were present almost all through and some appeared only during late stages of rice cultivation. Species of *Gloeotrichia, Phormidium* and *Hapalosiphon* colonised in latter phases, while the species of *Cylindrospermum* and *Porphyrosiphon* colonised in earlier stages.

A total fortytwo (42) genera and 240 Taxa of BGA are listed in the Table 3.3 Among the different genera highest number of species, varieties and forma were reported from the genera like *Oscillatoria* (29), *Phormidium* (24), *Anabaena* (21), *Lyngbya* (17), *Scytonema* (16), *Nostoc* (14), Calothrix (13), *Cylindrospermum* (9) *Gloeotrichia* (8), *Aphanocapsa* (8), *Gloeocapsa* (7), *Aulosira* (7) and Schizothrix (7).

Table 3.3: Detailed Checklist of Rice-field BGA Species

Orders	*Families*	*Genus*	*Species*
I. CHROOCO-CCALES	(A) Chrooco-ccaceae	1. *Microcystis* (4)	*M. elongata, M. aeruginosa, M. flos-aquae, M. bengalensis.*
		2. *Chroococcus* (3)	*C. minutus, C. pallidus, C. cohaerens*

(Contd...)

Orders	Families	Genus	Species
		3. *Gloeocapsa* (7)	*G. polydermatica, G. calcarea, G. decorticans, G. rupestris, G. montana, G. luteofusca, G. crepidinum*
		4. *Gloeothece* (4)	*G. samoensis, G. membranacea, G. rupestris, G. rupestris* var. *maxima.*
		5. *Aphanocapsa* (8)	*A. roeseana, A. roeseana* forma *major, A. banaresensis A. biformis, A. littoralis, A. pulchra, A. themalis, A. grevillei*
		6. *Aphanothece* (5)	*A. bullosa, A. microscopica, A. pallida, A. stagnina, A. naegelii.*
		7. *Merismopedia* (2)	*M. convoluta, M. tenuissima*
	(B) Entophysalidaceae	8. *Johannes-baptistia* (1)	*J. pellucida*
		9. *Chlorogloea* (1)	*C. fritschii*
II. CHAMAESIPHONALES	(A) Dermocarpaceae	10. *Stichosiphon* (1)	*S. sansibaricus*
III. PLEUROCAPSALES	(A) Pleurocapsaceae	11. *Myxosarcina* (1)	*M. spectabilis*
IV. NOSTOCALES	(A) Oscillatoriaceae	12. *Arthrospira* (5)	*A. jenneri, A. massartii, A. massartii* var. *indica, A. gomontiana, A. platensis v. tenuis*
		13. *Spirulina* (4)	*S. major, S. subsalsa, S. princeps, S. meneghiniana*
		14. *Oscillatoria* (29)	*O. annae, O. subbrevis, O. subbrevis* forma *major, O. acuta, O. brevis* var. *non-granulata, O. earlei, O. acuminata, O. schultzii, O. okeni, O. amphibia, O. obscura, O. princeps, O. curviceps, O. pseudogeminata, O. pseudogeminata*

(Contd...)

Orders	Families	Genus		Species
				var. *unigranulata*, *O. amphigranulata*, *O. proteus*, *O. quadripunctulata*, *O. anguina*, *O. anguina* forma, *O. proboscidea*, *O. chalybea*, *O. amoena*, *O. formosa*, *O. splendida*, *O. animalis*, *O. tenuis*, *O. raoi*, *O. limosa*
		15. *Trichodesmium*	(1)	*Trichodesmium thiebautii.*
		16. *Crinalium*	(1)	*C. magnum*
		17. *Porphyrosiphon*	(1)	*P. notarisii*
		18. *Phormidium*	(24)	*P. corium*, *P. corium var. capitatum*, *P. coriumu. major* *P. ambiguum*, *P. ambiguum* var. *major*, *P. fragile*, *P. bohneri*, *P. jadinianum*, *P. molle*, *P. rotunda* *P. laminosum*, *P. tenue*. *P. purpurascens*, *P. valderianum*, *P. stagnina*, *P. papyraceum*, *P. subfuscum*, *P. angustissimum*, *P. foveolarum*, *P. anomala*, *P. favosum*, *P. retzii*, *P. lucidum*, *P. uncinatum*.
		19. *Lyngbya*	(17)	*L. laxespiralis*, *L. spiralis*, *L. lagerheimii*, *L. lagerheimii* forma *lacerata*, *L. birgei*, *L. ceylanica*, *L. ceylanica* var. *bankurensis*, *L. confervoides* *L. majuscula*, *L. martensiana*, *L. hieronymusii*, *L. allorgei*, *L. cryptovaginata*, *L. stagnina*, *L. major*, *L. perelegans*, *L. aerugineo-coerulea*.
		20. *Schizothrix*	(7)	*S. telephoroides*, *S. friesii*, *S. muelleri*, *S. penicillata*, *S. lacustris*, *S. arenaria*, *S. lamyi*.

(Contd...)

Orders	Families	Genus		Species
		21. *Symploca*	(1)	*S. muralis*
		22. *Microeoteus* (5)		*M. chthonoplastes, M. subtorulosus, M. vaginatus, M. lacustris, M. paludosus.*
		23. *Hydrocoleum* (1)		*H. lyngbyaceum*
	(B) Nostocaceae	24. *Cylindrospermum*	(9)	*C. muscicola, C. muscicola* var. *longispora, C. stagnale, C. licheniforme, C. majus, C. indicum, C. alatosporum, C. michailouskoense, C. sphaerica*
		25. *Wollea*	(1)	*W. bharadwajae*
		26. *Nostoc*	(14)	*N. humifusum, N. calcicola, N. linckia,* var. *arvense N. paludosum, N. piscinale, N. punctiforme, N. hatei, N. muscorum, N. ellipsosporum, N. comminutum, N. depressum, N. verrucosum, N. commune, N. spongiaeforme*
		27. *Anabaena*	(21)	*A oryzae, A. khannae, A. naviculoides, A. fertilissima, A. ambigua, A. utermohlii, A. oscillarioides, A. iyengarii, A. iyengarii* var. *tenuis, A. iyengarii* var. *attenuata, A. orientalis, A. torulosa, A. fuellebornii, A. volzii, A. spiroides, A. gelatinicola, A. vaginicola, A. vaginicolas f. fertilissima A. variabilis, A. variabilis* var. *ellipsospora, A. circinalis* var. *crassa*
		28. *Pseudanabaena*	(1)	*P. schmidlei*
		29. *Nodularia*	(1)	*N. spumigena*
		30. *Aulosira*	(7)	*A. aenigmatica, A. pseudoramosa, A. prolifica, A. fritschii, A. fertilissima, A.*

(Contd...)

Orders	Families	Genus	Species
			fertilissima var. *tenuis*, *A. implexa*
	(C) Scytonem-ataceae	31. *Plectonema* (3)	*P. notatum*, *P. tomasinianum*, *P. nostocorum*.
		32. *Scytonema* (16)	*S. simplex*, *S. coactile*, *S. fritschii*, *S. hofmanni*, *S. fremyii*, *S. mirabile*, *S. schmidtii*, *S. caldarium*, *S. javanicum*, *S. leptobasis*, *S. burmanicum*, *S. guyanense*, *S. stuposum*, *S. cincinnatum*, *S. myochrous*, *S. tolypothrichoides*.
		33. *Tolypothrix* (2)	*T. nodosa*, *T. tenuis*.
	(D) Micro-chaetaceae	34. *Microchaete* (2)	*M. uberrima*, *M. tenera*.
	(E) Rivula-riaceae	35. *Homoeothrix* (1)	*H. juliana*
		36. *Calothrix* (13)	*C. fusca*, *C. fusca* var. *crassa*, *C. parietina*, *C. ghosei*, *C. brevissima*, *C. brevissima* var. *moniliformis*, *C. braunii*, *C. scytonemicola*, *C. castellii*, *C. castellii* var. *somastipurense*, *C. elenkinii*, *C. membranacea*, *C. marchica*.
		37. *Gloeotrichia* (8)	*G. natas*, *G. raciborskii*, *G. raciborskii* var. *kashiense*, *G. intermedia*, *G. intermedia* var. *kanwaensis*, *G. pisum*, *G. pilgeri*, *G. ghosei*.
V. STIGONE-MATALES	(A) Mastigo-cladaceae	38. *Mastigocladus* (1)	*M. laminosus*
	(B) Stigonem-ataceae	39. *Hapalosiphon* (4)	*H. welwitschii*, *H. fontinalis*, *H. luteolus*, *H. intricatus*
		40. *Westiellopsis* (1)	*W. prolifica*
		41. *Camptylonema* (1)	*C. indicum*
		42. *Stigonema* (1)	*S. ocellatum*

Taxonomic Descriptions
(With Keys)

Class : CYANOPHYCEAE Sachs

Order : CHROOCOCCALES Wettstein

Family : CHROOCOCCACEAE Nägeli

Key to the Genera

1. Cells arranged in regular transverse and longitudinal rows	..	*Merismopedia*
1. Cells not so arranged	..	2
2. Cells without individual envelop or sheath, mostly planktonic	..	*Microcystis*
2. Cells with or without individual envelops or sheaths, may be subaerial, aerial, or planktonic	..	3
3. Cells spherical	..	4
3. Cells elongate or ellipsoidal	..	6
4. Sheath vesicular	..	*Gloeocapsa*
4. Sheath not vesicular	..	5
5. Thallus mostly microscopic sometimes macroscopic, sheath of individual cells generally distinct	..	*Chroococcus*
5. Thallus mostly macroscopic, sheath of individual cells generally indistinct	..	*Aphanocapsa*
6. Colonial mucilage not homogeneous individual sheath vesicular	..	*Gloeothece*
6. Colonial mucilage generally homogeneous, individual sheath not vesicular	..	*Aphanothece*

Genus: MICROCYSTIS Kützing
Key to the Species

1. Colony many times longer than broad	..	3
1. Colony round or slightly longer than broad	..	2
2. Colony clathrate, colonial mucilage distinct, hyaline	..	*M. aeruginosa*
2. Colony not clathrate, colonial mucilage indistinct	..	*M. flos-aqueae*
3. Sheath thick and stratified	..	*M. bengalensis*
3. Sheath hyaline	..	*M. elongata*

1. *Microcystis elongata* Desikachary (Plate 1, figs. 5 & 8) Cyanophyta, 89, pl. 18, f. 7 & 8, 1959

Colony many times longer than broad, colonial mucilage hyaline, sometimes lamellated, cells spherical 3.5.-5 μ diam., with gasvacuoles.

From stagnant water of paddy field, Onda (Bankura)

Specimen No. Ban 34

Reported earlier from Madras (Desikachary, 1959) Patiala (Sarma & Kanta, 1978) & W.B. (Sinha & Mukhejee 1975).

2. *Microcystis aeruginosa* Kütz. (Plate 1, figs, 3 & 7 and Plate 38A) Tab. Phycologicae, 1: 8, pl. 8, f. 1, 1846, Geitler, 137, f. 59d, 1932; Desikachary, 93, pl. 17, f. 1, 2, 6; pl. 18, f. 10, 1959.

Colony round or slightly longer than broad, clathrate, colonial mucilage hyaline, cells spherical 3.5-4.5 μ diam., with gas-vacuoles.

From stagnant water of paddy fields, 24-Parganas (South) and Bankura. Specimen No. 273, Ban 39, DH 9

Previously reported from Berhampur (Rao, C.B. 1938); Bhubaneswar (Mohanty 1984); Banaras (Bharadwaja 1935); from paddy fields in Lucknow and neighbourhood (Prasad & Mehrotra 1980); Allahabad (Pandey & Pandey 1982, Chadha & Pandey 1983a, b); Meerut (Bendre & Kumar 1975); Delhi (Rao, C.S., 1940); Maharashtra (Dixit 1936, Kamat, 1963); Karnataka (Gonzalves & Kamat 1958, Bongale & Bharati 1980); Jammu (Anand 1979); Punjab (Pandhol & Grover 1976, Sarma & Kanta 1978); Bihar (Nasar & Munshi 1976, Jha et al. 1986, Saha & Pandit 1987); North East India (Reddy et al. 1986); W.B. (Biswas 1926, 1927, Pal & Santra 1982, Sinha & Mukherjee 1984).

3. *Microcystis flos-aquae* (Wittr.) Krichner (Plate 1. fig. 1) In Engler & Prantl, Natürlichen Pflanzenfam., I. (1a): 56, f. 49 (N) 1898; Geitler, 138, f. 59e, f, 1932, Desikachary, 94, pl. 17, f. 11 & pl. 18. f. 11, 1959.

Colony spherical to ellipsoidal, not clathrate, colonial mucilage indistinct, cells spherical 3.5-5 μ diam., with gas-vacuoles.

From paddy fields of Bagnan (Howrah) and Joyrambati (Bankura). Specimen nos. H7, Ban20

Reported earlier from U.P. (Bharadwaja 1935, Bendre & Kumar 1975, Pandey & Pandey 1982, Chadha & Pandey 1983a, b); M.P. (from river Narmada Srivastava & Sen, 1987); Maharashtra (Dixit 1936, Kamat 1963, Barhate & Tarar 1983, Kolte & Goyal 1985); Tamil Nadu (Fremy 1942); Karnataka (Gonzalves & Kamat 1958, Somashekar 1983); Jaipur (Tridevy 1982); Bihar (Rao 1939); Assam (Bruhl & Biswas, 1926); Ludhiana (Pandhol & Grover 1976); W.B. (Banerjee 1936, Biswas 1942, Sinha & Mukherjee 1975b).

4. *Microcystis bengalensis* Banerji (Plate 1 fig. 9)

Myxo. of Lower Bengal, I, J. Indian bot. Soc., 15: 295, f. 2A & B, 1936; Desikachary, 89, pl. 19, f. 5 & 6, 1959.

Colony longer than broad, sheath thick and stratified, cells spherical 3.5-5.5 μ diam., with gas-vacuoles.

From paddy fields, Diamond Harbour (South 24-Parganas).

Specimen No. DH 12

Previously reported from W.B. (Banerjee 1936, Maity & Santra 1985, Santra et al. 1988).

Genus: CHROOCOCCUS Nägeli
Key to the Species

1.	Thallus microscopic, in groups of 2-4	..	*C. minutus*
1.	Thallus large	..	2
	2. Cells without shealth 4.5-6 μ in diam.	..	*C. pallidus*
	2. Cells without sheath 2-3.5 μ diam.	..	*C. cohaerens*

5. *Chroococcus minutus* (Kütz.) Näg. (Plate, 2, figs. 1 & 2)

Gatt. einzell. Algen 46, 1849; Geitler, 232, f. 112a, 1932; Prescott, 449, pl. 100, f. 9, 1951; Desikachary, 103, pl. 24, f. 4 & pl. 26, f. 4, 15, 1959.

Cells single or in groups of 2-4; pale blue-green in colour; sheath hyaline not lamellated; cells spherical, with sheath 9-14 μ diam. and without sheath 6-10.5 μ diam.

From wet soil of the rice fields; Contai, Garhbetta (Midnapore) and Falta (South 24-Parganas).

Specimen nos. G. 103, C 107, 61.

Previously reported from M.P. (from river Narmada, Srivastava & Sen 1987); Jammu (Anand 1979); Punjab (Pandhol & Grover 1976, Sarma et al. 1979); Maharashtra (Kamat 1963, Barhate & Tarar 1983, Kolte & Goyal 1985); Rajasthan (Gupta 1972); Karnataka (Gonzalves & Kamat 1958, Bongale & Bharati 1980, Somashekar 1983, 1984a,b, Hegde & Bharati 1983); Bhubaneswar (Mohanty 1984), W.B. (Sinha & Mukherjee 1984, Maity & Santra 1985, Pal & Santra 1985).

6. *Chroococcus pallidus* Nag. (Plate 1, fig. 2)

Gatt. einzell, Algen, 46, pl. 1, f. 2, 1849; Geitler, 238, f. 116b, 1932; Desikachary, 108, pl. 26, f. 5, 1959.

Thallus gelatinous, cells single or in groups of 2, sometimes 4, with sheath 6.5-8 μ diam. when single, and up to 12 μ diam. when in groups of two, without sheath each cells 4.5-6 μ diam., sheath unlamellated. From paddy fields of Uluberia (Howrah).

Specimen No. H 18.

Previously reported from U.P. (Rao 1937, Pandey & Chaturvedi 1979, from Rohilkhand, Prasad & Mehrotra 1980 from Lucknow, Chadha & Pandey 1983 a,b, from Allahabad); M.P. (Srivastava & Sen 1987); Punjab (Pandhol & Grover 1976); Maharashtra (Dixit 1936, Barhate & Tarar 1983); Karnataka (Bongale & Bharati 1980); Bhubaneswar (Mohanty 1984), A.P. (Rao 1938).

7. *Chroococcus cohaerens* (Breb.) Näg. (Plate 1, fig. 6)

Gatt. einzell, Algen., 46, 1849; Geitler, 238, f. 116e, 1932; Desikachary 111, pl. 26, f. 3 & 9, 1959.

Thallus gelatinous, cells in groups of 2-4 forming flakes like structures, cells without sheath 2-3.5 (-4) µ diam., with sheath 6-8 µ diam. when in groups two cells and 7.5-11 µ diam when in groups of four cells, sheath unlamellated.

From paddy fields of Moynapur (Bankura).

Specimen no. Ban 14

Previously reported from Maharashtra (Nandkar et al. 1983); Karnataka (Bongale & Bharati 1980, Somashekar 1983); W.B. (Sinha & Mukherjee, 1975b).

Genus: GLOEOCAPSA Kützing

Key to the Species

1.	Sheath colourless	..	2
1.	Sheath yellowish to brown coloured	..	5
	2. Sheath unlamellated, thallus with calcium impregnation	..	*Gl. calcarea*
	2. Sheath lamellated and without calcium impregnation	..	3
3.	Cells without sheath 7.5-8 µ broad	..	*Gl. decorticans*
3.	Cells without sheath upto 5 µ broad	..	4
	4. Cells without sheath 2-5 µ broad and with sheath 4-8 µ broad cell contents homogeneous	..	*G. montana*
	4. Cells without sheath 3-5 µ broad and with sheath 10-15 µ diam.	..	*Gl. polydermatica*
5.	Brackish water form, cells very closely grouped	..	*Gl. crepidinum*
5.	Fresh water form, not so closely arranged	..	6
	6. Cells without sheath 6-11 µ broad	..	*Gl. rupestris*
	6. Cells wihout sheath 3.5-4.5 µ broad	..	*Gl. luteofusca*

8. *Gloeocapsa polydermatica* Kütz. (Plate 2, fig. 4)

Tab. Phycologicae, 1: 15, pl. 20, f. 3, 1846; Geitler, 185, f. 82c, 83c, 1932; Desikachary, 114, pl. 25, f. 1, 1959.

Thallus mucilaginous, dark blue green colour, cells spherical single or in groups of 2-6, cells without sheath 3-5.5 µ diam. and with sheath 10-15 µ (-20 µ) diam., sheath colourless very thick, distinctly lamellated. From paddy field soils, Kolaghat (Midnapore), and Midnapore Sadar.

Specimen no. K5, 393.

Previously reported from Meerut, (Bendre & Kumar, 1975); Karnataka (Bongale & Bharati 1980, Somashekar 1984b); Ghaziabad (Pal 1975), W.B. (Sinha & Mukherjee 1975b).

9. *Gloeocapsa calcarea* Tilden (Plate 2, fig. 7)

Minn. Bot. Studies, ser. 2, 1; 29, 1898; Geitler, 187, f. 83a, 1932; Desikachary, 115, pl. 24, f. 6, 1959.

Thallus with distinct calcium incrustation; cells with or without individual sheath, without sheath 7-8.5μ diam., sheath thin.

From paddy field soils, Nurpur (South 24-Parganas) & Kolaghat (Midnapore)

Specimen nos. 49, K5.

Previously it was reported from Madras (Ganapathi 1940); Karnataka (Bongale & Bharati 1980) and W.B. (Maity & Santra 1985, Santra et al. 1988).

10. *Gloeocapsa decorticans* (A.Br.) Richter (Plate 2, fig. 5)

ex Wille, Nyt. Mag. Naturvid., 62; 186, 1925; Geitler, 184, f. 83b, 1932; Desikachary, 114, pl. 24, f. 9, 1959.

Thallus blue-green, cells spherical, solitary or in groups of 2; cells with sheath 15.5-18 μ broad and 18-22 μ long and without sheath 6-8 μ broad and 7-10 μ long; when in groups of two 17-20 μ broad and 28-31 μ long; sheath colourless and lamellated.

From paddy field, soils of Bishnupur (Bankura) and Bagnan (Howrah).

Specimen no. Ban 12, H.23.

Previously reported from U.P. (Mitra 1951, Pandey 1965, Pal 1975, Chadha & Pandey 1983a,b); Punjab (Pandhol & Grover 1976); Karnataka (Bongale & Bharati 1980, Somashekar 1983); W.B. (Gupta 1975).

11. *Gloeocapsa rupestris* Kütz. (Plate 2, fig. 12)

Tab. Phycologicae, 1; 17, pl. 22, f. 2, 1846; Geitler, 194, f. 88e, 89, 1932; Prescott, 452, pl. 107, 1951; Desikachary, 117, 1959.

Thallus brownish, colonies of 2-4 cells; cells without sheath 7-11μ diam.; sheath lamellated.

From paddy field soils, Roychawk (South 24-Parganas), Taki (24-Parganas, North)

Specimen No. 23.

Earlier reported from U.P. (Pal 1975, Pandey 1982); Karnataka (Bongale & Bharati 1980); Maharashtra (Bhoge & Ragothaman 1986); Coimbatore & Berhampur (Laloraya & Mitra 1974); W.B. (Banerjee 1936, Pal & Santra 1982, Maity & Santra 1985, Santra et al. 1988).

12. *Gloeocapsa montana* Kütz. (Plate 2, fig. 6)

Phyc. gene., 173, 1843; Tab. phycologicae, 1: 15, pl. 19, f. 11, 1846; Desikachary, 123, pl. 24, f. 14, 1959.

Thallus yellowish to blue-green, mucilaginous, cells spherical with sheath 4-10.5 µ broad and without sheath 3-4.5 µ broad, colonies 15-27 µ broad, sheath lamellated, cell contents homogeneous.

From paddy field soils, Nurpur (South 24-Parganas); Jalpaiguri, Darjeeling & Cooch Behar.

Specimen no. 50.

Previously recorded from Assam (Biswas 1934); North East India (Reddy et al. 1986); Maharashtra (Barhate & Tarar 1983); Bhoge & Ragothaman 1986); Karnataka (Gonzalves & Kamat 1958); W.B. (Banerji 1936, Maity & Santra 1985, Santra et al. 1988).

13. *Gloeocapsa luteo-fusca* Martens (Plate 2, fig. 8).

ex wille, Vorarbeiten zu einer Monographie der Chroococcaceen, Nyt. Mag. Naturvid., 62: 191, pl. 11, f. 25, 1925; Desikachary, 119, pl. 24, f. 16, 1959.

Thallus mucilaginous, cells spherical to subspherical, with slight lamellation; cells without sheath 3.5-4.5 µ broad and 5-7.5 µ long. From paddy field soils, Islampur (West Dinajpur), Beldanga (Murshidabad).

Specimen nos. W6, M38.

Here the alga is slightly broader the type specimen.

Previously reported from Darjeeling (Das et al. 1974) and Murshidabad (Pal & Santra 1982).

14. *Gloeocapsa crepidinum* Thuret (Plate 1, fig. 4)

in Bornet and Thuret, Notes Algologiques, 1 : 1, pl. 1, f. 1-3, 1932; Desikachary, 117, pl. 27, f. 4,5, 1959.

Thallus gelatinous, cells in groups 2-4 and more; colonies of 2 cells 7.5-10 µ diam. and colonies of 4 cells 7.5-15 µ diam., single cell 5-8 µ diam., sheath unlamellated.

From paddy fields, Kakdwip and Frazerganj (South 24-Parganas).

Specimen nos 58, 70.

Previously reported from Poona (Schmidle 1900); Karnataka (Bongale & Bharati 1980); Lucknow and its neighbourhood (Prasad & Mehrotra 1980); Nainital (Singh 1974); W.B. (Mukhopadhyaya & Chatterjee 1981).

Genus: GLOEOTHECE Nägeli

Key to the species

1. Thallus expanded, gelatinous, cells without sheath 4-5.5 µ broad	..	*Gl. membranacea*
1. Thallus not so expanded	..	2
2. Cells without sheath upto 10 µ long	..	*Gl. samoensis* var. *major*
2. Cells longer	..	3

3. Cells without sheath 7.5-9 μ broad .. *Gl. rupestris* var. *maxima*

3. Cells without sheath 4.5-6 μ broad .. *Gl. rupestris*

15. *Gloeothece samoensis* var. *major* Wille (Plate 2, fig. 9)

in Rechinger, Bot. and Zool., Ergebn. Samoa und Solomoninselin, Susswass-eralgen, 6, pl. 1, f. 4, 1915; Geitler, 220, 1932; Desikachary, 128, pl. 23, f. 6, 1959.

Cells ellipsoid, mostly single, sometimes 2-6 cells in a colony, cells without sheath 5-6.5 μ broad, 7.5-10 μ long; sheath unlamellated, or faintly lamellated colourless.

From paddy field soils of Falta (South 24-Parganas).

Specimen no. 26.

Previously reported from Madras (Laloraya & Mitra 1974); Patiala (Sarma et al. 1979); Maharashtra (Barhate & Tarar 1983); U.P. (Pandey 1982); and W.B. (Maity & Santra 1985, Santra et al. 1988).

16. *Gloeothece membranacea* (Rabenh.) Bronet (Plate 2, fig. 14)

Alg. de Schousb., Mem. Soc. nat. Sci. nat., Cherbourg, 28: 175, 1892; Geitler, 220, 1932; Desikachary, 128, 1959.

Thallus blue-green, expanded, gelatinous; cells ellipsoidal, without sheath 4-6.5 μ broad and 7.5-9.5 μ long; sheath colourless.

From paddy field soils of Mathurapur (South 24-Parganas), Memary (Burdwan), Bagmundi (Purulia).

Specimen no. 30, 130, 166.

Previously it was reported from Madras (Fremy 1942); Patiala (Sarma & Kanta 1978); Karnataka (Bongale & Bharati 1980); U.P. (Prasad & Mehrotra 1980); Kolhapur (Kamat 1963) and W.B. (Maity & Santra 1985, Santra et al. 1988).

The present alga comes closer to *G. membranacea* (Rabenh.) Bornet in most of the characters but differs from it in possessing slightly broader (4-6.5 μ instead of 4.5-5.5 μ) and longer cells (7.5-9.5 μ long instead of 7.5-8.8 μ).

17. *Gloeothece rupestris* (Lyngb.) Bornet (Plate 2, fig. 10)

Alg. de Schousb., Mem. Soc. Sci. Nat. Math. de Cherbourg, 28: 177, 1892; Geitler, 221, 1932; Desikachary, 127, pl. 25, f. 4, 1959.

Cells ellipsoidal, 2-4 in a spherical colony; colony 28-35 μ diam; cells with envelope 8-11 μ diam., 10-15 μ long, without envelop 4.5-6 μ broad, 9-12 μ long, envelope colourless.

From paddy fields of Joynagar (24-Parganas, South), Howrah, Hooghly, Bankura, Midnapore & Purulia.

Specimen nos. 175, 173, 181, 185, 192, 195.

Previously reported from Tamilnadu (Fremy 1942, Anand & Revathi 1987); U.P. (Prasad & Mehrotra 1980, Pandey 1982); Patiala (Sarma & Kanta 1978): J & K (Goyal et al. 1984); Maharashtra (Barhate & Tarar 1983, Bhoge & Ragothaman 1986).

18. *Gloeothece rupestris* (Lyngb.) Bornet var. *maxima* West (Plate 2, fig. 13)

J. Roy. micr. Soc: 743, 1892.

Desikachary, 127, 1949.

Thallus yellowish to blue-green, cells elongated, without sheath 7.5-9 μ broad and 14-17.5 μ long; mucilage thick; cells with sheath 16-21 μ broad and 25-30 μ long.

From paddy field soils, Kolaghat (Midnapore), Burdwan & Birbhum.

Specimen nos. K8, 101, 190.

Previously reported from Karnataka (Gonzalves & Kamat 1958, Bongale & Bharati 1980); Ghaziabad (Pal 1975); Maharashtra (Barhate & Tarar 1983). Kodaikanal of Tamil Nadu (Suxena 1983).

Genus: APHANOCAPSA Näeg.

Key to the species

1.	Thallus gelatinous, distinct sheath at the periphery of the thallus	..	*A. thermalis*
1.	Thallus gelatinous or mucilaginous, sheath not seen at the periphery of the thallus	..	2
	2. Cells broader than 5 μ	..	3
	2. Cells smaller than 5 μ	..	7
3.	Cells arranged singly	..	4
3.	Cells arranged singly or in twos	..	6
	4. Sheath thick, cells 5-6 μ in diameter	..	*A. banaresensis*
	4. Sheath mucilaginous	..	5
5.	Cells 5.5-8 μ diam.	..	*A. roeseana*
5.	Cells 7.6-10 μ diam.	..	*A roeseana* f. *major*
	6. Cells 6.3-7 μ diam.	..	*A. biformis*
	6. Cells 4.5-6.5 μ diam	..	*A. littoralis*
7.	Cells 3-4.5 μ diam., arranged singly or in twos	..	*A. pulchra*
7.	Cells 4-5 μ diam., arranged singly	..	*A. grevillei*

19. *Aphanocapsa roeseana* de Bary (Plate 3, figs. 1 & 2)

Hedwigia, 9: 74, 1870; Geitler, 157, 1932; Desikachary, 131, 1959.

Thallus large, blue-green colour; cells spherical , 5.5-8 μ in diam., sheath mucilaginous and homogeneous.

From paddy fields of Beldanga (Murshidabad), Silda, Contai (Midnapore), Canning (South 24-Parganas) in floating condition.

Specimen nos. M33, Sil 70a, C 117, Ca 14

Earlier this alga was reported from Madurai, Coimbatore & Hyderabad (Laloraya & Mitra 1974); U.P. (Chaturvedi & Pandey 1976, Prasad & Mehrotra 1980); Karnataka (Gonzalves & Kamat 1958, Bongale & Bharati 1980); Maharashtra (Kamat 1968, Bhoge & Ragothaman 1986); Patiala (Sarma & Kanta 1978); W.B. (Pal & Santra 1982, Pal et al. 1986, Santra et al. 1988, Sinha & Mukherjee 1984, Maity and Santra 1985).

20. *Aphanocapsa roeseana* de Bary forma *major* Laloraya et Mitra (Plate 3, fig. 3)

Nova Hedwigia, 268, pl. 1, f. 12, 1974.

Thallus dark blue-green colour, large; cells spherical, 7.6-10 μ in diam., mucilage sheath homogeneous.

From paddy field soils, Jhargram and Garhbeta (Midnapore).

Specimen nos. Jh C, G 124a, and G 124b.

The present alga resembles *Aphanocapsa roeseana* f. *major* Laloraya et Mitra in all respects but differs from the same in having slightly larger cells (7.6-10 μ diam. instead of 7-9 μ diam.).

Earlier, the alga was reported from paddy field soils of Coimbatore and Calcutta (Laloraya & Mitra 1974).

21. *Aphanocapsa banaresensis* Bharadwaja (Plate 3, fig. 4; and Plate 38B)

Myxophyceae of the United Provinces, Proc. Indian Acad. Sci., B, 2: 96, f. 1B, 1935; Desikachary, 133, pl. 22, f. 8, 1959.

Thallus light blue-green, mucilaginous, colony macroscopic sometimes colourless, upto 1.5 cm. in diam., cells spherical or oval, sometimes with individual sheath; cells without sheath 5-6 μ in diam. and with sheath 7.5-9 μ in diam., rarely cells with sheath upto 11.5 μ diam., sheath unstratified, hyaline.

From paddy field soils, Contai and Junpat (Midnapore).

Specimen nos. C9, J40, 389.

The present alga differs from *Aphanocapsa banaresensis* Bharadwaja in having slightly larger cells.

Previously it was reported from Benaras (Bharadwaja 1935); paddy fields of Lucknow and its neighbourhood (Prasad & Mehrotra 1980); paddy field soils of Berhampur of Orissa (Laloraya *et* Mitra 1974); Karnataka (Bongale & Bharati 1980, Somashekar, 1983); Punjab (Sarma *et al.*, 1979).

22. *Aphanocapsa biformis* A. Br. (Plate 3, fig. 6; and Plate 38C)

in Rabenhorst, Fl. Eur. Alg., 2: 246, 1865; Forti, 68, 1907; Geitler, 158, f. 70, 1932; Desikachary, 134, pl. 21, f. 3, 4, 1959.

Thallus expanding, gelatinous, very light blue-green colour; cells spherical, loosely arranged, 6.3-7 µ diam. sometimes two to four cells in a common sheath forming 11-16 µ diam. colony.

From moist soil of paddy field, Kolaghat (Midnapore).

Specimen no. K3

Earlier recorded from Allahabad (Chadha & Pandey 1983a, b); paddy fields of U.P. (Prasad & Mehrotra 1980); M.P. (Mishra & Purohit 1979); Karnataka Bongale & Bharati 1980, Somashekhar 1983); Vidarbha region of Maharashtra (Kolte & Goyal 1985); Nagpur (Barhate & Tarar 1982); Kolhapur (Kamat 1963); North East India (Reddy et al. 1986); Orissa (Mohanty 1982); Travancore (Parukutty 1940) and W.B. (Pal & Santra 1987).

23. *Aphanocapsa littoralis* Hansg. (Plate 3, fig. 7)

Beitr. Zur Kenntnis der Meeresalg. und Bact. Fl., Sitzb. K. Boehm. Ges. Wiss., Math. Na Cl., 229, 189. Geilter, 153, f. 66b, 1932; Desikachary, 131, pl. 21, f. 1, 1959.

Thallus amorphous, mucilaginous without having any definite shape, blue-green; cells 4.5-6.5 µ diam. spherical to subspherical; arranged singly or in twos.

From paddy fields of Asansol (Burdwan) on moist soil.

Specimen no. Bur 13.

Here cells are broader (4.5-6.5 µ diam. instead of 4-6 µ).

Earlier reported from Trivandrum (Parukutty 1940); Indian Ocean (Karsten 1907); Cape Comarin (Venkataraman 1957); Karnataka (Somashekar 1984b).

24. *Aphanocapsa pulchra* (Kutz.) Rabenh. (Plate 3, fig. 5).

Fl. Eur. Alg., 2: 49, 1865; Geitler, 155, f. 69g, 1932; Desikachary, 132, pl. 21, f. 2, 1959.

Thallus gelatinous, blue-green, cells spherical, 3-4.5 (-5) µ diam., single or in twos.

From paddy field soils, Diamond Harbour (24-Parganas) & Islampur (West Dinajpur); Jalpaiguri, Cooch Behar.

Specimen nos. DH 4, W9, 199, 201.

Previously reported from Maharashtra (Hof & Fremy 1932-33, Bhoge & Ragothaman 1986); U.P. (Rao, C.B. 1937, Pandey 1965, Singh 1974, Chadha & Pandey 1983a,b); Karnataka (Gonzalves & Kamat 1958, Somashekar 1983, 1984b); Punjab (Pandhol & Grover 1976, Sarma & Kanta, 1978; Tamil Nadu (Anand & Revathi 1987); Jaipur (Trivedy 1982); Puri (Rao, C.B. 1938) & W.B.

(Pal & Santra 1982, Sinha & Mukherjee 1984, Maity & Santra 1985, Santra et al. 1988).

25. *Aphanocapsa thermalis* (Kutz.) Brügg. (Plate 3, fig. 8)

Bundn. Alg., Jahresber, Naturf. des Graubundens, n.f., 8: 244, 1863; Geitler, 159, 1932; Desikachary, 135, 1959.

Thallus gelatinous, cell spherical to slightly ellipsoidal, 3-5 μ broad, cells densely arranged, mucilage homogeneous distinct sheath at the periphery of the thallus.

From paddy field soils, Diamond Harbour (24-Parganas), Birbhum, Malda.

Specimen nos. DH 7, 177, 181.

Present alga resembles *Aphanocapsa thermalis* (Kütz.) Brügg, but differs from the same in having slightly broader cells (3-5 μ instead of 2.5-4.2 μ).

Reported earlier from Bombay (Gonzalves, 1947); W.B. (Maity & Santra 1985, Santra et al. 1988).

26. *Aphanocapsa grevillei* (Hass.) Rabenh. (Plate 3, fig. 9 and Plate 38D)

Fl. Eur. Alg., 2: 50, 1865; Geitler, 159, f.71, 1932; Desikachary, 134, pl. 21, f.9, 1959.

Thallus gelatinous, pale blue-green; cells spherical, 4-5μ diam., with granulated contents.

From paddy field soils, Garhbeta (Midnapore).

Specimen no. G 15, 173.

Previously reported from Calcutta (Bruhl & Biswas 1923, Banerjee 1936); Benaras (Rao, C.B. 1937); Allahabad (Pandey & Pandey 1982); Karnataka (Gonzalves & Kamat 1958; Somashekar 1983); Panhala (Kamat 1961-62); Kolhapur (Kamat 1963); U.P. (Pandey 1965, Chaturvedi & Pandey 1976, Prasad & Mehrotra 1980); Punjab (Sarma et al. 1979); Bhubaneswar (Mohanty 1982, 1984).

Genus: APHANOTHECE Nägeli

Key to the Species

1.	Thallus spherical and large upto few cm. in diam.	..	4
1.	Thallus small	..	2
	2. Sheath sometimes distinct	..	3
	2. Sheath not distinct, cells 4-5.5 μ diam.	..	*A. naegelii*
3.	Sheath distinct only in the peripheral part of the thallus, cells without sheath 6-8 (-9) μ broad	..	*A. pallida*
3.	Sheath of the inner cells sometimes distinct, cells without sheath 4-5 μ broad	..	*A. microscopica*

4. With or without individual sheath, cells cylindrical, thallus greenish yellow .. *A. bullosa*
4. Without individual sheath, cells ovoid to cylindrical, thallus brownish colour .. *A. stagnina*

27. *Aphanothece bullosa* (Menegh.) Rabenh. (Plate 1, fig. 10)

Fl. Eur. Alg., 2: 65, 1865; Geitler, 172, 1932; Desikachary, 142, pl. 22, f. 12, 1959.

Thallus large, spherical, 5-10 cm. diam., greenish yellow colour, cells cylindrical, with or without individual sheath, without sheath 4-5 μ broad and 6-11 μ long.

From paddy fields of Sankrail (Howrah).

Specimen no. H 15, 157, 159, 166.

Previously reported from Benaras (Rao, C.B. 1937); Ballia (Pandey 1965); Allahabad (Mitra 1951; Chadha & Pandey 1983a,b); Berhampur of Orissa (Rao C.B. 1938); River Kapila of Karnataka (Somashekar 1984b); Mahabaleshwar of Maharashtra (Kamat 1962-63b).

28. *Aphanothece microscopica* Näg. (Plate 1, fig. 11).

Gatt. einzell. Algen, 59, pl. 1, H, 1949; Geitler, 172, f. 79, 1932; Desikachary, 142, pl. 22, f. 4, 5, 9, 1959.

Thallus gelatinous, small, blue-green colour, cells cylindrical, 4-5 μ broad and 7.5-10 μ long.

From paddy fields of Harmasra (Bankura); Diamond Harbour (South 24-Parganas); 24-Parganas North, Nadia, Murshidabad.

Specimen nos. Ban 40, DH 18, 172, 169, 145, 139

29. *Aphanothece pallida* (Kütz.) Rabenh. (Plate 3, fig. 10)

Fl. Eur, Alg., 2: 64, 1865; Geitler, 171, f. 78,1932; Desikachary, 140, pl. 22, f. 3, 1959.

Thallus blue-green to brownish colour, gelatinous, soft; cells elliptical or oblong, sometimes cells with sheath; sheath diffluent in the inner part; cells without sheath 6-8 (-9) μ broad and 10-12 μ long, and cells with sheath 8.5-11.5 μ broad and 10-15 μ long.

From paddy field soils of Jhargram, Midnapore Sadar and Asansol (Burdwan), Diamond Harbour (South 24-Parganas).

Specimen nos. J. 38, 393, A9, 71.

Earlier the alga was reported from Orissa (Rao, 1938b, Mohanty 1984); A.P. (Rao 1938a); Maharashtra (Gonzalves & Joshi 1946, Kamat 1961-62, 1963, Kamat & Harankhedkar 1976, Kolte & Goyal 1985); U.P. (Pandey 1965, Pandey & Chaturvedi 1979, Prasad & Mehrotra 1980); Bihar (Rao C.S. 1939); Karnataka (Gonzalves & Kamat 1958, Bongale & Bharati 1980, Somashekar 1984b); Punjab (Grover & Pandhol 1975, Sarma et al. 1979); Jammu &

Kashmir (Goyal et al. 1984); Tamil Nadu (Anand & Revathi 1987); W.B. (Sinha & Mukherjee 1975b, Sengar & Pandey 1985, Pal & Santra 1985, Pal et al. 1986).

30. *Aphanothece stagnina* (Spreng.) A. Braun (Plate 3, fig. 11)

In Rabenhorst, Fl. Eur. Alg., 2: 66, 1865; Geitler, 164,f. 72, 75a, b, 1932; Desikachary, 137, pl. 21, f. 10, 1959).

Thallus blue-green when young, becomes brownish when matured, irregularly lobed, 1-5 cm. diam., gelatinous; cells oblong 4.5-6.5 μ broad and 5.5-9 μ long, without individual sheath, densely arranged; mucilage homogeneous.

From paddy fields of Midnapore Sadar in free floating condition, Bankura, Howrah.

Specimen nos. 79, 121, 129.

Previously it was reported from Paddy field soils of Patna, Hyderabad, Berhampur (Laloraya & Mitra 1974); Karnataka (Bongali & Bharati 1980, Somashekar 1983); Bihar (Jha et al. 1986); Tamil Nadu (Anand & Revathi 1987); Punjab (Grover & Pandhol 1975, Pandhol & Grover 1976, Sarma et al. 1979) & W.B. (Banerji 1936, Sinha & Mukherjee 1975b).

31. *Aphanothece naegelii* Wartm. (Plate 3, fig. 12)

in Rabenhorst, Fl. Eur. Alg., 2: 65, 1865; Geitler, 172, 1932; Desikachary, 141, pl. 22, f. 7, 1959.

Thallus pale blue-green; cells 4-5.5 μ diam., and 6.5-8.5 μ long, sheath diffluent.

From paddy fields of Raghunathganj (Murshidabad), Uluberia (Howrah).

Specimen nos. M 7, H 10.

Here cells are slightly longer (4-5.5 μ x 6.5-8.5 μ instead of 3.5-4.5 x 6.5-8 μ).

Reported earlier from U.P. (Rao 1936, Pandey 1965, Prasad & Mehrotra 1980, Pandey 1982, Chadha & Pandey 1983a,b); Nagpur (Barhate & Tarar 1982); Kolhapur (Kamat 1963); Punjab (Pandhol & Grover 1976, Sarma et al. 1979); Bihar (Jha et al. 1986); Karnataka (Bongale & Bharati 1980); W.B. (Pal & Santra 1982, Sengar & Pandey 1985).

Genus: MERISMOPEDIA Meyen

Key to the Species

1. Thallus convolute, cells oblong, 4-5 μ broad .. *M. convoluta*
1. Thallus not convolute, cells sub-spherical, 1.5-2 μ broad .. M. *tenuissima*

32. *Merismopedia convoluta* Breb. (Plate 2, fig. 11)

in Kützing, Species Algarum, 472, 1849; Geitler, 262, 1932; Desikachary, 152, pl. 29, f. 8, 12 & 13, 1959.

Thallus blue-green, flat, often convolute, cells oblong, 4-5 μ broad, 6.5-9 μ long, arranged in single plane, in transverse and longitudinal rows.

From paddy fields, Bishnupur (Bankura), Purulia.

Specimen no. Ban 40, 127.

Previously reported from Maharashtra (Bhoge & Ragothaman 1986); Jammu (Anand 1979); Allahabad (Pandey & Pandey 1982); Karnataka (Somashekar & Ramaswamy 1982, Somashekar 1983, 1984b) & W.B. (from Hooghly river—Biswas 1942, and Bankura - Sinha & Mukherjee 1975b).

33. *Merismopedia tenuissima* Lemm. (Plate 2, fig. 3 and Plate 39A)

Beitr. Kennt. Planktonalg., II, Bot. Centralbl. 76, 154, 1898; Geitler, 263, f. 129a, b, 1932; Desikachary, 154, pl. 29, f. 7 & pl. 30, f. 8, 9.

Cells closely packed, 32-80 cells in a colony; cells subspherical, 1.5-2 μ broad.

From paddy fields of 24-Parganas (South & North), Nadia, Malda, Murshidabad.

Specimen nos. 107, 113, 122, 135, 139.

Previously reported from U.P. (Bendre & Kumar 1975); Maharashtra (Kamat 1963, Nandkar et al. 1983, Barhate & Tarar 1983, Bhoge & Ragothaman 1986); Jaipur (Tridevy 1982); Orissa (Rao 1938b); Delhi (Rao 1940); W.B. (Sinha & Mukherjee 1984).

Family: ENTOPHYSALIDACEAE Geitler

Key to the Genera

1. Cells arranged in a single row, thallus cylindrical .. *Johannesbaptistia*
1. Cells arranged in one or more radial rows, thallus not cylindrical .. *Chlorogloea*

Genus: JOHANNESBAPTISTIA J. de Toni

34. *Johannesbaptistia pellucida* (Dickie) Taylor et Drouet (Plate 4, fig. 1) Bull. Torrey bot. Cl., 65: 285, 1938; Desikachary, 165, pl. 32, f. 14-19, 1959.

Thallus filamentous, cells discoid, arranged in single row 4-6 μ broad and 2.5-4.5 μ long; mucilage homogeneous, firm, hyaline.

From paddy field soils, cannning (South 24-Parganas).

Specimen no. SP 14.

Cells of present alga are slightly larger than those of the type species (4-6 μ broad and 2.5-4.5 μ long instead of 3.9-5.2 μ x 2.6-3.9 μ).

Reported earlier from estuarine regions of Madras (Iyenger and Desikachary 1946, Rao, C.B. 1947); Karnataka (Bongale & Bharati 1980); and W.B. (Pal & Santra 1982, Maity & Santra 1985, Santra et al. 1988).

Genus: CHLOROGLOEA Wille

35. *Chlorogloea fritschii* Mitra (Plate 1, fig. 12)

Two new algae from Indian Soils, Ann. Bot., Lond., n.s., 14, 457, f. 25-43, 1950; Desikachary, 163, pl. 31, f. 1-16, 1959.

Cells rounded or slightly angular, arranged vertically and horizontally, single or in groups of 2 to many cell. Each cells 5-8 μ in diam.

From paddy field soils, Falta (South 24-Parganas), 24-Parganas (North), Midnapore, Howrah, Hooghly, Nadia.

Specimen nos. SP 22, 116, 123, 134, 145, 152.

Previously reported from Orissa (Rao C.B. 1938); U.P. (Rao C.B. 1936, Pandey 1965, Chaturvedi & Pandey 1976, Prasad & Mehrotra 1980, Pandey 1982, Pandey & Pandey 1982); Patiala (Sarma & Kanta 1978); Maharashtra (Schmidle 1900, Kamat 1963, Barhate & Tarar 1983, Bhoge & Ragothaman 1986); Bihar (Saha & Pandit 1987); Karnataka (Bongale & Bharati 1980, Somashekar 1983, Somashekar & Ramaswamy 1983) & W.B. (Sinha & Mukherjee 1975b, Pal et al. 1986).

Order : CHAMAESIPHONALES R.v. Wettstein

Family : DERMOCARPACEAE Geitler

Genus : *STICHOSIPHON* Geitler

36. *Stichosiphon sansibaricus* (Hieron.) Drouet et Daily (Plate 1, fig. 14)

Butler Univ. Studies, 10: 223, 1952; Rev. Coccoid Myxophyceae, 125, f. 253, 1956; Geitler, Kryptogamenflora, 412, f. 240, 1932, as *S. filamentosus* (Ghose) Geitler; Desikachary, 176, pl. 32, f. 9-13 & pl. 33, f. 8-10, 1959.

Plant microscopic, attached to the host, erect or curved; at first clavate but later on cylindrical, 6-7.5 μ broad, 20-90 μ long, sometimes upto 150 μ long; sheath thin, cell content divides into a series of endospores; spores spherical 4-5 μ (-5.5 μ) diam.

Epiphytic on a species of *Cladophora* growing on the shell of a snail in a paddy field, Midnapore Sadar.

Specimen no. 602.

Previously reported from Karnataka (Gonzalves & Kamat 1958 as *S. indica* Rao); Orissa (Rao 1938); U.P. (Gupta 1956, Pandey & Pandey 1982); Gujarat (Kamat 1967); Himachal Pradesh (Vasishta 1968); Tamil Nadu (Fremy 1942, Suxena 1983); Andaman and Nicobar (Prasad & Srivastava 1986); W.B. (Maity & Santra 1985).

Order : PLEUROCAPSALES Geitler

Family : PLEUROCAPSACEAE Geitler

Genus : MYXOSARCINA Printz.

37. *Myxosarcina spectabilis* Geitler (Plate 1, fig. 13)

Arch. Hydrobiol. (u. Planktonk.) 12(4); 624, 1923; Desikachary, 178, pl. 30, f. 1-5 & pl. 31, f. 17-22, 1959.

Plant in three dimensional colonies, cells 5-10 μ diam. colonial sheath hyaline.

From paddy fields of Digha (Midnapore), Diamond Harbour (South-24-Prga).

Specimen nos. C 38, DH 17.

Previously reported from U.P. (Prasad & Mehrotra 1980, Pandey & Pandey 1982, Chadha & Pandey 1983a,b); Karnataka (Somashekar 1983, 1984b); Patiala (Sarma et al. 1979) and W.B. (Mukhopadhyay & Chatterjee 1981, Santra et al. 1988).

Order : NOSTOCALES Geitler

Family : OSCILLATORIACEAE Kirchner

Key to the Genera

1. Trichomes flattened .. *Crinalium*
1. Trichomes not so .. 2
    2. Trichomes generally many in a sheath .. 3
    2. Trichomes without a sheath or single within a sheath .. 6
3. Trichomes twisted like a rope within the sheath .. *Microcoleus*
3. Trichomes not so twisted .. 4
    4. Sheath slimy, end cell with a calyptra .. *Hydrocoleum*
    4. Sheath more or less firm, end cell without a calyptra .. 5
5. Trichomes many, rarely single within the sheath, sheath not lamellated .. *Schizothrix*
5. Trichomes single, rarely two within the sheath, sheath lamellated and reddish brown .. *Porphyrosiphon*
    6. Trichomes without sheath .. 9
    6. Trichomes with a sheath .. 7
7. Sheath firm .. 8
7. Sheath more or less firm, partly diffluent, filaments forming gelatinous or leathery stratum .. *Phormidium*
    8. Filaments not in bundles .. *Lyngbya*
    8. Filaments forming tufts of fascicles .. *Symploca*
9. Trichomes more or less straight, not spirally coiled .. 11
9. Trichomes spirally coiled .. 10

10. Showing many coils, cross walls if present not clearly seen .. *Spirulina*
10. Showing comparatively fewer coils, cross walls clearly seen .. *Arthrospira*

11. Forming planktonic bundles .. *Trichodesmium*
11. Not in bundles .. *Oscillatoria*

Genus: ARTHROSPIRA Stizenberger
Key to the Species

1. Spirals narrower .. 4
1. Spirals broader .. 2
   2. Distance between two spirals generally less than 50 μ .. 3
   2. Distance greater .. *A. massartii*
3. Trichome 4.5-5.5 μ broad, end cell with calyptra .. *A. massartii* v. *indica*
3. Trichomes 5.5-6.5 μ broad, end cell rounded without calyptra .. *A. platensis* v. *tenuis*
   4. Breadth of spirals 6-8 μ, trichome 4-5 μ broad, cells 2-3.5 μ long .. *A. gomontiana*
   4. Breadth of spirals 10-15 μ, trichome 5-7 μ broad, cells 4-6 μ long .. *A. jenneri*

38. *Arthrospira jenneri* Stizenb. ex Gomont (Plate 4, fig. 2)

Gomont, Monogr. Oscillariees, 247, pl. 7, f. 26, 1892; Desikachary, 192, pl. 35, f. 3, 1959.

Thallus blue-green, trichome spirally coiled, at a regular distance, apex not attenuated, trichome unconstricted at the cross wall, 5-7 μ broad; breadth of spirals 10-15 μ; distance between two spirals 22-28 μ, cells are approximately as long as broad or slightly shorter, 4-6 μ, cross walls granulated.

From paddy field soils, Midnapore Sadar.

Specimen no. 396

Previously the alga was reported from Ghaziabad (Pal 1975); Allahabad (Pandey & Pandey 1982); Maharashtra (Barhate & Tarar 1983); Punjab (Grover & Pandhol 1975, Pandhol & Grover 1976); Bihar (Jha et al. 1986) W.B. (Pal & Santra 1987).

39. *Arthrospira massartii* Kuffareth (Plate 4, fig. 6)

Contribution a l'etude Flore Algologique du Luxembourg meridional, II, Ann. biol. lacustre, 7: 264, f. 16, 1914; Desikachary 191, pl. 35, f. 9, 10, 1959.

Trichome loosely coiled, spirals 25-28 μ broad, distance beween two spirals 75-90 μ; cells 5-6 μ broad and 2-4 μ long cross wall not constricted.

From paddy field soils, Kalyani (Nadia).

Specimen no. N 7.

Earlier the species was reported from W.B., Jaipur (Tridevy 1982).

40. *Arthrospira massartii* var. *indica* Desikachary (Plate 4, fig. 4)

Desikachary, Cyanophyta, 191, pl. 35, f. 7 & 8, 1959.

Trichome 4.5-5.5 μ broad, spirals 35-40 μ broad, distance between two spirals 40-45 (-50) μ, cross wall not constricted, cells 2-3 μ long, end cell with calyptra.

From paddy fields Onda (Bankura).

Specimen no. Ban 11.

Previously reported from Madras (Desikachary 1959); W.B. (Sinha & Mukherjee 1975b).

41. *Arthrospira gomontiana* Setchell (Plate 4, fig. 5)

Bull. Torrey bot. Cl., 22: 430, 1895; Desikachary, 192, 1959.

Trichome regularly spirally coiled; distance between spirals 10-16.5 μ; spirals 6-8 μ broad, cells 4-5 μ broad and 2-3.5 μ long.

From paddy field soils, Diamond Harbour (South 24-Parganas).

Specimen no. 45.

Reported earlier from Ghaziabad (Pal 1975); from paddy fields of U.P. (Prasad & Mehrotra 1980) and W.B. (Maity & Santra 1985). The present alga has many resemblances with *Arthrospira gomontiana* Setchell, and it comes close to the var. *crassa* Gupta

42. *Arthrospira platensis* (Nordst.) Gomont

Monogr. Oscillariees, 247, pl. 7, f. 27; 1892.

var. *tenuis* (Rao, C.B.) Desikachary (Plate 4, fig. 3)

Desikachary, Cyanophyta, 190, pl. 35, f. 4 & 11, 1959.

Trichome 5.5-6.5 μ broad, spirals 32-41 μ broad and distance between two spirals 38-48 μ, cross wall not constricted end cell rounded, cells 2-3 μ long.

From paddy fields soils of Haldia (Midnapore).

Specimen no. Hal 11

Previously reported from U.P. (Rao C.B. 1937, Pandey & Chaturvedi 1979); Bihar (Rao C.S. 1939); Karnataka (Somashekar 1983). The present alga has many resemblances with the specimen of Darbhanga (Rao C.S. 1939).

Genus: SPIRULINA Turpin em. Gardner

Key to the Species

1. Spirals irregular .. *S. meneghiniana*

1. Spirals regualar .. 2
   2. Trichome 1.5-2 μ broad .. 3
   2. Trichome 4.5-5 μ broad .. *S. princeps*
3. Spirals 3.8-5 μ away from each other .. *S. major*
3. Spirals densely coiled, not more than 2 μ away from each other .. *S. subsalsa*

43. *Spirulina major* Kütz. ex Gomont (Plate 4, fig. 7)

Kutzing, Phyc. gene. 183, 1843; Gomont, Monogr. Oscillariees, 251, pl. 7, f. 29; 1892; Geitler, 930, f. 595, 1932; Desikachary, 196, pl, 36, f. 13, 1959.

Trichome 1.5-2 μ broad; spirals 3-4.5 μ broad and 3-5 μ distant; regularly coiled and blue-green colour.

From paddy field soils, Haldia and Kolaghat (Midnapore), Kalyani (Nadia).

Specimen nos. Hal 1, Hal 10, Hal 11, K 14, N 16.

The present alga differs from the type species in having slightly broader spirals.

Earlier the alga was reported from Maharashtra (Kamat 1963, Bhoge & Ragothaman 1986); U.P. (Rao C B, 1937, Gupta 1956, Bendre & Kumar 1975, Pal 1975, Chadha & Pandey 1983a); W.B. (Biswas 1926, Banerjee 1938, Sinha & Mukherjee 1975a, Pal & Santra 1985, Santra & Pal 1988, Santra et al. 1988); Bihar (Saha & Pandit 1987); A.P. (Rao C.B. 1938, Laloraya & Mitra 1975); Ahmedabad (Kamat 1962-63); Kolhapur (Kamat 1963); Nagpur (Nandkar et al. 1983); Cauvery of Karnataka (Somashekar 1983), Karnataka (Somashekar 1984a, b, Somashekar & Ramaswamy 1983).

44. *Spirulina subsalsa* Oerst. ex Gomont (Plate 4, figs, 8 & 9; Plate 39B)

Gomont, Monogr. Oscillariees, 253, pl. 7, f. 32, 1892; Geitler, 927, f. 593a, 1932; Desikachary, 193, pl. 36, f. 3, 9, 1959.

Trichome 1.5-2 μ broad, densely spirally coiled, sometimes loosely coiled, spirals 4.5-5.5 μ broad; colour blue-green.

From paddy field soils, Haldia and Narajole (Midnapore).

Specimen nos. Hal 14, Hal 7, 43.

Here spirals slightly broader.

Previously the species was recorded from W.B. (Banerji 1938); Orissa (Rao C.B. 1938); Assam (Parukutty 1939); Maharashtra (Kamat 1975); Karnataka (Bonagale & Bharati 1980, Somashekar 1984); A.P. (Rao C.B. 1938); Ghaziabad (Pal 1975); Bihar (Sinha & Srivastava 1980, Jha et al. 1986, Saha & Pandit 1987); North East India (Reddy et al. 1986); Punjab (Pandhol & Grover 1976).

45. *Spirulina princeps* W. et G.S. West (Plate 4, fig. 10)

A contribution to the fresh water algae of Ceylon, Trans. Linn. Soc. (Lond.), Bot., 2 ser., 6: 205, 1902; Geitler, 931, f. 593d, 1932; Desikachary, 197, pl. 36, f. 7, 1959.

Trichome blue-green, regularly spirally coiled, trichome 4-5.5 μ broad, spirals 10.5-12.5 μ broad and 10-11 μ distant.

From paddy field soils, Narajole (Midnapore)

Specimen no. 43.

Previously it was reported from Jammu (Anand 1979); Patiala (Sarma & Kanta 1978); M.P. (Roy & Sen 1985); U.P. (Bendre & Kumar 1975, Pal 1975, Prasad & Mehrotra 1980); Maharashtra (Kamat 1968, 1975; Kolte & Goyal 1985, Barhate & Tarar 1983); W.B. (Maity & Santra 1985, Santra et al. 1988).

46. *Spirulina meneghiniana* Zanard. ex Gomont (Plate 4, fig. 16)

Gomont, Monogr. Oscillariees, 250, pl. 7, f. 28, 1892; Geitler, 928, f. 593d, 1932; Desikachary, 195, pl. 36, f. 8, 1959.

Trichome 1.5-2 μ broad, irregularly coiled, spirals 3-5 μ broad, distance between spirals ranges from 3-5 μ.

From paddy fields, Onda (Bankura)

Specimen no. 80

Previously reported from Maharashtra (Kamat 1968, 1975; Nandkar 1983); Ahmedabad (Kamat 1962-63); Ghaziabad (Pal 1975); Allahabad (Pandey & Pandey 1982); Tamil Nadu (Anand & Revathi 1987); Cauvery of Karnataka (Somashekar 1983); River Kapila of Karnataka (Somashekar 1984b); Bihar (Sinha & Srivastava 1980) & W.B. (Banerji 1938).

Genus : OSCILLATORIA Vaucher
Key to the species

1. Trichomes constricted at the cross walls .. 2
1. Trichomes not constricted at the cross walls .. 10
    2. Trichomes attenuated .. 3
    2. Trichomes not distinctly attenuated .. 6
3. Trichomes 4-5 μ broad at the middle, gradually attenuated, cross wall slighly constricted and granulated, calyptra present .. *O. amoena*
3. Trichomes attenuated only at the end, calyptra absent .. 4
    4. Trichome 2-3 μ broad, cross-wall granulated .. *O. schultzii*
    4. Trichome broader .. 5
5. Cross-wall not granulated, 5-6 μ broad, cells 2-3.5 μ long .. *O. okeni*
5. Cross-wall granulated, 4-6 μ broad, cells 2.5-5 μ long .. *O. formosa*

6. Trichomes broader than 5 μ, cells subquadrate .. 7
6. Trichomes narrower, 1.5-2 μ broad, cells longer than breadth, 4-5 μ long, one gas vacuole on either side of the septum .. *O. amphigranulata*

7. End cell rounded .. 8
7. End cell rounded conical, sometimes trichome may be very slightly attenuated at ends, 6-7 μ broad, cells 2.5-4 μ long .. *O. proteus*
8. Cross-wall slightly constricted, septa granulated, end cell with outer thickened membrane, trichome 5-8 μ broad .. *O. tenuis*
8. Cross-wall distinctly constricted, septa not granulated, end cell without outer thickened membrane .. 9
9. Trichome 8-12 μ broad, cells 3.5-5.5 μ long, end cell slightly bent .. *O. chalybea*
9. Trichome 7-8.5 μ broad, cells 2-3.5 μ long .. *O. annae*
10. End cell capitate .. 11
10. End cell not capitate .. 15
11. Trichomes upto 8 μ broad, ends distinctly attenuated .. 12
11. Trichomes broader .. 14
12. Trichomes 6-8 μ broad, cells upto 1/3 as long as broad, end cell will thickened membrane or with calyptra .. 13
12. Trichomes 2-3 μ broad, cells longer than broad, trichome gradually attenuated at the ends, ends screw like, sickle shaped .. *O. splendida*
13. Not constricted at the cross-wall .. *O. anguina*
13. Slightly constricted at the cross-wall .. *O. anguina* forma

14. Ends of trichomes distinctly attenuated, end cells flatly rounded, with thick membrane trichome 11-14.5 μ broad .. *O. proboscidea*
14. Ends of trichomes slightly attenuated, trichome 24-48 μ broad .. *O. princeps*
15. Trichome not attenuated at the apex .. 16
15. Trichome attenuated at the apex .. 24

16. Cells shorter than broad	..	17
16. Cells longer than broad	..	21
17. Cells upto 1/3 as long as broad	..	18
17. Cells longer 2.5-5 μ long, trichome 5-6 μ broad, cross-wall indistinct but marked by the presence of granules on either side of the septum	..	*O. raoi*
18. End cell flatly rounded with outer thickened membrane, cross-wall granulated	..	*O. limosa*
18. End cell not so	..	19
19. Trichome more or less straight cross-wall not granulated	..	20
19. Trichomes curved, ends rounded trichomes 10.5-13 μ broad, cells 2.5-4.5 μ long, cross-wall sometimes granulated	..	*O. curviceps*
20. Trichomes 6-8 μ broad, cells 1.5-2 μ long	..	*O. subbrevis*
20. Trichomes 10-11 μ broad, cells 2.5-3 (-4) μ long	..	*O. subbrevis* f. *major*
21. Trichomes 1.5-2 μ broad, cross-wall indistinct, maked by a pair of granules on either side	..	*O. quadripunctulata*
21. Trichomes broader, cross-wall not so	..	22
22. Trichomes 1.5-3 μ broad, cells upto 4 μ long, cross-wall thick	..	23
22. Trichomes 2.5-3.5 μ broad, cells 4-7 μ long, cross-wall not so thick, one granule on either side of the septum	..	*O. amphibia*
23. Cross-wall not granulated cells 1.5-2.5 μ broad & 2-3 μ long	..	*O. pseudogeminata*
23. One granule on either side of septum, cells 2-3 μ broad, 3-4 μ long	..	*O. pseudogeminata* v. *unigranulata*
24. End cells rounded, apex of the trichome slightly attenuated, 4-5 μ broad, cells 1.5 μ long	..	*O. obscura*
24. End cells conical or subobtuse, trichome distinctly attenuated	..	25
25. Cells quadrate to shorter than broad	..	26
25. Cells quadrate to longer than broad	..	28

26. Cells upto 1/2 as long as broad; trichome 4.5-5 µ broad, cells 2-2.5 µ long, end cell conical and bent .. *O. brevis* var. *non-granulata*

26. Cells may be longer .. 27

27. Trichomes sometimes aggregated into bundles, arranged in parallel, trichomes 4-5.5 µ broad, cells 2.3-3.5 µ long, ends subobtuse, straight or bent aside .. *O. acuta*

27. Trichomes not in aggregated bundles, 3-4 µ broad, cells 2-4.5 µ long, ends conical, pointed and bent .. *O. animalis*

28. Trichomes 2.5-3 µ broad, cells 2.5-5 µ long, end cell prominently pointed .. *O. earlei*

28. Trichomes 3-5 µ broad, cells 5-8 µ long, end cell mucronate .. *O. acuminata*

47. *Oscillatoria annae* van Goor (Plate 4, figs. 12 & 13 and Plate 39C)

Zur Kenntnis des Oscillatoriaceen, Rec. Trav. bot. neerland., 15: 289, pl. 2, f. la-c, 1918, Geitler, 943, 1932; Desikachary, 203, pl. 38, f. 13, 1959.

Thallus dull blue-green; trichome straight, slightly constricted at the cross-walls, 7-8.5 µ broad; cells 2-3.5 µ long, endcell rounded, calyptra absent.

From soils and stagnant water of paddy field, Panskura (Midnapur). Specimen no. 255.

The present alga resembles *Oscillatoria annae* van Goor but differs from the same in having slightly broader trichomes (7-8.5 µ instead of 7.5-8 µ).

Previously reported from Maharashtra (Kamat 1963, 1968, Kolte & Goyal 1985); Karnataka (Somashekar 1984b); Punjab (Sarma & Kanta 1978); U.P. (Chadha & Pandey 1983a, Khan 1985); Ahmedabad (Kamat 1962-63) & W.B. (Pal & Santra 1985, Santra et al. 1989).

48. *Oscillatoria subbrevis* Schmidle (Plate 4, fig. 11)

Engler's Bot. Jahrb., 30: 243, pl. 4, f. 7, 1901; Geitler, 949, f. 601b, 1932; Prescott, 491. pl. 107, f. 23, 1951; Desikachary, 207, pl. 37, f. 2 & pl. 40, f. 1, 1959.

Trichomes solitary, 6-8 µ broad, not attenuated at the apices; cells 1.5-2 µ (-3 µ) long, not granulated at the cross-walls, end cell rounded.

From water logged paddy field, Diamond Harbour (South 24-Parganas), Balurghat (West Dinajpur), Jalpaiguri.

Specimen no. 27, Bol 9, Jal 12.

Reported earlier from Maharashtra (Barhate & Tarar 1982, Nandkar et al. 1983, Kolte & Goyal 1985); J & K (Goyal et al. 1984); Punjab (Grover & Pandhol 1975, Pandhol & Grover 1976, Sarma & Kanta 1978); Karnataka (Bongale & Bharati 1980, Somashekar & Ramaswamy 1983, Somashekar 1984b); Tamil Nadu (Anand & Revathi 1987); Bihar (Jha et al 1986; Saha &

Pandit 1987); Trichur of Kerala, Patna of Bihar & Mangalore of Karnataka (Laloraya & Mitra 1974); Assam (Kanta & Pandey 1984); Meerut (Bendre & Kumar 1975); Allahabad (Chadha & Pandey 1983a); Nainital (Singh 1974), Rohilkand of U.P. (Pandey 1982); W.B. (Laloraya & Mitra 1974, Pal & Santra 1985, Maity & Santra 1985, Santra & Pal 1988, Santra et al. 1988).

49. *Oscillatoria subbrevis* Schmidle forma *major* West (Plate 4, figs, 14 & 15)

J. Linn. Soc. (Lond.), Bot., 38: 78, 1907; Rao, C.B. 171, 1936; Desikachary, 209, 1959.

Thallus blue-green, trichome 10-11 μ broad, more or less straight, cross-walls not constricted; cells 2.5-3 (-4) μ long and cell rounded, calyptra absent.

From water logged paddy field of Haldia (Midnapore) and Purulia.

Specimen nos. Hal 3, Hal 13, Pur 1.

The present alga differs from the type in having broader trichome and longer cells. Regarding cell length the present alga resemble forma *crassa* Dixit in which cells 2.5-3 μ long; while in comparison to trichome diameter it comes close to forma *major* West, G.S. in which cells are 10-10.5 μ broad.

Earlier, the alga was recorded from Madhya Pradesh and Uttar Pradesh (Laloraya & Mitra 1974).

50. *Oscillatoria acuta* Bruhl et Biswas, Orth. mut. Geitler (Plate 5, fig. 1)

Bruhl & Biswas, Indian Bark Algae, J. Dept. Sc., Cal. Univ., 5: 3, pl. 1, fig. 6a and b, 1922; Geitler, 978, 1932; Desikachary, 240, pl. 39, f. 5, 8, 1959.

Trichomes free and sometimes aggregated into bundles, arranged in parallel unconstricted at the cross-walls, straight but attenuated at the apices, apices subobtuse, non-capitate, straight or abruptly bent aside, trichome about 600-1000 μ long; cells 4-5.5 μ broad and 2.3-3.5 μ long, contents granular.

From water logged paddy field of Jhargram, Tamluk, Silda and Midnapore Sadar (Midnapore).

Specimen nos. J75, Sil 51, T 12, D 34.

The present form is similar to *O. acuta* Bruhl et Biswas, orth. mut Geitler but differs in having shorter cells (2.3-3.5 μ long instead of 3-4 μ) and longer trichomes (600-1000 μ long instead of 70-400 μ).

Previously reported from Karnataka (Bongale & Bharati 1980, Somashekar 1983, 1984b); Maharashtra (Kamat 1963, Barhate & Tarar 1983); U.P. (Das et al. 1974, Bendre & Kumar 1975, Prasad & Mehrotra 1980); Madurai & Gaya (Laloraya & Mitra 1974); Sagar of M.P. (Mishra & Purohit 1979); W.B. (Biswas 1936).

51. *Oscillatoria brevis* (Kutz.) Gomont. var. *non-granulata* Kamat (Plate 5, fig. 6).

Kamat, 283, pl. 15, f. 88, 1963.

Thallus blue-green, trichomes long, not constricted and not granulated at the cross walls, tapering and bent towards apices; cells 4.5-5 μ broad and 2-2.5 μ long; end cell conical and bent.

From stagnant water of paddy field, Tamluk, Contai (Midnapore).

Specimen nos. T 90, C 1.

Previously reported from Maharashtra (Kamat 1963, Ashtekar & Kamat 1980).

52. *Oscillatoria earlei* Gardner (Plate 5, fig. 4)

New Myxophyceae from Puerto Rico, Mem. N.Y. bot. Gdn., 7: 36, pl. 8, f. 70, 1927; Geitler, 976, f. 615a. 1932; Desikachary, 238, pl. 38, f. 15, 1959.

Trichomes blue-green, straight, curved and attenuated at the ends, 2.5-3 μ broad, cross-walls not granulated, not constricted; cells quadrate to longer than broad 2.5-5 μ long, end cells pointed.

From water logged paddy field of Tamluk, Silda, Keshpur (Midnapore).

Specimen nos. T 10, K 36, Sil 51.

The present alga resembles *O. earlei* Gardner in all respects but differs from the same in having slightly larger cells.

Reported earlier from Maharashtra (Kamat 1968, Bhoge & Ragothaman 1986); Meerut (Bendre & Kumar 1975); Allahabad (Chadha & Pandey 1983a); Pachmarhi of M.P. (Das et al. 1974); Tamil Nadu (Anand & Revathi 1987).

53. *Oscillatoria acuminata* Gomont (Plate 5, fig. 2)

Monogr. Oscillariees, 227, pl. 7, f. 12, 1892; Geitler 978, f. 603d, 1932; Desikachary, 240, pl. 38, f. 7 & pl. 40, f. 13, 1959.

Trichomes blue-green, straight, not constricted at the cross-walls, cells 3-5 μ broad, 5-8 μ long; ends tapering pointed, bent and generally mucronate, calyptra absent.

From paddy fields, Midnapore Sadar.

Specimen no. D 86.

Previously reported from Karnataka (Bongale & Bharati 1980, Somashekar 1984b); Maharashtra (Dixit 1936); U.P. (Bendre & Kumar 1975 from Meerut, Pandey 1982 from Rohilkand, Khan 1985 from Sultanpur) & W.B. (Sinha & Mukherjee 1975b, Mukhopadhyay & Chatterjee 1981).

54. *Oscillatoria schultzii* Lemm. (Plate 5, fig. 5)

Forschb. Biol. stat., Plon, 12: 145, pl. 4, f. 14 & 15, 1905; Geitler 970, 1932; Desikachary, 232, pl. 41, f. 11. 1959.

Thallus blue-green, trichome distinctly constricted at the cross-walls, cross-wall granulated gradually attenuated at the apices; cells quadrate to barrel shaped, 2-3 μ broad and 1.5-2.5 (-3) μ long, end cell conical.

From paddy field soils of Ghatal, Contai (Midnapore).

Specimen nos. Gh 11, C4.

Earlier this species was reported from Maharashtra (Gonzalves & Gangla 1949, Bhoge & Ragothaman 1986); Karnataka (Bongale & Bharati 1980);

Punjab (Grover & Pandhol 1975, Pandhol & Grover 1976); U.P. (Pal 1975; Chadha & Pandey 1983a), W.B. (Pal & Santra 1985).

55. *Oscillatoria okeni* Ag. ex Gomont (Plate 5, fig. 10)

Gomont, Monogr. Oscillariees, 232, pl. 7, f. 18, 1892; Geitler, 969, f. 608a, 1932; Desikachary, 231, pl. 38, f. 17. 1959.

Thallus pale blue-green, trichomes fragile and distinctly constricted at the cross-walls, 5-6 μ broad, attenuated at the ends, ends slightly curved, subconical, without calyptra; cells 2-3.5 μ long.

From water logged paddy field, Panskura (Midnapore).

Specimen no. Pa 32.

The present alga differs from *Oscillatoria okeni* Ag. ex Gonot in having slightly shorter cells (2-3.5 μ long, instead of 2.7-4.5 μ).

Previously it was reported from U.P. (Rao 1937b, Bendre & Kumar, 1975, Chaturvedi & Pandey 1976, Pandey & Chaturvedi 1979, Chadha & Pandey 1983a); Kolhapur (Kamat 1963); Vidarbh (Kamat 1975); Punjab (Pandhol & Grover 1976, Sarma & Kanta 1978); Orissa (Rao,1938b, Mohanty 1982); Kerala (Parukutty 1940); Delhi (Rao 1940); W.B. (Sinha & Mukherjee 1975a).

56. *Oscillatoria amphibia* Ag. ex Gomont (Plate 5, fig. 8)

Gomont, Monogr. Oscillariees, 221, pl. 7, f. 4-5, 1892; Geitler, 966, 1932; Desikachary, 229, pl. 37, f. 6, 1959.

Thallus dark blue-green; trichomes straight or curved, not constricted at the cross walls, two granules at the septa; cells 2.5-3.5 μ broad and 4-7 μ long; end cell rounded.

From paddy field soils Onda (Bankura).

Specimen no. Ban 40

Earlier recorded from Assam (Bruhl & Biswas 1926); Kolhapur (Kamat 1963); Ghaziabad (Pal 1975); river Cauvery of Karnataka (Somashekar 1983) & W.B. (Bruhl & Biswas 1922, Biswas 1925, 1926, 1942, Banerji 1938, Sinha & Mukherjee 1975b, Mukhopadhyay & Chatterjee 1981, Pal & Santra 1982).

57. *Oscillatoria obscura* Bruhl et Biswas (Plate 5, fig. 13)

Algae of the Bengal Filter Beds, J. Dept. Sci., C.U., 4: 6, pl. 2, f. 9, 1922; Geitler, 945, 1932; Desikachary, 207. 1959.

Thallus blue-green, trichome 4-5 μ broad, not constricted at the cross-walls, cells granulated, slightly attenuated at the apex; cells 1.5 μ long, end cell rounded.

From paddy field soils of Onda (Bankura), Balurghat, Raiganj (West) Dinajpur.

Specimen nos. Ban 42, WD 4, WD 9

Previously reported from U.P. (Rao 1937b, Pandey 1965, Bendre & Kumar 1975, Prasad & Mehrotra 1980, Pal 1975, Chadha & Pandey 1983a); Bihar (Laloraya & Mitra 1974, Jha et al. 1986); Orissa (Rao 1936b), Laloraya & Mitra 1974, Mohanty 1982, 1984); A.P. (Laloraya & Mitra 1974); Tamil Nadu (Anand & Revathi 1987); Karnataka (Bongale & Bharati 1980, Somashekar & Ramaswamy 1983, Somashekar 1984b); Punjab (Grover & Pandhol 1975, Pandhol & Grover 1976); Jammu & Kashmir (Goyal et al. 1984); W.B. (Laloraya & Mitra 1974, Sinha & Mukherjee 1975b, Pal & Santra 1982).

58. *Oscillatoria princeps* Vaucher ex Gomont (Plate 5, fig. 11 and Plate 39D)

Gomont, Monogr. Oscillariees, 206, pl. 6, f. 9, 1892; Geitler, 947, f. 598a, 1932; Desikachary 210, pl. 37, f. 1, 10, 11, 13, 14, 1959.

Thallus blue-green, trichomes straight, not constricted at the cross-walls, 24-48 μ broad; apices straight or slightly bent, slightly attenuated; cells 4-6.5 μ long; end cells capitate.

From paddy field soils, Salboni, Haldia (Midnapore).

Specimen nos. 127, 17.

Previously reported from all over India (from Hyderabad of A.P, Madhurai of Tamil Nadu, Trichur of Kerala - Laloraya & Mitra 1974); Karnataka (Bongale & Bharati 1980, Hedge & Bharati 1983, Somashekar 1983, 1984a, b); Maharashtra (Kamat 1963, Kolte & Goyal 1985); Orissa (Mohanty 1982, 1984); Tamil Nadu (Anand & Revathi 1987); Jammu & Kashmir (Anand 1979, Goyal et al. 1984); North East India (Reddy et al. 1986); U.P. (Pandey 1982); Punjab (Pandhol & Grover 1976); W.B. (Laloraya & Mitra 1974, Sinha & Mukherjee 1975a, Pal & Santra 1985, Santra et al. 1988, 1989).

59. *Oscillatoria curviceps* Ag. ex Gomont (Plate 6, fig. 5)

Monogr. Oscillariees, 213, pl. 6, f. 14, 1892, Geitler, 947, f. 598e, 1932; Desikachary, 209, pl. 38, f. 2, 1959.

Thallus dark blue-green, trichomes curved straight, not attenuated at the apex, not constricted at the cross-walls, cross-wall sometimes granulated, 10.5-13 μ broad, cells 2.5-4.5 μ long; ends without calyptra, rounded.

From paddy field soils, Kakdweep (South 24-Parganas), Cooch Behar.

Specimen nos. SP 22, CB 4.

Previously reported from Karnataka (Bongale & Bharati 1980, Somashekar 1983, 1984a,b, Somashekar & Ramaswamy 1983); U.P. (Bendre & Kumar 1975, Chadha & Pandey 1983a); Maharashtra (Nandkar et al. 1983, Barhate & Tarar 1983); Punjab (Pandhol & Grover 1976, Sarma & Kanta 1978); Jammu (Anand 1979); Bihar (Jha et al. 1987, Saha & Pandit 1987); Orissa (Mohanty 1982, 1984) & W.B. (Mukhopadhyay & Chatterjee 1981, Santra & Pal 1988, Santra et al. 1988, 1989).

60. *Oscillatoria pseudogeminata* G. Schmid (Plate 5, fig. 3)

Ber, dtsch. bot. Ges, 32: 124, f. 4, 1914; Geitler, 966, f. 616, 1932; Desikachary, 228, pl. 41, f. 10, 1959.

Thallus dirty blue-green, trichomes curved or straight cells 1.5-2.5 μ broad, 2-3 μ long, cross-wall not constricted end cell rounded, without calyptra.

From paddy fields of Jalpaiguri, Cooch Behar.

Specimen nos. Jal 7, CB 8.

Previously reported from Assam (Biswas 1934); Benaras (Rao, C.B. 1937); Meerut (Bendre & Kumar 1975); Cuttack (Rao, C.B. 1938); Bhubaneswar (Mohanty 1982); Madhubani (Rao 1939); Bihar (Jha et al. 1986); North East India (Reddy et al. 1986); Kolhapur (Kamat 1963); Panhala of Maharashtra (Kamat 1961-62); Vidarbha of Maharashtra (Kamat 1975); Khandesh of Maharahstra (Barhate & Tarar 1983); River Cauvery (Somashekar 1983); River Kapila (Somashekar 1984b) & W.B. (Mukhopadhyay & Chatterjee 1981).

61. *Oscillatoria pseudogeminata* var. *unigranulata* Biswas (Plate, 5, fig. 7)

Malayan aquatic Biology, J. Fed. Malay Stat. Mus, 14: 409, Pl. 9, f. 7, 1929; Geitler, 966, 1932; Desikachary, 229, pl. 39, f. 19, pl. 41, f. 17, 1959.

Trichome 2-3 μ broad, straight or curved; cells 3-4 μ long, cross-wall not constricted, thick, with one granule on either side of cross-wall.

From paddy fields of Bishnupur (Bankura).

Specimen no. B53

Previously reported from Delhi (Rao C.S. 1940); Maharashtra (Bhoge & Ragothaman 1986); W.B. (Sinha & Mukherjee 1975a).

Here, cells are slightly longer (3-5 μ long instead of 2.5-4 μ).

62. *Oscillatoria amphigranulata* van Goor (Plate, 5, fig. 9)

Zur Kenntnis der Oscillatoriaceen, Rec. trav. bot. neerland. 15: 255, pl. 2, f. 2, 1918; Geitler, 964, 1932; Desikachary, 226, pl. 37, f. 4, 1959.

Trichome straight, cells 1.5-2 μ broad and 4-5 μ long, cross-wall constricted, one gas vacuole on either side of the septa, end cell rounded without calyptra.

From paddy fields of Bishnupur (Bankura).

Specimen no. B55.

Reported earlier from W.B. (Sinha & Mukherjee 1975b); Kolhapur (Kamat 1963); Vidarbh (Kamat 1975); River Kapila (Somashekar 1984b); Cauvery (Somashekar 1983).

63. *Oscillatoria proteus* Skuja (Plate 7, fig. 8)

Zur Süsswasseralgen flora Burmas, Nov. Acta Reg. Soc. Upsal., ser. IV, 14: 48, pl. 8, f. 11-13, 1949; Desikachary, p. 221, pl. 41, f. 15, 16, 18, 1959.

Trichome, straight or curved, cells (5-) 6-7 μ broad and 2.5-4 (-4.5) μ long, cross-wall distinctly constricted; end cell rounded conical, without calyptra.

From stagnant water of paddy fields, Jamtala (Midnapore), Onda (Bankura).

Specimen nos. M 18, Ban 39.

Previously reported from Sagar (Mishra & Purohit 1979); Jaipur (Tridevy 1982); Maharashtra (Nandkar et al. 1983, Barhate & Tarar 1983); Karnataka (Somashekar 1984); Patiala (Sarma & Kanta 1978), W.B. (Sinha & Mukherjee 1975b, Pal & Santra 1985).

64. *Oscillatoria quadripunctulata* Bruhl et Biswas (Plate 6, figs, 8 & 9)

Algae of Bengal Filter Beds, J. Dept. Sci. C.U., 4: 5, pl. 1, f. 6, 1922; Geitler, 966, 1932; Desikachary, 227, pl. 37, f. 5, 1959.

Thallus blue green colour, trichomes curved or straight, closely entangled, cells 1.5-2 μ broad and 3-4 μ long, not constricted at the cross-wall, cross-wall indistinct, with two granules on either side.

From paddy fields of Jairambati (Bankura).

Specimen no. Ban 19.

Previously reported from W.B. (Sinha & Mukherjee (1975b, Mukhopadhyay & Chatterjee 1981); U.P. (Chaturvedi & Pandey 1976).

65. *Oscillatoria anguina* (Bory) Gomont (Plate 5, fig. 12)

Gomont, Monogr. Oscillariees, 214, pl. 6, f. 16, 1892; Geitler, 948, f. 599b, 1932; Desikachary, 210, pl. 38, f.11, 1959.

Thallus dark blue-green colour, trichomes 6-6.5 μ broad, not constricted at the cross wall; cells 2-4 μ long, ends attenuated and curved, end capitate.

From water logged paddy fields of Jhilimili (Bankura).

Specimen no. Ban 22.

Previously reported from Ballia & Ghazipur of U.P. (Pandey 1965); Orissa (Mohanty 1982); West Bengal (Mukhopadhyay & Chatterjee 1981, Sinha & Mukherjee 1975a); Bihar (Rao 1939); Delhi (Rao 1940); Bombay (Gonzalves and Joshi 1949).

66. *Oscillatoria anguina* (Bory) Gomont forma (Plate 6, fig. 15)

Skuja, Zur Süssawasseralgenflora Burmas, 46, pl. 8, f. 1, 2, 1949; Desikachary, 211, pl. 41, f. 5, 6, 1959.

Filaments straight or curved, trichomes 6-8 μ broad, slightly constricted at the cross-wall, cells 2-3.5 μ long, end cell with calyptra.

Previously reported from Bankura (W.B.) Sinha & Mukherjee, 1984.

67. *Oscillatoria proboscidea* Gomont (Plate 6, figs. 1 & 2 and Plate 11, fig. 15)

Monogr. Oscillariees, p. 209, pl. 6, f. 10 & 11, 1892; Geitler 948, f. 598b, 1932; Desikachary, 211, pl. 38, f. 9, 1959.

Trichomes straight or curved, attenuated at the apex, 11-14.5 μ broad, not constricted at the cross-walls, cells 2.5-4 μ long, apex flatly rounded, capitate and with thick membrane.

From paddy fields of Moynapur (Bankura)

Specimen no. Ban 88.

Previously reported from U.P. (Singh 1939a,b, Pandey 1965, 1982, Chaturvedi & Pandey 1976, Chadha & Pandey 1983a); Bihar (Sinha & Srivastava 1980); Karnataka (Bongale & Bharati 1980); Delhi (Rao 1940); W.B. (Sinha & Mukherjee 1975a).

68. *Oscillatoria chalybea* (Mertens) Gomont (Plate 6, fig. 3 & 4)

Gomont, Monogr. Oscillariees, 232, pl. 7, f. 19, 1892; Geitler, 956, f. 608b, 1932; Desikachary, 218, pl. 38, f. 3, 1959.

Thallus dark blue, green filaments straight or curved trichomes 8-12 μ broad; cross-wall constricted; cells 3.5-5.5 μ long, end cell obtuse, without calyptra.

From paddy fields of Midnapore, Howrah and Hooghly.

Specimen nos. M11, H20, Hg15.

Previously reported from A.P. (Rao 1938); Orissa (Rao 1938b, Mohanty 1982); Delhi (Rao 1940); Bombay (Gonzalves & Joshi 1946); Jalgaon district (Bhoge & Ragothaman 1986); Kolhapur (Kamat 1963); U.P. (Rao 1937b, Pandey & Chaturvedi, 1979, Pandey & Pandey 1982); Bhagalpur (Saha & Pandit 1987); Karnataka (Bongale & Bharati 1980, Somashekar 1983, 1984a,b, Somashekar & Ramaswamy 1983); W.B. (Biswas 1926, Sinha & Mukherjee 1975a, Mukhopadhyay & Chatterjee 1981, Pal & Santra 1985, Santra et al. 1988).

69. *Oscillatoria amoena* (Kütz.) Gomont (Plate 6, fig. 11).

Gomont, Monogr. Oscillariees, 225, pl. 7, f. 9, 1892; Geitler, 969, f. 603a, 611k, 1932; Desikachary, 230, pl. 40, f. 12, 1959.

Thallus blue-green, trichomes straight or slightly curved; trichomes 4-5 μ broad at the middle and graudally attenuated at the ends, narrowed upto 2.5 μ. Cross-wall slightly constricted, septa granulated, cells 2.5-5 μ long, end cell broadly conical, calyptra present.

From paddy fields of Canning (South-24-Parganas).

Specimen no. SP57

Previously reported from U.P. (Chaturvedi & Pandey 1976, Prasad & Mehrotra 1980); Karnataka (Bongale & Bharati 1980, Somashekar 1983, 1984, Somashekar & Ramaswamy 1983); J. & K. (Goyal et al. 1984); Sagar of M.P. (Mishra & Purohit 1979); Patiala (Sarma et al. 1979); W.B. (Sinha & Mukherjee 1975b, Mukhopadhyay & Chatterjee 1981).

70. *Oscillatoria formosa* Bory ex Gomont (Plate 6, fig. 7)

Gomont, Monogr. Oscillariees, 230, pl. 7, f. 16, 1892; Geitler, 970, f. 619b, 1932; Desikachary, 232, pl. 40,f. 15, 1959.

Filaments straight; trichomes 4-6 μ broad; cross-wall constricted and granulated, apex attenuated and slightly bent; cells 2.5-5 μ long; end cell without calyptra.

From paddy fields of Jhargram (Midnapore).

Specimen no. J37.

Previously reported from U.P. (Gupta 1956, Pandey 1965, Pandey & Chaturvedi 1979, Bendre & Kumar 1975, Pandey & Pandey 1982, Rao 1937b); A.P. (Rao 1938a); Karnataka (Bongale & Bharati 1980, Somashekar 1983, 1984a,b); Ahmedabad (Kamat 1962-63); Patiala (Sarma et al. 1979); Kolhapur (Kamat 1963); Nagpur (Nandkar et al. 1983); Bhagalpur (Saha & Pandit 1987); W.B. (Biswas 1926, Sinha & Mukherjeee 1975b, Pal & Santra 1985, Santra et al. 1988).

71. *Oscillatoria splendida* Grev. ex Gomont (Plate 7, fig. 10)

Gomont, Monogr, Oscillariees, 224, pl. 7, f. 7 & 8, 1892; Geitler, 972, f. 611 m-o, 620d-f, 1932; Desikachary, 234, pl. 37, f. 7 & 8, pl. 38, f. 10, & pl. 40, f. 11, 1959.

Filaments straight or curved; trichomes 2-3 μ broad; gradually attenuated at the ends; not constricted at the cross-wall, cross-wall granulated; cells 4-8 μ long; ends sickle shaped or screw like, end cell capitate and rounded.

From paddy fields of Jhilimili (Bankura), Howrah, Purulia.

Specimen nos. Ban 28, H 14, Pur 17.

Previously reported from Ahmedabad (Kamat 1962-63); Maharashtra (Ashtekar & Kamat 1980); Bhagalpur (Nasar & Munshi 1976); River Cauvery of Karnataka (Somashekar 1983); W.B. (Bruhl & Biswas 1922, Banerjee 1938, Sinha & Mukherjee 1975b).

72. *Oscillatoria animalis* Ag. ex Gomont (Plate 5, fig. 14)

Gomont, Mongr. Oscillariees, 227, pl. 7, f. 13, 1892; Forti, 178, 1907; Fremy, 223, f. 193. 1929; Geitler 978, f. 603a, 1932; Desikachary, 239, pl. 40, f. 14, 1959.

Thallus dark blue green colour, filaments straight, trichomes 3-4 μ broad, not constricted at the cross-wall, end cell bent, conical and pointed. Cells 2-4.5 μ long.

From paddy field of Purulia, Bankura and Midnapore.

Specimen nos. 121, Ban 45, Pur 31.

Previously reported from Nainital (Singh 1974); Madurai of Madras, Asansol of W.B., Trichur of Kerala (Laloraya & Mitra 1974b); U.P. (Das et al. 1974, Bendre & Kumar 1975, Pandey & Chaturvedi 1979, Chadha & Pandey 1983a); Maharashtra (Kamat 1963, Ashtekar & Kamat 1980, Chaporkar & Gangawane 1984, Nandkar et al. 1983); Ahmedabad (Kamat 1962-63); Karnataka (Somashekar 1983, 1984a,b); Bihar (Saha & Pandit 1987); Orissa (Mohanty 1982).

73. *Oscillatoria tenuis* Ag. ex Gom. (Plate 7, fig. 4)

Gomont, Monogr. Oscillariees, 220, pl.7, f. 2-3, 1893; Geitler, 959, f. 611 f-g, 1932; Desikachary, 222, pl. 42, f. 15, 1959.

Thallus blue-green; trichomes straight, slightly constricted at the cross-wall, sometimes appears unconstricted cross-wall granualted, 5-8 μ broad; cells 2.5-4 μ long, end cell rounded, with thickened outer membrane.

From paddy fields, Onda (Bankura), Falta (South 24-Parganas).

Specimen nos. Ban 37, 113.

Previously reported from North East India (Reddy et al. 1986); Karnataka (Somashekar 1983, 1984a & b); Orissa (Rao 1938b, Mohanty 1982, 1984); Maharashtra (Gonzalves & Joshi 1946, Kamat 1963, Ashtekar & Kamat 1980, Barhate & Tarar 1983, Nandkar et al. 1983); U.P. (Rao 1937b, Bendre & Kumar 1975); Bihar (Rao 1939); W.B. (Biswas 1925, 1926, Banerjee 1938, Sinha & Mukherjee 1984, Santra & Pal 1988).

74. *Oscillatoria raoi* De Toni (Plate 6, fig. 6)

Diagnoses Algarum novarum, I, cent. 6: 560, 1939; Desikachary, 223, pl. 42, f. 16-19, 1959.

Thallus dull blue-green, trichomes straight, long, sometimes curved, 5-6 μ broad, cross-wall indistinct but marked by the presence of granules of either side; cells 2.5-5 μ long, end cells rounded, sometimes slightly constricted at the base of the end cell, calyptra absent.

From paddy fields of Chinsura (Hooghly), Bishnupur (Bankura).

Specimen nos. H 17, Ban 37.

Previously reported from Karnataka (Somashekar 1983, 1984a,b); Meerut (Bendre & Kumar 1975); Ludhiana (Pandhol & Grover 1976); Patiala (Sarma & Kanta 1978); Jaipur (Tridevy 1982); Maharashtra (Barhate & Tarar 1983, Bhoge & Ragothaman 1986); W.B. (Sinha & Mukherjee 1984).

75. *Oscillatoria limosa* Ag. ex Gom. (Plate 6, fig. 10)

Gomont, Monogr. Oscillariees, 210, pl. 6, f. 13, 1892; Geitler, 944, f. 598d, 1932, Desikachary, 206, pl. 42, f. 11, 1959.

Thallus dark blue-green colour; trichomes generally straight, not constricted at the cross-wall, cross-wall granulated, 12-16 (-20) μ broad, cells 4-5 μ long, end cell flatly rounded, having outer thickened membrane.

From paddy fields of Midnapore, Howrah and Burdwan.

Specimen nos. 173, How 13, Bur 23.

Previously reported from Bihar (Jha et al. 1986); Orissa (Rao 1938b, Mohanty 1982); Maharashtra (Ashtekar & Kamat 1980, Patil & Satav 1986); U.P. (Prasad & Mehrotra 1980, Pandey 1982, Pandey & Pandey 1982); Tamil Nadu (Ganapathi 1940a); Shillong & Kohima (Reddy et al. 1986), Patiala (Sarma & Kanta 1978); J & K (Anand 1979, Goyal et al. 1984); Karnataka (Bongale & Bharti 1980, Hegde & Bharati 1983); River Cauvery (Somashekar 1983); W.B. (Prain 1905, Biswas 1926, Sinha & Mukherjee 1984, Santra et al. 1988, 1989).

Genus: TRICHODESMIUM Ehrenb.

76. *Trichodesmium thiebautii* Gomont (Plate 6, fig. 12)

Gomont, J. de Bot., 4: 356, 1890; Mongr. Oscillariees, 197, pl. 6, f. 2-4, 1892; Desikachary, 243, pl. 42, f. 4 & 5, 1959.

Trichome 7-12.5 μ broad, cells upto 2 times as long as broad.

From water logged rice field, Diamond Harbour (South 24-Parganas), Basirhat (North 24-Parganas).

Specimen nos. 105, 193.

Reported earlier from South Indian Coast (Iyengar & Desikachary 1944); W.B. (Pal & Santra 1982, Santra et al. 1988, 1989).

Genus: CRINALIUM Crow

77. *Crinalium magnum* Fritsch et John (Plate 9, fig. 1)

Ann. Bot. Lond., 6: 393, f. 8A, 1942; Desikachary, 247, pl. 46, f. 7 & 8, 1959.

Trichome straight, short, flattened, slightly constricted at the cross-wall, sheeth diffluent and closed at both ends, end cells convex, cells 15-19 μ broad and 2-3 μ long.

From paddy field soils of Falda (South 24-Parganas).

Specimen no. 69.

Previously reported from Tamil Nadu (Mitra 1951); Andaman & Nicobar (Prasad & Srivastava 1986); West Bengal (Santra et al. 1988).

Genus: PORPHYROSIPHON Kütz.

78. *Porphyrosiphon notarisii* (Menegh) Kutz. ex Gomont (Plate 7, fig. 6 and Plate 40A)

Tab. Phycologicae, 2: 7, pl. 27, f. 1, 1850; Gomont, 331, 1892; Geitler, 986, f. 631, 1932; Desikachary, 248, pl. 47, f. 9, 1959.

Filaments densely arranged forming expanded, reddish brown thallus, sheath firm, lamellated, orange-red or colourless; trichomes unconstricted or slightly constricted at the cross-walls; cells 10-12.5 μ broad and 6-8.5 μ long.

From paddy field soil, Garbeta, Kharagpur, Keshpur (Midnapore), Alipur Duar (Jalpaiguri) and Asansol (Burdwan).

Specimen nos. G 72, G 121, K 38, A 42, Jal 29

Previously this alga was reported from Chilka (Biswas 1924); M.P. (Das et al. 1974; Misra & Purohit 1979); Nainital (Singh 1974); Allahabad (Chadda & Pandey 1983a); W.B. (Sinha & Mukherjee 1975a; Sengar & Pandey 1985).

Genus: PHORMIDIUM Kütz.

Key to the Species

1.	Trichomes distinctly constricted	.. 2
1.	Trichomes not distinctly constricted	.. 8

2. Trichomes 7-8 μ broad, cells 2-2.5 μ long, cross-wall granulated, end cell with a calyptra .. *P. lucidum*

2. Trichomes upto 5 μ broad .. 3

3. Trichomes upto 2.5 μ broad .. 4

3. Trichomes broader .. 7

4. Cells upto 3 μ long .. 6

4. Cells may be longer .. 5

5. Cells 3-5 μ long, 1-2 μ broad, end cell conical .. *P. tenue*

5. Cells 1.5-5 μ long, 0.5-1 μ broad .. *P. angustissimum*

6. Trichomes 1.5-2.5 μ broad, cells quadrate to slightly longer, 2.5-3 μ long, end cell conical .. *P. fragile*

6. Trichomes 1.5-2.3 μ broad, cells quadrate to subquadrate, 1-1.5 (-2) μ long, end cell rounded .. *P. foveolarum*

7. Trichomes 4-5 μ broad, cells subquadrate, 2-3.5 μ long, end cell rounded or slightly conical .. *P. jadinianum*

7. Trichomes 2.5-4 μ broad, cells quadrate to cylindrical 3-8 μ long, end cell rounded .. *P. molle*

8. Trichomes 1-1.5 (-2) μ broad, cross-wall granulated, cells 2-4 μ long, end cell conical .. *P. laminosum*

8. Trichomes broader .. 9

9. End cell with a calyptra .. 10

9. End cell without calyptra .. 13

10. Cells upto 4.5 μ long .. 11

10. Cells longer .. 12

11. Cells 1-2 μ long, trichomes 8-9.5 μ broad, filaments 13-14.5 (-15) μ broad, end cell round with calyptra .. *P. stagnina*

11. Cells 2.5-4.5 μ long, trichomes 7.5-10 μ broad, end cells conical .. *P. subfuscum*

12. End cell with a hemispherical calyptra, trichomes gradually attenuated at the apex, straight, truncated, 5-7 μ broad, cells 4-6 μ long .. *p. favosum*

12. End cell with a depressed calyptra, apex briefly attenuated and slightly curved, capitate, trichomes 6-8 μ broad, cells 3-6 μ long .. *P. uncinatum*

13. Trichomes upto 6 μ broad .. 14

13. Trichomes broader .. 21

14. Cells upto 2.5 μ long .. 15

	14. Cells may be longer	..	16
15.	Trichomes 5-6 μ broad, cells 1.5-2.5 μ long, and cell rounded	..	*P. ambiguum*
15.	Trichomes 2.5-3.5 μ broad, cells 1.8-2.5 μ long, end cell rounded	..	*P. bohneri*
	16. Cross-walls with granules end cells rounded, trichomes 2-3 μ broad	..	17
	16. Cross-walls without granules, end cells round to obtuse conical or conical, trichomes may be broader	..	18
17.	Cells 4.5-6.5 μ long, single granule on either side of the septum	..	*P. valderium*
17.	Cells 2.5-4 (-5) μ long, two granules on either side of the septum	..	*P. purpurascens*
	18. Cells 2.5-4 μ long, trichomes 4-5 μ broad,	..	*P. papyraceum*
	18. Cells longer	..	19
19.	Trichomes 2.7-4 μ broad, 4-7 μ long	..	*P. corium*
19.	Trichomes broader	..	20
	20. Trichomes 5-6 μ broad, cells 4.5-6 μ long, end cells round to obtuse conical, with thick outer membrane	..	*P. corium* v. *capitatum*
	20. Trichome 4.5-6 (-7) μ broad, cells 3-6 μ long, and cell conical, without thick outer membrane	..	*P. corium* f. *major*
21.	End cell rounded	..	22
21.	End cell truncated, sometimes with a thick outer membrane, trichomes 5.5-8 μ broad, cells 4.5-8 μ long	..	*P. retzii*
	22. Cells 4-10 μ long, trichomes 7.5-9 μ broad	..	*P. rotunda*
	22. Cells shorter in length	..	23
23.	Cells 1-2 μ long, trichomes 8-10.5 μ broad, cross-walls not granulated	..	*P. anomala*
23.	Cells 2-3.5 μ long, trichomes 8-9.5 μ broad, cross-walls granulated	..	*P. ambiguum* v. *major*

79. *Phormidium corium* (Ag.) Gomont (Plate 7, fig. 1)

Gomont, J. de Bot. Fr., 4: 355, 1890; Monogr. Oscillariees, 172, pl. 5, f. 1 & 2, 1892; Geitler, 1018, f. 649, b,c; 1932; Desikachary, 269, pl. 44, f. 10-11, 1959.

Thallus expanded membranous; trichomes densely entangled, with thin sheath, not constricted at the cross-walls, ends straight, slightly conical on obtuse conical; cells 2.7-4 μ broad, and 4-7 μ long; calyptra absent.

From paddy field soils, Silda, Godapiasal (Midnapore).

Specimen nos. Sil. S1, Goda 133

Here the alga differs from the type species in having 2.7-4 μ broad trichomes instead of 3-4.5 μ.

Previously reported from Shillong (Parukutty 1939); Chilka (Biswas 1924); Cuttack (Rao 1938); Maharashtra (Kamat 1963, Ashtekar & Kamat 1980); U.P. (Prasad & Mehrotra 1980; Pandey & Pandey 1982, Chadha & Pandey 1983a); Karnataka (Somashekar 1984b).

80. *Phormidium corium* var. *capitatum* Gardner (Plate 9, fig. 2)

Univ. Calif. Publ. Bot., 14: 4, pl. 1, f. 1, 1927; Geitler, 1019, 1932; Desikachary, 271, pl. 43, f. 11-12, 1959.

Trichome straight, 5-6 μ broad, not constricted at the cross-wall, sheath thin; cells 4.5-6 μ long; end cells round to obtuse conical with thick outer membrane.

From paddy field soils of Simlipal (Bankura), Siliguri (Darjeeling).

Specimen nos. Ban 41, Dar 19.

Previously reported from Orissa (Rao 1939); U.P. (Bendre & Kumar 1975); W.B. (Sinha & Mukherjee 1984).

81. *Phormidium corium* forma *major* Santra et Pal (Plate 7, fig. 7 and Plate 40B)

Plant mass blue-green, trichomes long, curved, slightly constricted at the cross-walls; not granulated at the cross-wall; end cell conical, without calyptra; cells 4.5-6 (-7) μ diam. and 3-6 μ long.

On moist soil paddy field, Salboni (Midnapore).

Collection nos. 330, 67, Specimen kept in the personal collection of the senior author.

Planta massa caeruleo-viridis; trichomata longa; curvata, palulum constricta ad septa transversa, non granulata ad septa transversa, cellula terminalis conica, absque calyptra; cellala 4.5-6 (-7) μ diam., 3-6 μ longae.

In solo madido, in oryzetis ad Salboni (Midnapore). Collectione sub numero 330, 67, typus positus in herbrio herbario auctoris senioris.

The present alga differs from the type in having broader cells 4.5-6 (-7) μ broad instead of 3-4.5 μ and slight constriction at the cross-walls.

82. *Phormidium ambiguum* Gomont (Plate 6, figs. 13 & 14 and Plate 40C)

Monogr. Oscillariees, 178, pl. 5, f. 10, 1892; Geitler, 1015, f. 647e, 1932; Desikachary, 266, pl. 44, f. 16, pl. 45, f. 5-8, 1959.

Thallus blue green to yellowish green; filaments densely entangled trichomes very slightly or not constricted at the cross-walls; sheath thin, sometimes thick; filament 6.5-7.5 μ broad; cells 5-6 μ broad and 1.5-2.5 μ long, end cell rounded, calyptra absent.

From paddy field soils, Panskura (Midnapore), Mukutmonipur (Bankura), Jhalda (Purulia).

Specimen nos. 19, Ban 15, Pur 22.

The alga differs from the type species in having slightly longer cells, and no constriction at the cross-wall, though the type species are described to be constricted.

The present species was also reported from Maharashtra (Kamat 1963, Ashtekar & Kamat 1980); Karnataka (Bongale & Bharati 1980, Somashekar 1983, 1984b); U.P. (Prasad & Mehrotra 1980, Chadha & Pandey 1983a); North East India (Reddy et al. 1986); Assam (Kant & Pandey 1984); W.B. (Sinha & Mukherjee, 1975b, Pal & Santra 1985).

83. *Phormidium ambiguum* var. *major* Lemmermann (Plate 6, figs. 16-18 and Plate 40D)

Kryptogamenflora der Mark Brandenburg, 3: 127, 1910; Geitler, 1015, 1932; Desikachary, 266, 1959.

Trichomes straight 8-9.5 μ broad, cells 2-3.5 μ long cross-walls granulated, cross-wall indistinct but becomes prominent by the presence of granules.

From paddy field soils of Bagmundi (Purulia), Suri (Birbhum).

Specimen nos. Pur 9, Bir 10.

Previously reported from Bombay (Dixit 1936); Nagpur (Barhate & Tarar 1982).

84. *Phormidium fragile* (Meneghini) Gomont (Plate 7, fig. 9)

Gomont, Monogr. Oscillariees, 163, pl. 4, f. 13-15 1892; Geitler, 999, f. 636b, 1932; Desikachary, 253, pl. 44, f. 1-3, 1959.

Thallus yellowish blue-green, mucilaginous, sheath diffluent, trichomes entangled, distinctly constricted at the cross-walls, attenuated at the apex, 1.5-2.5 μ broad; cells 2.5-3 μ long, end cell conical, calyptra absent.

From paddy field soils, Digha (Midnapore), Panagarh (Burdwan).

Specimen nos. 450, Bur 19.

The present form has slightly broader cells (1.5-2.5 μ broad instead of 1.2-2.3 μ).

Earlier, the alga was reported from Chilka Lake (Biswas 1924); Delhi (Rao 1940); Bombay (Schmidle 1900); Jalgaon of Maharashtra (Bhoge & Ragothaman 1986); U.P. (Pandey 1965, Prasad & Mehrotra 1980, Chadha & Pandey 1983a); Patiala (Sarma & Kanta 1978); Karnataka (Somashekar 1984b); W.B. (Banerjee, 1938, Sinha & Mukherjee 1975b, Mukhopadhyay and Chatterjee 1981, Santra & Pal 1988, Santra et al. 1988).

85. *Phormidium bohneri* Schmidle (Plate 7, fig. 11)

Beitr, z. Algenfl. Afrikas, Engler's Bot. Jahrb., 30: 59, pl. 2, f. 11, 1901; Forti in De Toni, Sylloge Algarum, 222, 1907; Geitler, 1008, 1932; Desikachary, 261, 1959.

Thallus blue-green, mucilaginous; trichomes straight, not constricted at the cross-walls, 2.5-3.5 μ broad, not attenuated at the apex; sheath thin; cells 1.8-2.5 μ long, end cell rounded.

From paddy field soils, Narajole (Midnapore), Howrah, Hoogly.

Specimen nos. 131, How 7, Hoog 17.

The present forms have slightly broader cells than the type species (2.5-3.5 μ broad instead of 1.7-2 μ), and in this respect resemble Rao's form (Rao, 368, 1937b).

Earlier the alga was reported from U.P. (Rao 1937b); A.P. (Rao 1938); Madurai of Madras (Laloraya & Mitra 1974); Karnataka (Bongale & Bharati 1980); Maharashtra (Kamat 1963, Kamat & Harankhedkar 1976, Tarar & Shewale 1984); Punjab (Sarma et al. 1979).

86. *Phormidium jadinianum* Gomont (Plate 7, fig. 5)

Bull. Soc. Bot. Fr., 40: 161, 1893; Geitler, 1002, f. 640, 1932; Desikachary, 256, pl. 55, f. 9, 1959.

Thallus dark-green; filaments more or less parallel; sheath thin; trichomes 4-5 μ broad, constricted at the cross-walls cells generally shorter than broad, sometimes quadrate, 2-3.5 μ long, end cell rounded, sometimes slightly conical, calyptra absent.

From paddy field soils, Salboni, Contai (Midnapore); Ratua and Gazol (Malda), Raigunj (West Dinajpur).

Specimen nos. C85, S 135, Mal 13, 17, WD 03.

The present form is similar to *P. jadinianum* Gomont in having trichome characters, but differs from the type species due to showing end cells rounded; in this respect it resembles Rao's specimen from Benaras (Rao, 173, 1936).

Previously the alga was recorded from Karnataka (Bongale & Bharati 1980); Patiala (Sarma & Kanta 1978).

87. *Phormidium molle* (Kütz.) Gomont (Plate 7, fig. 3)

Monogr. Oscillariees, 163, pl. 4, f. 12, 1892; Geitler, 1000, 1932; Desikachary, 255, pl. 59, f. 8, 1959.

Thallus mucilaginous, thin, blue-green, sheath colourless, diffluent; trichomes straight or bent, distinctly constricted at the cross-walls, not attenuated at the ends, 2.5-4 μ broad, cells quadrate to cylindrical, 3-8 μ long, end cell rounded, calyptra absent.

From paddy field soils, Midnapore Sadar, Howrah, 24-Parganas (South & North).

Specimen nos. D 127, How 13, 103, 239.

Here cells are slightly broader and longer than the type species (2.5-4 μ broad instead of 2.7-3.3 μm and 3-8 μ long instead of 3-7.8 μ).

Previously it was recorded from Ahmedabad (Kamat 1962-63); Karnataka (Bongale & Bharati 1980, Somashekar 1984b); Patiala (Sarma & Kanta 1978); Maharashtra (Kamat 1963, Kolte & Goyal 1985); Jaipur (Tridevy 1982); U.P. (Mitra 1951, Prasad & Mehrotra 1980, Pandey 1982); W.B. (Prain 1905, Sinha & Mukherjee 1975a, Mukhopadhyay & Chatterjee 1981).

88. *Phormidium rotunda* Santra et Pal (Plate 8, figs. 14-16 and Plate 41A)

Plant-mass blue-green, sheath diffluent, trichomes usually curved, not tapering towards the apices, not constricted at the cross-walls, cells 7.5-9 μ broad, 4-10 μ long, end cell rounded calyptra absent.

On paddy field soil, Garbeta (Midnapore).

Collection no. 308; specimen kept in the personal collection of senior author.

Phormidium rotunda Santra et Pal (Plate 8, figs. 14-16 and Plate 41A).

Planta massa caeruleo viridis, vagina diffluens, trichomata vulgo curvata, non fastigata ad apices, non constricta ad septa transversa, cellulae, 7.5-9 μ diam, 4-10 μ longae, cellula terminalis rotundata, absque calyptra.

In solo oryzetorum, Garbeta (Midnapore).

Specimen no. 308; typus positus in herbario auctoris senioris.

89. *Phormidium laminosum* Gomont (Plate 8, figs. 12 & 13).

Essai class. Nostoc. homo., J. de Bot., 4: 355, 1890; Monogr. Oscillariees, 167, pl. 4, f. 21 & 22, 1892; Geitler, 1005, f. 642c, 1932; Desikachary, 259, pl. 44, f. 6, 1959.

Thallus light blue-green, membranous, filaments densely arranged; sheath thin diffluent, trichomes 1-1.5 (-2) μ broad, not constricted at the cross-wall cross-wall granulated, cells 2-4 μ long, end cell conical, calyptra absent.

From paddy fields of Santiniketan (Birbhum), Susunia (Bankura).

Specimen no. Ban 44, Bir 84.

Previously reported from U.P. (Prasad & Mehrotra 1980); Maharashtra (Barhate & Tarar 1983, Choporkar & Gangawane 1987); Punjab (Sarma et al., 1979); Karnataka (Bongale & Bharati 1980); W.B. (Sinha & Mukherjee 1975b).

90. *Phormidium tenue* (Menegh.) Gomont (Plate 8, figs. 4 & 5)

Monogr. Oscillariees, 169, pl. 4, f. 23-25, 1892; Geitler, 1004, f. 642d, e, 1932; Desikachary, 259, pl. 43, f. 13-15; pl. 44, f. 7-9, 1959.

Thallus light blue-green, membranous, filaments densely entangled; sheath thin diffluent, trichomes 1-2 μ broad, constricted at the cross-wall; cells 3-5 μ long, end cell conical, without calyptra.

From paddy fields of Egra (Midnapore), Simlipal (Bankura); Jhalda (Purulia)

Specimen nos. 97, 35, Ban 5, Pur 9.

Previously reported from Tamilnadu (Anand & Revathi 1987); Maharashtra (Kamat & Harankhedkar 1976, Barhate & Tarar 1983); Sagar of M.P. (Mishra & Purohit 1979); U.P. (Pandey 1965, Prasad & Mehrotra 1980, Pandey 1982, Chadha & Pandey 1983a, Das et al. 1974); Trichur of Kerala & Patna of Bihar (Laloraya & Mitra 1974b); Karnataka (Bongale & Bharati 1980; Somashekar 1983) and W.B. (Sinha & Mukherjee 1975b, Mukhopadhyay & Chatterjee 1981).

91. *Phormidium purpurascens* (Kütz.) Gomont (Plate 8, fig. 6).

Essai class. Nostoc. homo., J. de Bot., Fr., 4: 355, 1890; Monogr. Oscillariees, 166, pl. 4, fig. 19, 1892; Forti 223, 1907; Geitler, 1009, fig. 644c 1932; Desikachary, 262, pl. 44, f. 4 & pl. 45, f. 1-4, 1959.

Thallus leathery; trichomes densely arranged, sheath diffluent, trichomes 2-3 μ broad, cells 2.5-4 (-5) μ long, cross-wall not constricted, with 2 granules on either side of the cross-wall; end cell rounded.

The present alga has similarity with Rao's specimen (Rao, 168, 1938b).

From paddy field soils of Murshidabad, Malda, West Dinajpur, Darjeeling, Jalpaiguri.

Specimen nos. Mur 15, 31, Mal 3, 8, 18, Dar 3, 9, Jal 7, 12, WD 2,6,9.

Previously reported from Maharashtra (Ashtekar & Kamat 1980, Barhate & Tarar 1983, Bhoge & Ragothaman 1986); Karnataka (Bongale & Bharati 1980, Somashekar 1983, 1984b); Orissa (Rao 1938); U.P. (Gupta 1956); Bihar (Saha & Pandit 1987).

92. *Phormidium valderianum* (Delp.) Gomont (Plate 8, fig. 17)

Monogr. Oscillariees, 167, pl. 4, f. 20, 1892; Geitler, 1011, f. 645c, 1932; Desikachary, 263, pl. 44, f. 5, 1959.

Thallus pale blue-green; trichomes densely entangled, flexuous, 2-3 μ broad, not constricted at the cross-wall, sheath diffluent, cells cylindrical, 4.5-6.5 μ long, single granule on either side of the cross-wall, end cell rounded without calyptra.

From paddy fields of Nadia, 24-Parganas (North), Hooghly, Burdwan, Birbhum.

Specimen nos. N3, 9, 12; Hoog 14, 21, 26; Bur 1, 7, 8; Bir 3, 7, 9, 15, 18; S 3, 7.

Previously reported from A.P. (Rao 1938a); Ahmedabad (Kamat 1962-63); W.B. (Sinha & Mukherjee 1975b).

93. *Phormidium stagnina* Rao, C.B. (Plate 8, fig. 3)

The Myxophyceae of the Madras Presidency, I, J. Indian bot. Soc., 17: 93, f. 4-7, 1938; Desikachary, 265, pl. 45, f. 16-18, 1959.

Thallus soft, light blue-green colour, filaments 13-14.5 (-15) μ broad, sheath diffluent, hyaline; trichomes 8-9.5μ broad, not constricted at the cross-walls, cells 1-2 μ long, end cell rounded, calyptra present.

From paddy fields of Bankura, 24-Parganas, Howrah, Hooghly.

Specimen nos. Ban 1, 4, 9; S -3, 7, 12, 16; How 1, 5, 7; Hoog 2, 7, 14.

Previously reported from Karnataka (Bongale & Barhati 1980), Maharashtra (Kamat & Harankhedkar 1976); W.B. (Sinha & Mukherjee 1975b, Mukhopadhyay & Chatterjee 1981, Santra et al. 1988).

94. *Phormidium papyraceum* (Ag.) Gomont (Plate 8, figs. 7 & 8)

Essai class Nostocacees homocystees, Morots J. Bot., 4: 355, 1890; Monogr. Oscillariees, 173, pl. 5, f. 3 & 4, 1892; Desikachary, 271, 1959.

Thallus dark blue-green expanding, leathery, filaments long, flexuous entangled with other filaments, sheath thin, sometimes diffluent; Trichomes 4-5 μ broad; not constricted at the cross-wall, cells quadrate to subquadrate, 2.5-4 μ long, cross-wall not granulated; end cell round to obtuse conical, calyptra absent.

From paddy fields of 24-Parganas (South), Howrah.

Specimen nos. S 31, 49, 50; How 19, 25, 27

Previously reported from Puri (Rao 1939b); Karnataka (Bongale & Bharati 1980); Patiala (Sarma & Kanta 1978); Maharashtra (Kamat 1963, Kamat & Harankhedkar 1976, Ashtekar & Kamat 1980); culture of paddy field soils of Madurai (Laloraya & Mitra 1974); West Bengal (Biswas 1941).

95. *Phormidium subfuscum* Kütz. ex Gomont (Plate 8, fig. 1 & 2)

Kützing, Phyc. gene., 195, 1843; Gomont, Monogr. Oscillariees, 182, pl. 5, f. 17-20, 1892; Geitler, 1022, f. 652d-g, 1932; Desikachary, 273, pl. 44, f. 22, 23, 1959.

Thallus dark blue-green colour, filaments straight, sometimes parallel, sheath diffluent, trichomes 7.5-10 μ broad, not constricted at the cross-walls, cross wall granulated, cells 2.5-4.5 μ long, ends briefly attenuated and capitate and cell conical.

From paddy fields of Bankura, 24-Parganas (North), Purulia, Birbhum.

Specimen nos. Ban 15, 18, 31, S 7, 8, 9; Pur 7, 13, 17; Bir 12, 19, 23.

Previously reported from U.P. (Rao 1937b); West Bengal (Sinha & Mukherjee 1975b, Mukhopadhyay & Chatterjee 1981).

96. *Phormidium angustissimum* W. et G.S. West (Plate 7, fig. 2)

Welwitsch's African freshwater Algae, J. Bot., Lond., 35: 298, 1897; Geitler, 996, 1932; Desikachary, 253, 1959.

Thallus pale blue-green colour, membranous; trichomes densely entangled, 0.5-1 μ broad, constricted at the cross-wall, sheath thin, diffluent, cells cylindrical, 1.5-5 μ long.

From paddy field soils of Darjeeling, 24-Parganas (South).

Specimen nos. Dar 7, 11, 13, 15; S - 1, 3, 5, 18.

Previously reported from U.P. (Pandey 1975, Prasad & Mehrotra 1980, Chadha & Pandey 1983a, Singh 1974); Karnataka (Bongale & Bharati 1980); Ahmedabad (Kamat 1962-63); Maharashtra (Barhate & Tarar 1983); Jaipur (Tridevy 1982); Patna, Mangalore, Madurai & Berhampur (Laloraya & Mitra 1974); West Bengal (Laloraya & Mitra 1974, Mukhopadhyay & Chatterjee 1981).

97. *Phormidium foveolarum* (Mont.) Gom. (Plate 9, figs, 3 & 4).

Monogr. Oscillariees, 164, pl. 4, f. 16, 1892; Geitler, 999, f. 636, c, d, 1932; Desikachary, 254, 1959.

Thallus dark blue-green colour, expanded and mucilagenous, trichomes entangled, 1.5-2.3 μ broad, constricted at the cross-wall, sheath thin and diffluent, cells quadrate to shorter than broad, 1-1.5 (-2) μ long, end cell rounded and calyptra absent.

From paddy field soils of Darjeeling, Bankura and Cooch Behar.

Specimen nos. Dar - 18, 22, 27; Ban - 9, 14; Cooch - 11, 17, 21.

Previously reported from U.P. (Mitra 1951, Pandey 1965, Pandey & Chaturvedi 1979, Prasad & Mehrotra 1980, Chadha & Pandey 1983a); Maharashtra (Kamat & Harankhedkar 1976, Ashtekar & Kamat 1980, Kolte & Goyal 1985); Patiala (Sarma & Kanta 1978); J & K (Goyal et al., 1984); M.P. (Das et al. 1974); W.B. (Laloraya & Mitra 1974, Das et al. 1974, Sinha & Mukherjee 1984).

98. *Phormidium anomala* Rao, C.B. (Plate 9, fig. 5)

The Myxophyceae of the United Provinces, III, Proc. Indian Acad. Sci., B. 6: 371, f. 7F -I, 1937; Desikachary, 266, pl. 45, f. 11-13, 1959.

Thallus dark blue-green colour, expanded and mucilaginous; trichomes 8-10.5 μ broad; cross-wall not constricted, not granulated, sheath thin; cells 1-2 μ long, end cell rounded, calyptra absent.

From paddy fields of Nadia & Malda.

Specimen nos. N - 1, 7, 12, 16; Mal - 3, 8, 9.

Previously reported from Bihar (Jha et al. 1986); U.P. (Chaturvedi & Pandey 1976, Prasad & Mehrotra 1980, Khan 1985); Patiala (Sarma & Kanta 1978); Karnataka (Bongale & Bharati 1980); Maharashtra (Kamat 1963, Barhate & Tarar 1982, Kolte & Goyal 1985); Tamil Nadu (Anand & Revathi 1987).

99. *Phormidium favosum* (Bory) Gom. (Plate 9, figs, 6 & 7)

Monogr. Oscillariees, 180, pl. 5, f. 14-15, 1982; Geitler, 1024, f. 652a,b, 1932; Desikachary, 275, pl. 44, f. 20, 21, 1959.

Thallus dark blue green colour, expanded, trichomes entangled, 5-μ broad, not constricted at the cross-wall, gradually attenuated at the apex; cells quadrate to shorter than broad, 4-6 μ long, end cell obtuse truncated with a hemispherical calyptra.

From paddy field soils of 24-Parganas (N), Howrah, Hooghly and Bankura.

Specimen nos. S - 17, 31, 45; How - 7, 13, 17; Hoog - 12, 16, 18; Ban - 27.

Previously reported from Karnataka (Bongale & Bharati 1980); Hyderabad (Laloraya & Mitra 1974); Orissa (Rao 1938); U.P. (Rao 1937, Prasad & Mehrotra 1980), W.B. (Sinha & Mukherjee 1984).

100. *Phormidium retzii* (Ag.) Gom. (Plate 9, figs. 8 & 9)

Essai class. Nostoc. homo., J. de Bot., Fr., 4: 355, 1890; Monogr. Oscillariees, 175, pl. 5, f. 6-9, 1892; Geitler, 1012, f. 647, 1932; Desikachary, 268, pl. 44, f. 13, 15, 1959.

Thallus blue-green colour, trichomes 5.5-8 μ broad, straight or curved, not constricted at the cross-walls, cross-wall distinct, sheath diffluent, cells 1.5-8 μ long, end cell truncated, sometimes with a thick outer membrane, calyptra absent.

Reported from rice field of Bankura, Howrah & Hooghly.

Specimen nos. Ban 7, 12, 14; How - 21, 23; Hoog -3, 13.

Previously reported from A.P. (Rao 1938); Orissa (Rao 1938); Maharashtra (Barhate & Tarar 1983); U.P. (Rao 1937, Prasad & Mehrotra 1980); Bihar (Rao 1939); Assam (Parukutty 1939); W.B. (Chatterjee et al. 1980, Sinha & Mukherjee 1984, Santra et al. 1988).

101. *Phormidium lucidum* Kütz. ex Gom. (Plate 9, fig. 10)

Kützing, Phyc. gene., 194, 1843; Gomont, Monogr. Oscillariees, 179, pl. 5, f. 11-12, 1892; Geitler, 1025, f. 648f-h, 1932; Desikachary 275, pl. 44, f. 17, 18, 1959.

Thallus pale blue-green colour, trichomes flexuous, 7-8 μ broad, sheath diffluent, sometimes slightly attenuated at the apex; cross-wall constricted and granulated; cells 2-2.5 μ long, end cell with a calyptra.

From paddy field soils of Bankura, Burdwan.

Specimen nos. Ban - 3, 9, 19; Bur- 2, 8, 25.

Previously reported from W.B. (Sinha & Mukherjee 1984).

102. *Phormidium uncinatum* (Ag.) Gom. (Plate 9, figs. 11 & 12)

Essai class. Nostocacees Homocystees, Morot's J. de Bot., 4: 355, 1890; Monogr. Oscillariees, 184, pl. 5, f. 21 & 22, 1892; Geitler, 1025, f. 652 h, i, 1932; Desikachary, 276, pl. 43, f. 1, 2 & pl. 45, f. 9, 10.

Thallus dark blue-green expanded; trichomes straight or curved, sheath diffluent; trichomes not constricted at the cross-wall, 6-8 μ broad, briefly attenuated at the apex, ends slightly curved; cells quadrate to subquadrate, 3-6 μ long; end cell with a depressed calyptra.

From paddy field soils of Jalpaiguri, Darjeeling and Bankura.

Specimen nos. Jal - 13, 17; Dar - 3, 6, 9; Ban - 7, 13

Previously reported from U.P. (Singh 1974); Tamilnadu (Mitra 1951); Maharashtra (Kamat 1963); Jammu (Anand 1979); Karnataka (Bongale & Bharati 1980); W.B. (Sinha & Mukherjee 1984).

Genus: LYNGBYA Ag.

Key to the species

1.	Filaments more or less spirally coiled	.. 2
1.	Filaments not so spirally coiled	.. 5
	2. Filaments curved, sometimes spirally coiled, may be straight also	.. 4
	2. Filaments distinctly spirally coiled	.. 3
3.	Trichomes 9-10 μ broad, cells 4.5-7.5 μ long	.. *L. laxespiralis*
3.	Trichomes 5-7.5 μ broad, cells 1.5-2.5 μ long,	.. *L. spiralis*
	4. Trichomes 2-2.3 μ broad, sheath thin, cells 1.5-2.5 μ long, cross-wall not constricted	.. *L. lagerheimii*
	4. Trichomes 4-5 μ broad, sheath thick hyaline with lacerated margin, cells 3-5 μ long, cross-wall slightly constricted	.. *L. lagerheimii* f. *lacerata*
5.	Filaments form a thin felt	.. 6
5.	Filaments not so	.. 8
	6. Trichomes slightly constricted at the cross-wall, 5-6 μ broad, cells 4-5.5 μ long, end cell rounded	.. *L. cryptovaginata*
	6. Trichomes not constricted at the cross-wall	.. 7
7.	Trichomes 4.5-6 μ broad, cells subquadrate 2-3 μ long, end cell with thick outer membrane	.. *L. aerugineo-coerulea*
7.	Trichomes 5-5.5(-6.5) μ broad, cells 3.5-6 μ long, end cell rotund, without cap or calyptra	.. *L. allorgei*
	8. Sheath not lamellated	.. 9
	8. Sheath lamellated, at least when old	.. 12
9.	Filaments 1.5-2.5 μ broad, cross-wall with a single granule on either side of septation	.. *L. perelegans*
9.	Filaments broader	.. 10
	10. Sheath colourless	.. 11
	10. Sheath red when old, filaments 12-14 μ broad, cells quadrate to subquadrate, end cells rotund, cross-walls not granulated	.. *L. ceylanica*

11.	Filaments 11-15 μ broad, cells 2-5(-6.5) μ long, end cells rotund, cross-walls somctimes granulated	..	*L. martensiana*
11.	Filaments 11-12.5 μ broad, cells 4-6.5 (-9) μ long, end cells rounded, cross-walls granulated and with vacuoles	..	*L. hieronymusii*
	12. Sheath distinctly lamellated	..	13
	12. Sheath lamellated when old	..	15
13.	Cells upto 4 μ long, sheath yellowish when mature; filaments 12-15 μ broad, trichomes 10-12 μ broad, end cell rounded	..	*L. stagnina*
13.	Cells may be longer, sheath colourless	..	14
	14. Filaments 17-20 μ broad, trichomes 12-15(-16) μ broad, cells 2.5-5(-7) μ long, and cell rounded, sometimes with thickened membrane	..	*L. major*
	14. Filament 40-62 μ broad, trichomes 30-41 μ broad, cells 5-10.5 μ long, end cell rotund	..	*L. majuscula*
15.	Cells 2-4 μ long, thallus yellowish brown, filaments 15-20 (-25) μ broad, trichomes 10-15 (-20) μ broad	..	*L. confervoides*
15.	Cells longer	..	16
	16. Thallus blue-green to yellowish green, filaments 23-26 (-32) μ broad, trichomes 17.5-19 (-25) μ broad, cells 4.5-8 μ long, end cell rounded	..	*L. birgei*
	16. Thallus pale-blue-green, filaments 12-16 μ broad, trichomes 9-12 μ broad, cells 5-9 μ long, end cell with thickened membrane	..	*L. ceylanica* v. *bankurensis*

103. *Lyngbya laxespiralis* Skuja (Plate 8, figs. 9 & 11)

Zur Süsswasseralgenflora Burmas, 53, pl. 10, f. 10-11, 1949; Desikachary, 289, pl. 50, f. 4, 1959.

Thallus blue-green, intermixed with other algae (*Schizothrix friesii*) filaments spirally arranged, sometimes more or less straight distance between two spirals 130-200 μ; sheath thin, firm, colourless, not lamellated; filaments 10-14.5 μ broad; trichomes 9-10 μ broad, slightly constricted at the cross-walls; cells shorter than broad, 4.5-7.5 μ long, apical cell rounded.

From rice field soils, Garhbeta (Midnapore), Onda (Bankura), Jhalda (Purulia) and Asansol (Burdwan).

Specimen nos. G 72, Ban - 12, Pu 25, Bur 18.

The present form differs from the type specimen in having broader filament (10-14.5 μ broad, instead of 9.5-11 μ), shorter cells (4.5-7.5 μ long instead of 5-8 μ) and longer distance between two spirals (130-200 μ instead of 100-110 μ).

Previously reported from Karnataka (Somashekar 1984b) and Punjab (Sarma & Kanta 1978).

104. *Lyngbya spiralis* Geitler (Plate 10, fig. 1)

Geitler, 1042, f. 659, 1932; Desikachary, 289, pl. 48, f. 1, 1959.

Thallus leathery, blue-green; filaments 5-7.5 μ broad, entirely or partly spirally coiled; sheath not lamellated, colourless; trichomes not attenuated, not capitate 5-7.5 μ broad; cells 1.5-2.5 μ long, end cell rotund or rounded, calyptra absent.

From paddy fields of Asansol (Burdwan), Kotulpur (Bankura).

Specimen nos. Bur 22, Ban 14.

Previously reported from Ahmedabad (Kamat 1962-63); Nagpur (Barhate & Tarar 1982, Tarar & Shewale 1984); Karnataka (Bongale & Bharati 1980); Marathwada of Maharashtra (Ashtekar & Kamat 1980); Bihar (Jha et al. 1986); W.B. (Sinha & Mukherjee 1975a, Mukhopadhyay & Chatterjee 1981, Sengar & Pandey 1985, Pal et al. 1986).

105. *Lyngbya lagerheimii* (Mob.) Gomont (Plate 11, fig. 14).

Essai class. Homo., J. de. Bot., 4: 354, 1890; Monogr. Oscillariees, 147, pl. 4, f. 6, 7, 1892; Geitler, 1044, f. 661g, 1932; Desikachary, 290, pl. 48, f. 6 & pl. 53, f. 2, 1959.

Filaments irregularly entangled; straight or curved; trichomes 2-2.3 μ broad sheath thin; cells 1.5-2.5 μ long, cross-wall not constricted, end cell rounded.

From water logged paddy field, Bishnupur (Bankura).

Specimen no. Ban 31

Previously reported from U.P. (Prasad & Mehrotra 1980, Pandey, D.C. 1965, Pandey 1982); Karnataka (Bongale & Bharati 1980, Somashekar 1983, 1984b); Maharashtra (Ashtekar & Kamat 1980, Bhoge & Ragothaman 1986); Kodaikanal of Tamilnadu (Suxena 1983) & W.B. (Sinha & Mukherjee 1975b).

106. *Lyngbya lagerheimii* forma *lacerata* Prasad et al. (Plate 8, fig. 10) Prasad et al., 64, f. 3, 6 & 6, 1976.

Filaments single or in groups straight or curved; sheath thick, hayaline, margin lacerated; trichomes 4-5 μ broad, slightly constricted at the cross-walls; cells nearly quadrate or slightly shorter, 3-5 μ long, end cell rounded or slightly attenuated, calyptra absent.

From paddy field soil, Narajole (Midnapore).

Specimen nos. M 22, 139

Previously reported from U.P. (Prasad et al. 1976)

107. *Lyngbya birgei* Smith G.M. (Plate 10, figs. 2, 3, 5 and 6)

New or interesting Algae from the Lakes of Wisconsin, Bull. Torrey bot. 61; 43: 482, pl. 26, f. 28, 1916

Geitler, 1048, f. 663, 1932; Desikachary, 296, pl. 50, f. 7, 8, 1959.

Thallus blue-green to yellowish green, filaments straight, free floating, 23-26(-32) μ broad sheath firm, colourless, unlamellated, but lamellated in the older filaments, 2.5-5 μ thick; trichomes not constricted but some filaments having lamellated sheath show clear constrictions at the cross-walls; trichomes 17.5-19 (-25) μ broad; cells 4.5-8 μ long, sometimes with gas vacuoles, end cell rounded not capitate.

From paddy fields, Garhbeta (Midnapore), Onda (Bankura).

Specimen nos. 468, M 43, Ban 28.

The present form has slightly broader filaments 23-26 (-32)μ instead of 20-24 μ, and longer cells (4.5-8μ) long instead of 2-2.5 μ in this respect it resemble Gupta's form.

Previously reported from Allahabad (Mitra 1951, Chadha & Pandey 1983); from paddy fields of U.P. (Prasad & Mehrotra 1980); Karnataka (Bongale & Bharati 1980); Punjab (Pandhol & Grover 1976, Sarma et al. 1979) & W.B. (Pal et al. 1986).

108. *Lyngbya ceylanica* Wille (Plate 10, fig. 7)

Denkschr. Akad. Wiss. Wien. Math. nat. Kl., 91: 161, pl. 3, f. 27-29, 1914; Geitler, 1054; f. 668a, 1932; Desikachary 299, pl. 54, f. 4, 1959.

Filaments 12-14 μ broad, straight; sheath; becomes red when colourless but old trichomes unconstricted at the cross-wall, not attenuated at the apices; 8.5-10 μ broad; generally cells 4.5-7.5 μ long, sometimes quadrate, end cell rotund, calyptra absent.

From paddy fields, Falta, Diamond Harbour (South 24-Parganas).

Specimen nos. 33, 70, 71.

Previously reported from Namdur of A.P. (Rao 1936, 1938); U.P. (Rao 1937, Prasad & Mehrotra 1980, Pandey & Pandey 1982, Chadha & Pandey 1983); Nagpur (Kamat & Harankhedkar 1976, Barhate & Tarar 1982); Karnataka (Bongale & Bharati 1980, Somashekar 1984); W.B. (Sinha & Mukherjee 1984, Maity & Santra 1985, Santra et al. 1988).

109. *Lyngbya ceylanica* Wille var. *bankurensis* Sinha & Mukherjee (Plate 9, fig. 13)

Sinha & Mukherjee, Phykos, vol. 23, 148, 1984.

Thallus pale blue-green colour, filaments straight or curved, 12-16 μ broad; sheath thin when immatured, but become thick and lamellated in mature filaments; trichomes 9-12 μ broad, not constricted at the cross-wall; cells 5-9 μ long, end cell round with thickened membrane.

From paddy fields of Moynapur, Bishnupur (Bankura), Garhbeta (Midnapore).

Specimen nos. Ban 12, 26; M 23.

Previously reported from W.B. (Sinha & Mukherjee 1984).

110. *Lyngbya confervoides* C. Ag ex Gomont (Plate 11, fig. 1 and Plate 41B)

Agardh, Syst. Alg., 73, 1824; Gomont, Monogr. Oscillariees, 156, pl. 3, f. 5 & 6, 1892; Geitler, 1061, f. 672b, 1932; Desikachary, 314, pl. 49, f. 9 & pl. 52, f. 13, 1959.

Thallus yellowish brown; filaments straight or slightly curved, 15-20 (-25) μ broad; sheath colourless, unlamellated but becomes lamellated when old; trichomes not constricted at the cross-walls, granulated, trichomes 10-15 (-20) μ broad; cells 2-4 μ long, end cell rotund.

From paddy fields, Haldia, Digha (Midnapore), Bak khali (South 24-Prgs.).

Specimen nos. 8, 14, 370, 375.

Reported earlier from A.P. (Rao 1938a); Punjab (Gupta 1950); Orissa (Biswas 1932b, Rao 1938b, Mohanty 1982); Bombay (Boergesen 1935, Dixit 1936); Karnataka (Bongale & Bharati 1980, Somashekar 1983); W.B. (Maity & Santra 1985, Santra & Pal 1988, Santra et al. 1988, 1989).

111. *Lyngbya majuscula* Harvey ex Gomont (Plate 10, fig. 4 and Plate 41C) Harvey in Hooker, English Flora, 5(Pt. 1): 370, 1833, Gomont, Monogr. Oscillariees, 151, pl. 3, f. 3, 4, 1892; Geitler, 1060, f. 672c, d; 1932; Desikachary 313, pl. 48, f. 7, pl. 49, f. 12 & pl. 52, f. 10, 1959.

Thallus expanded, blue-green to yellowish brown; filaments very long; sheath colourless, lamellated, 5-10 μ thick, outside rough; filaments 40-62 μ broad; trichomes not constricted at the cross-walls, 30-41 μ broad, not attenuated at the ends; cells 5-10.5 μ long, end cell rotund calyptra absent.

From paddy fields, Falta, Diamond Harbour (South 24-Parganas) and Tamluk (Midnapore).

Specimen nos. 46, 29, T 32.

The present forms have longer cells than that of type species (6.5-10.5 μ long instead of 2-4 μ).

Previously reported from U.P. (Gupta 1956, Pandey & Chaturvedi 1979, Prasad & Mehrotra 1980, Pandey & Pandey 1982); Hyderabad (Laloraya & Mitra 1974); Maharashtra (Kamat 1963, Barhate & Tarar 1981, 1982, Tarar & Shewale 1984, Bhoge & Ragothaman 1986); Karnataka (Somashekar 1983, 1984b) and W.B. (Chatterjee et al. 1980, Santra et al. 1988, 1989).

112. *Lyngbya martensiana* Menegh. ex Gomont (Plate 11, figs. 2-4)

Meneghini, Conspectus Algologiae euganeae, 12, 1837; Gomont, Monogr. Oscillariees, 145, f. 17, pl. 3, 1892; Geitler, 1064, f. 676, 1932; Desikachary, 318, pl. 52, f. 6, 1959.

Thallus blue-green; filaments entangled, 11-15 μ broad, sheath colourless, thick, not lamellated; trichomes not constricted at the cross-walls sometimes granulated, 9 12 μ broad, ends not attenuated; cells 2-5(-6.5) μ long, end cell rotund without calyptra.

From paddy fields, Keshpur (Midnapore), Asansol, Burdwan and Onda (Bankura).

Specimen nos. K 15, H 38, Bur 16, Ban 34.

The present alga comes closer to *L. martensiana* Menegh. ex Gomont in most of the character but differs from the type species due to the presence of longer cells (2-5(-6.5) μ long instead of 1.75-3.3μ) and slightly broader filaments. In Indian material the breadth of the filament is 9.5-16 μ, a range in which the filaments of the present alga can be included.

Previously reported from Manipur (Biswas 1930); U.P. (Rao 1937, Prasad & Mehrotra 1980, Pandey 1982, Pandey & Pandey 1982, Chadha & Pandey 1983); Ahmedabad (Kamat 1962-63); North East India (Reddy et al. 1986); Bihar (Jha et al. 1986); Punjab (Singh 1941, Sarma & Kanta 1978); Karnataka (Bongale & Bharati 1980, Somashekar 1983, 1984); Maharashtra (Kamat 1963, Ashtekar & Kamat 1980, Nandkar et al. 1983, Bhoge & Ragothman 1986); W.B. (Banerjee 1938, Laloraya & Mitra 1974, Maity & Santra 1985, Santra et al. 1988).

113. *Lyngbya hieronymusii* Lemm. (Plate 11, figs. 5 & 6)

Forschbers., Plon, 12: 146, pl. 4, f. 12, 13, 1905; Geitler, 1047, f. 656a, 1932; Desikachary, 297, pl. 48, f. 4, 1959.

Thallus blue-green, free-floating; filaments straight; 11-12.5 μ broad; sheath firm, hyaline; trichomes 8.5-10 μ broad, not constricted at the cross-walls, granulated with vacuoles, cells 4-6.5(-9) μ long, end cell rounded.

From paddy fields, Panskura, Dharma (Midnapore).

Specimen nos. Pa 18, Pa 19, Pa 20, D 101.

The present alga differs from the type species in having slightly barrower trichomes (8.5-10 μ broad instead of 11-13 μ) and longer cells (4-6.5 (-9) μ long instead of 2.5-4 μ), but in this respect it comes close to forma described by Laloraya & Mitra (1974, 282).

Previously reported from Maharashtra (Kamat & Harankhedkar 1976, Barhate & Tarar 1983, Bhoge & Ragothaman 1986); J. & K. (Goyal et al. 1984); Punjab (Pandhol & Grover 1976, Sarma & Kanta 1978); Assam (Kant & Pandey 1984); U.P. (Mitra 1951, Gupta 1956, Chaturvedi & Pandey 1976, Pandey & Chaturvedi 1979, Chadha & Pandey 1983a); Karnataka (Bongale & Bharati 1980); W.B. (Santra & Pal 1988, Santra et al. 1988, 1989).

114. *Lyngbya allorgei* Fremy (Plate 11, figs. 7 & 8)

Myxo. d' Afr. equat. franc., 189, f. 156, 1929; Geitler, 1059, f. 671, 1932; Desikachary, 313, pl. 54, f. 6. 1959.

Filaments solitary or entangled, long, forming a thin felt, filaments 6-7.5 μ broad, trichomes 5-5.5 (-6.5) μ broad, straight or bent, not constricted at the cross-walls, cells 3.5-6 μ long, nearly quadrate, or subquadrate end cell rounded, without cap or calyptra.

From paddy fields of Contai, Belda (Midnapore).

Specimen nos. C 2, C 6, C 8, C 120, 129

The present form differs from the type species in having broader trichomes (5-5.5 (-6.5) μ broad instead of 3.5-4 μ, in this respect it comes close to the form described by Dixit. Regarding the breadth of the filament, the present alga also resembles *Lyngbya dixitii* Kamat (1963, 295, pl. 17, f. 101) but differs in the length of the cells which are shorter in Kamat's species (3-4μ long).

Previously reported from Karnataka (Bongale & Bharati 1980, Somashekar 1983, Somashekar & Ramaswami 1983); U.P. (Chaturvedi & Pandey 1976); Punjab (Sarma et al. 1979); Maharashtra (Dixit 1936, Kamat & Harankhedkar 1976, Barhate & Tarar 1983).

115. *Lyngbya cryptovaginata* Schkorbatow (Plate 11, figs. 9 & 10)

Beobachtung über einige Cyanophy. d. Gouvern Charkov (Ukraine) Arch. Russ. Protistologia, 6(-4): 125, pl. 8, f. 14-16, 1927; Geitler, 12: 401, 1925; Desikachary, 297, pl. 50, f. 6, 1959.

Filaments free floating, blue-green, straight, 6.5-8.5 μ broad, sheath colourless, trichomes slightly constricted at the cross-walls, 5-6 μ broad, cells nearly quadrate 4-5.5 μ long, end cell rounded.

From paddy field soils, Narajole (Midnapore).

Specimen no. 446

Previously the alga was recorded from Maharashtra (Gonzalves & Joshi 1943); Karnataka (Bongale & Bharati 1980); U.P. (Bendre & Kumar 1975); Punjab (Pandhol & Grover 1976, Sarma & Kanta 1978); Bihar (Saha & Pandit 1987); W.B. (Pal & Santra 1987).

116. *Lyngbya stagnina* Kütz (Plate 11, fig. 11)

Species Algarum, 281, 1849; Tab. Phycologicae, 1: 48, pl. 87, f. 5, 1849; Geitler, 1066, 1932; Desikachary, 317, 1959.

Thallus pale blue-green, filaments 12-15 μ broad, sheath colourless, yellowish when mature; lamellated; trichomes 10-12 μ broad, not constricted at the cross-walls, granulated; cells 2-3.5 (-4) μ long, end cell rounded.

From paddy field soils of Burdwan, Birbhum, Malda, Nadia.

Specimen nos. Bur 7, Bir 9, Mal 14, 16, N 7, 13.

Earlier this was recorded from U.P. (Rao 1937); Punjab (Sarma et al. 1979); W.B. (Sinha & Mukherjee 1975).

117. *Lyngbya major* Menegh. ex Gomont (Plate 12, fig. 1-3)

Meneghini, Conspectus Algologiae euganeae, 12, 1837; Gomont, Monogr. Oscillariees, 144, pl. 3, f. 15; 1892; Geitler, 1066, f. 679a, 1932; Desikachary, 320, pl. 52, f. 11, 1959.

Filaments straight or curved, 17-20 μ broad, sheath thick, lamellated, upto 2.5 μ broad, trichomes 12-15 (-16) μ broad, not constricted at the cross-walls; septa granulated, cells 2.5-5(-7) μ long, end cell rounded, sometimes with thickened membrane.

From paddy fields of 24-Parganas (South), Howrah, Nadia, West Dinajpur, Jalpaiguri, Cooch Behar.

Specimen nos. S 17, 21, How 3, 9, N 7, 13, WD 7, 17 Jal 4, 9; Cooch 5

Previously reported from Orissa (Mohanty 1982, 1984); Tamilnadu (Suxena 1983); U.P. (Rao 1937b, Prasad & Mehrotra 1980); Punjab (Sarma et al. 1979); Karnataka (Somashekar 1983); Bihar (Jha et al. 1986); W.B. (Banerjee 1938, Sinha & Mukherjee 1975, Mukhopadhyay & Chatterjee 1981).

118. *Lyngbya perelegans* Lemm. (Plate 11, figs. 12 & 13)

Planktonalg., Ergebnisse einer Reise n.d. Pacific, Abh. nat. Ver. Bremen, 16: 355, 1899, Geitler, 1056, f. 66a-c, 1932; Desikachary, 309, pl. 48. f. 8 & pl. 53, f. 1, 1959.

Thallus light blue-green colour, filaments densely entangled straight or curved, 1.5-2.5(-3) μ broad; sheath thin, trichomes 1.2-2 μ broad, not constricted at the cross-wall with a single granule on either side of the septum; cells 2.5-5 μ long; end cell rotund.

From paddy fields of Bankura, Purulia and Birbhum.

Specimen nos. Ban 7, 10, Pur 15, 18, Bir 5, 11.

Previously reported from North India (Randhawa 1936c); U.P. (Pandey 1965); Maharashtra (Kamat 1963); Karnataka (Bongale & Bharati 1980); W.B. (Sinha & Mukherjee 1975a).

119. *Lyngbya aerugineo-coerulea* (Kutz.) Gom. (Plate 9, fig. 14)

Monogr. Oscillariees, 146, pl. 4, f. 1-3, 1892; Geitler, 1062, f. 670. 1932; Desikachary, 315, pl. 48, f. 9, 1959.

Thallus pale blue-green colour, expanded, filaments 4.5-6 μ broad, sheath thin; trichomes 4.5-6 μ broad, not constricted at the cross-wall, cells 2-3 μ long, end cell conical or rotund, with thick outer membrane.

From paddy fields Bishnupur (Bankura), Hooghly.

Specimen nos. Ban 7, 9; Hoog 12, 16.

Previously reported from Bihar (Jha et al. 1986); from paddy fields of U.P. (Pandey 1965, Prasad & Mehrotra 1980); Maharashtra (Kamat 1963, Kamat & Harankhedkar 1976, Barhate & Tarar 1982, 1983, Tarar & Shewale 1984); Punjab (Pandhol & Grover 1976, Sharma & Kanta 1978); Karnataka (Bongale & Bharati 1980, Somashekar 1984a,b, Somashekar & Ramaswamy 1983); W.B. (Sinha & Mukherjee 1984).

Genus: SCHIZOTHRIX Kutz.

Key to the species

1. Sheath generally coloured .. 2
1 Sheath colourless .. 4
    2. Cells quadrate to subquadrate, sheath colourless to golden yellow, trichomes 7.5-10 μ broad, cells 5-7.5 μ long, end cell obtuse .. *S. muelleri*
    2. Cells quadrate to cylindrical .. 3
3. Sheath golden-yellow, trichomes 3-4 μ broad, cells 4-7.5 μ long, end cell conical .. *S. lamyi*
3. Sheath red inside and colourless outside; trichomes 4-5.5 μ broad, cells 5-7.5 μ long, end cell rounded .. *S. telephoroides*
    4. Trichomes upto 3 μ broad .. 5
    4. Trichomes broader .. 6
5. Trichomes 1.3-2 broad, slightly constricted at the cross-wall cells 2-3.5 μ long, end cell slightly conical .. *S. lacustris*
5. Trichomes 2-3 μ broad, not constricted at the cross-wall, cells 4-5 μ long, end cell conical and pointed .. *S. arenaria*
    6. Sheath not lamellated, trichomes 3-5 μ broad, cells 4-8 μ long, unconstricted or slightly constricted at the cross-wall .. *S. penicillata*
    6. Sheath lamellated, trichomes 4-6 μ broad, cells 4-7 μ long, distinctly constricted at the cross-wall .. *S. friesii*

120. *Schizothrix telephoroides* (Mont.) Gomont (Plate 12, fig. 7; and Plate 41D) Monogr. Oscillariees, 319, pl. 10, f. 1-4, 1892; Geitler, 1101, f. 707, 1932; Desikachary, 330, pl. 57, f. 2, 7, 8, 1959.

Thallus reddish; filaments with thick lamellated sheath, sheath red inside and faintly coloured or colourless outside, outside of the sheath uneven, generally single trichome within the sheath, rarely two; trichomes 4-5.5 μ broad, slightly constricted at the cross-walls, cells 5-7.5 μ long, end cell rounded.

From paddy fields, Garhbeta & Panchkhuri (Midnapore), Onda (Bankura).

Specimen nos. Panch. 102, Panch 92, G 121, Ban 8.

The present form resembles *S. telephoroides* (Mont.) Gomont but differs from it not in having distinct constrictions.

Previously reported from North East India (Reddy 1986); W.B. (Sinha & Mukherjee 1975b).

121. *Schizothrix friesii* (Ag.) Gomont (Plate 12, fig. 9 and Plate 42A)

Monogr. Oscillariees, 316, pl. 9, f. 1, 2, 1892; Geitler, 1076, f. 685 & 686, 1932; Prescott, 507, pl. 114, f. 5, 1951; Desikachary, 328, pl. 57, f. 1, 6, 1959.

Thallus blackish green, filaments forming fascicles, parallel, straight, rarely curved; sheath usually colourless, lamellated, trichomes generally single, sometimes two to three within the sheath, blue-green, rarely slightly brown; trichomes distinctly constricted at the cross-walls, 4-6 μ broad, cells quadrate to slightly longer, 4-7 μ long, end cell obtuse or conical.

From paddy field soils, Keshpur (Midnapore), Purulia (Sadar).

Specimen nos. K 38, Pur 7 A

Previously reported from Maharashtra (Ashtekar & Kamat 1980, Kolte & Goyal 1985); Darjeeling (W.B.).

122. *Schizothrix muelleri* Näg. ex Gomont (Plate 12, figs. 8 & 10 and Plate 42B) Nägeli, in Kützing, species Algarum, 320, 1849; Gomont, Monogr. Oscillariees, 321, pl. 10, f. 5-7, 1892; Geitler, 1110, f. 7-15, 1932; Desikachary, 330, pl. 57, f. 9 & 10, 1959.

Thallus expanding, filaments arranged in creeping bundles, sheath colourless to golden yellow, gelatinizing, lamellated, generally single trichome within the sheath rarely a few; trichome 7.5-10 μ broad slightly constricted at the cross-walls, cells 5-7.5 μ long, end cell obtuse.

From paddy field soils, Garhbeta (Midnapore) & West Dinajpur.

Specimen nos. G 72a, W 32

Previously, this alga was reported from W.B. (Pal & Santra 1982).

123. *Schizothrix penicillata* (Kütz.) Gomont (Plate 12, figs. 4 &5)

Monogr. Oscillariees, 305, pl. 7, f. 8-10, 1892; Geitler, 1096, f. 702. 1932; Desikachary, 323, 1959.

Thallus dark-green; filaments unbranched; sheath not lamellated uneven outside, colourless, generally with single trichome, rarely two to four; trichomes unconstricted or slightly constricted at the cross-walls, 3-5 μ broad, cells 4-8 μ long, end cell obtuse or slightly conical.

From paddy field soils, Salboni (Midnapore), Diamond Harbour (South 24-Parganas).

Specimen nos. 101, 171, 392.

Previously reported from W.B. (Santra et al. 1988).

124. *Schizothrix lacustris* A. Br. ex Gomont (Plate 15, fig. 1)

A Braun in Kützing, Species Algarum, 320, 1849; Gomont, Monogr. Oscillariees, 301, pl. 6, f. 9-12, 1892; Geitler, 1092, f. 698, 699. 1932; Desikachary, 325, pl. 56, f. 6 & 10, 1959.

Filaments densely arranged, branched at the ends; sheath colourless; trichomes 2-5 in a sheath, 1.3-2 μ broad, slightly constricted at the cross-wall, cells 2-3.5 μ long.

From paddy fields of Burdwan

Specimen nos. Bur 18, 24

Previously reported from Karnataka (Somashekar 1984b); Maharashtra (Ashtekar & Kamat 1980); W.B. (Sinha & Mukherjee 1975b).

Here cells slightly broader (1.3-2 μ broad instead of 1-1.5 μ).

125. *Schizothrix arenaria* (Berk.) Gomont (Plate 9, fig. 15)

Monogr. Oscillariees, 312, pl. 8, f. 11 & 12, 1892; Geitler, 1085, f. 693d-e, 1932; Desikachary, 327, pl. 59, f. 12, 1959.

Trichomes 2-3 μ broad, sheath firm and pointed at the ends, not constricted at the cross-wall; 4-5 μ long, end cell pointed and conical.

From paddy field soil of Bankura, Malda & Murshidabad.

Specimen nos. Ban 2, 5; Mal 9, 19, Mur 3, 7.

Previously reported from paddy fields of U.P. (Prasad & Mehrotra 1980); Maharashtra (Gonzalves & Gangla 1949, Ashtekar & Kamat 1980); W.B. (Sinha & Mukherjee 1975b).

126. *Schizothrix lamyi* Gomont (Plate 12, fig. 6)

In Bornet, Algues du department de la Haute - Vienne et., Bull. Soc. bot., Fr., 38: 250. 1891; Monogr. Oscillariees, 323, pl. 11, f. 1-3, 1892; Geitler, 1114, f. 720, 1932; Desikachary, 330, pl. 56, f. 1, 4, 1959.

Filaments richly branched; sheath golden yellow, lamellated, outside uneven, pointed at the apex; trichomes 3-4 μ broad, distinctly constricted at the cross-walls, cells 4-7.5 μ long, end cell conical

From paddy field soils, Diamond Harbour (South 24-Parganas).

Specimen no. 71, 92.

Previously reported from W.B. (Maity & Santra 1985, Santra et al. 1988).

Genus: SYMPLOCA Kützing

127. *Symploca muralis* Kutz. ex Gomont (Plate 13, fig. 1 & 2).

Kutzing, Phyc. gene., 201, 1843, Gomont, Monogr. Oscillariees, 112. pl. 11, f. 10, 1892; Geitler, 1124, f. 732, 1932, Desikachary, 337, 1959.

Filaments very closely arranged forming dark blue-green tufts; sheath firm, thick, cells 3.5-4 μ broad, and 2.5-4 μ long, not constricted at the cross-walls, end cell conical, sometimes with thickened outer membrane.

From paddy field soils, Silda, Salboni (Midnapore).

Specimen nos. Sil 70, Sal 135

The present form resembles *S. muralis* Kütz. ex Gomont in many characters but here the end cell is not so thickened.

Previously reported from Allahabad (Gupta 1956); Maharashtra (Kamat 1963, Barhate & Tarar 1983).

Genus: MICROCOLEUS Desmazieres
Key to the species

1.	Trichomes not constricted at the cross-wall	.. 2
1.	Trichomes constricted at the cross-wall	.. 3
	2. Trichome with a conical calyptra, cells 3.5-5 μ broad, 3-6 μ long	.. *M. vaginatus*
	2. Trichome without calyptra, cells 3.8-5 μ broad, 4-6 μ long	.. *M. paludosus*
3.	End cell conical, pointed, trichome 2.5-3.5 μ broad, cell 4-7 μ long	.. *M. chthonoplastes*
3.	End cell may be slightly conical but not pointed, sometimes with rounded tip	.. 4
	4. Trichomes 4-5 μ broad, cells short barrel shaped 4-5.5 μ long, end cells slightly conical	.. *M. subtorulosus*
	4. Trichomes 4-5.5 μ broad, cells cylindrical, 6.5-10 μ long, end cells slightly conical with rounded tips	.. *M. lacustris*

128. *Microcoleus chthonoplastes* Thuret ex Gomont (Plate 13, fig. 3)

Thuret, Ann. Sci. nat. Bot. 6th ser., 1: 378, 1875; Gomont Monogr. Oscillariees, 353, pl. 14, f. 5-8, 1892;

Geitler, 1133, f. 739, 1932; Desikachary, 343, pl. 60, f. 7-9, 1959.

Filaments with many trichomes, sheath gelatinising, 16-30 μ broad, trichomes arranged in rope like bundles, distinctly constricted at the cross-walls, trichomes 2.5-3.5 μ broad; cells 4-7 μ long, end cell conical, pointed.

From paddy field soils, Digha (Midnapore).

Specimen no. 408.

The present alga comes closer to *Microcoleus chthonoplastes* Thuret ex Gomont in most of the characters and measurements.

The present form can also be compared to *M. lacustris* (Rabenh.)

Farlow because of its conical end cells, but in *M. lacustris* (Rabenh.) Farlow cells are 4-5.5 μ broad.

Previously reported from U.P. (Mitra 1951, from paddy fields - Pandey 1965, Prasad & Mehrotra 1980, Chadha & Pandey 1983a); Bihar (Jha et al. 1986); J.K.

(Goyal et al. 1984), Coimbatore, Hyderabad, Trichur of Kerala (Laloraya & Mitra 1974); Karnataka (Bongale & Bharati 1980); Maharashtra (Kamat 1963, Kolte & Goyal 1985); W.B. (Laloraya & Mitra 1974, Mukhopadhyay & Chatterjee 1981, Sengar & Pandey 1985, Santra & Pal 1988, Santra et al. 1988).

129. *Microcoleus subtorulosus* (Breb.) Gomont. (Plate 13, figs. 4 & 5)

Essai. class, Nostocacees homocystees, J. de. Bot., 4: 352, 1890; Monogr. Oscillariees, 369, pl. 14, f. 14-15, 1892; Geitler, 1143, f. 751, 1932; Desikachary, 345, pl. 56, f. 8,9, 1959.

Thallus blue-green sheath gelatinous, filaments 25-50 μ, in diameter, trichomes many, twisted but become parallel at the apex, gradually attenuated at the end, distinctly constricted at the cross-walls, trichomes 4-5 μ broad, cells 4-5.5 μ long, short barrel shaped, end cell slightly conical.

From paddy field soils Garhbeta, Silda (Midnapore).

Specimen nos. G 111, G 103, Sil 50, D 35.

The present form resembles *M. Subtorulosus* (Breb.) Gomont but differs from it in having narrower trichomes 4-5 μ instead of 6-10 μ) and cells are not much longer as described for the type species.

Previously reported from Punjab (Pandhol & Grover 1976); Karnataka (Bongale & Bharati 1980); W.B. (Pal & Santra 1982, Sinha & Mukherjee 1984).

130. *Microcoleus vaginatus* (Vaucher) Gomont (Plate 13, fig. 8)

Essai class. Nostocacees homocystees, Morot's J. de. Bot, 4: 353, 1890; Monogr. Oscillariees, 355, pl. 14, f. 12, 1892; Geitler, 1136, f. 741, 1932; Desikachary 343, pl. 56, f - 3, 1959.

Thallus dark blue-green colour, trichomes in together, 3.5-5 μ broad not constricted at the cross-wall, sheath hyaline, cells 3-6 μ long, end cell with a conical calyptra.

From paddy field soils Bishnupur (Bankura), Burdwan and West Dinajpur, Specimen no. Ban 16, Bur 5, WD 18.

Previously reported from Punjab (Pandhol & Grover 1976, Sarma et al.1976); Karnataka (Somashekar 1983); W.B. (Sinha & Mukherjee 1975b).

131. *Microcoleus lacustris* (Rabenh.) Farlow (Plate 13, fig. 6) Alg. es. Am. bor., nr. 227 bis., 1877; Geitler, 1142, f. 749, 750 a. 1932; Desikachary, 345, pl. 60, f. 4, 5, 1959.

Thallus dark blue-green; expanded trichomes many together, sheath gelatinous, colourless; trichomes distinctly constricted at the cross-wall, 4 - 5.5 μ broad; cells longer than broad, 6.5 - 10 μ long, end cell slightly conical with rounded tip.

From paddy fields of Nadia, Hooghly, Howrah, Bankura, Malda, Darjeeling Specimen nos. N 3, Hoog 1, How 14, Ban 31, Mal 8, Dar 12.

Previously reported from Karnataka (Somashekar 1983, 1984b); U.P. (Chadha & Pandey 1983a); Punjab (Sarma et al. 1979); Coimbatore (Laloraya & Mitra 1974); W.B. (Sinha & Mukherjee 1975b).

132. *Microcoleus paludosus* (Kütz.) Gomont (Plate 13, figs. 7 & 9)

Monogr. Oscillariees, 358, pl.14, f. 13; 1892; Geitler, 1114, f. 753, 1932; Prescott 505, pl. 113, f. 5, 1951; Desikachary, 344, pl. 56, f. 2, 1959.

Filaments form dark blue-green thallus, sheath gelatinous, with many rope-like twisted trichomes, trichomes more or less straight towards apex, trichomes not constricted at the cross-wall, 3.8-5 μ broad, cells as long as or slightly longer than broad, 4-6 μ long; end cell not capitate, slightly conical.

From paddy field soils, Haldia, Kolaghat, Contai, Tamluk, Silda (Midnapore).

Specimen nos. H 83, Ko 71, C 10, C 118, T 4, Sil 51.

Previously this alga was recorded from Punjab (Pandhol & Grover 1976, Sarma & Kanta 1978); Karnataka (Bongale & Bharati 1980, Somashekar 1984b); M.P. (Mishra & Purohit 1979); U.P. (Bendre & Kumar 1975; Pandey 1982); W.B. (Laloraya & Mitra 1974, Sinha & Mukherjee 1975b), Sengar & Pandey 1975, Mukhopadhyay & Chatterjee 1981, Pal & Santra 1985, Santra et al. 1989).

Genus: HYDROCOLEUM Kütz.

133. *Hydrocoleum lyngbyaceum* Kützing ex Gomont (Plate 14, figs. 3, 6 & 7) Kützing, Species Algarum, 259, 1849; Gomont. Monogr., Oscillariees, 337, pl. 12, f. 8-10, 1892; Geitler, 1150, f. 757, 1932; Desikachary, 347, pl. 46, f. 1, 2, 6, 1959.

Thallus gelatinous, dark blue-green, filaments unbranched or branched; sheath gelatinous, diffluent; trichomes many in the sheath or single when in branches; when single thick sheath may present around the trichome, trichome 7.5-14 μ broad, both narrow and broad types of trichomes are found within the same thallus, granulated at cross-walls, not constricted, cells 2-3 μ long, ends truncated attenuated, with calyptra and outer membrane of the end cell thickened.

From paddy fields, Geonkhali (Midnapore) & Diamond Harbour (South 24-Parganas).

Specimen no. 280, DH 76

The present form is similar to *H. lyngbyaceum* Kütz. ex Gomont but differs from it in having narrower trichomes (7.5-14 μ instead of 8-16 μ) and cells which are not much longer as described for the type species (2-3 μ long instead of 2.5-4.5 μ).

Previously it was reported from W.B. (Pal & Santra 1982, Santra et al. 1988).

Family: NOSTOCACEAE Kütz.

Key to the Genera

1. Trichomes with a sheath, heterocyst intercalary .. 2

1.	Trichomes without firm sheath	.. 3
	2. Sheath firm, cells not discoid, heterocysts and spores generally cylindrical	.. *Aulosira*
	2. Sheath hyaline, thin, cells discoid, heterocysts compressed, spores globose	.. *Nodularia*
3.	Without heterocysts	.. *Pseudanabaena*
3.	With heterocyst	.. 4
	4. Heterocyst terminal, spore next to the terminal heterocyst, trichome without sheath	.. *Cylindro-spermum*
	4. Heterocyst rarely terminal, generally intercalary	.. 5
5.	Filaments not free, variously curved or entangled in a definite colony	.. *Nostoc*
5.	Filaments or trichomes not so curved or entangled	.. 6
	6. Filaments free or in a formless gelatinous mass	.. *Anabaena*
	6. Thallus tubular, or cylindrical, soft, attached, filaments more or less parallel	.. *Wollea*

Genus: CYLINDROSPERMUM Kutz.
Key to the species

1.	Shape of spores spherical	.. *C. sphaerica*
1.	Shape of spores not spherical	.. 2
	2. Spore wall totally smooth	.. 3
	2. Spore wall not so	.. 7
3.	Heterocysts upto 6 μ broad, spores upto 30 μ long	.. 4
3.	Heterocysts broader 6-7 μ broad, 7-11 μ long, spores cylindrical, 11.5-13 μ broad, 30-34 μ long	.. *C. stagnale*
	4. Spores upto 13.5 μ broad	.. 5
	4. Spores broader, 12-15 μ broad, 19-27 μ long, heterocysts 4.5-5.5 μ broad and 6.5-8.5 μ long	.. *C. lichaniforme*
5.	Heterocysts oval or nearly spherical, 5-6 μ broad, 6.5-8 μ long, spores ellipsoidal, 8-12 μ broad, 18-25 μ long	.. *C. micheilo-skoense*
5.	Heterocysts oblong	.. 6
	6. Spores 8.5-11.5 μ broad, 14.5-20(-21) μ long; heterocysts sometimes slightly conical	.. *C. muscicola*
	6. Spores 10-13.5 μ broad, 22-28 μ long	.. *C. muscicola* v. *longispora*

7. Spores 8-10 μ broad, 16-20 μ long; outer membrane of spore smooth but with faint lines of striations .. *C. indicum*
7. Spores broader .. 8
   8. Spore wall with radial striations .. *C. alatosporum*
   8. Spore wall with papillae .. *C. majus*

134. *Cylindrospermum muscicola* Kütz. ex Born. et Flah. (Plate 14, figs. 8, 9 & 10) Born. et Flah, 1888, Revision des Nostocacees heterocystees 254, 1888; Geitler, 822, f. 520d; 1932; Prescott, 531, pl. 122, f. 16, 1951; Desikachary, 366, pl. 65, f. 3, 1959.

Thallus gelatinous; trichomes 3.5-5 μ broad, constricted at the cross-walls 300-350 μ long, cells 4-5.5 μ long, nearly quadrate, or slightly longer; heterocysts 4-5.5 μ broad and 5-9 μ long, oblong slightly conical, sometimes at the both ends, spores oblong, with rounded ends, 8.5-11.5 μ broad, 14.5-20 (-21) μ long, epispore smooth, yellowish brown.

From paddy field soils, Contai (Midnapore) and Onda (Bankura). Nadia.

Specimen nos. C 3, Ban 31, N 23.

The present form is near to *Cylindrospermum muscicola* Kutz. ex Born. et Flah. in most of the characters but differs from the type species in having slightly broader trichomes (3.5-5 μ instead of 3-4.7 μ), longer cells (4-5.5 μ instead of 4-5 μ), in this respect it comes close to the form described by Singh (1939a).

Regarding the breadth of the spore the present form can also be compared to the *C. muscicola* var. *longispora* Dixit, but spores are much longer in *C. muscicola* var. *longispora* Dixit.

Previously the alga was reported from Karnataka (Bongale & Bharati 1980; Somashekar 1983, 1984b); North East India (Reddy et al. 1986); Tamil Nadu (Anand & Revathi 1987); M.P. (Mishra & Purohit 1979); J. & K. (Goyal et al. 1984); U.P. (Prasad & Mehrotra 1980, Pandey & Pandey 1982, Chadha & Pandey 1983a); Maharashtra (Kamat 1962-63b, Kulti & Goyal 1985, 1986); W.B. (Laloraya & Mitra 1974, Sinha & Mukherjee 1975a, Mukhopadhyay & Chatterjee 1981).

135. *Cylindrospermum muscicola* var. *longispora* Dixit (Plate 15, figs. 9-11)

The myxophyceae of Bombay Province - I, Proc. Indian Acad. Sci., B, 3: 100, f. 3A, B, 1936; Desikachary, 367, pl. 64, f. 2, 10, 1959.

Trichomes 3.5-4.5 μ broad, cells 4.5-6.5 μ long; heterocysts 5-6 μ broad and 8-10.5 μ long; spores 10-13.5 μ broad, 22-28 μ long.

From paddy fields Simlipal (Bankura), Purulia.

Specimen nos. Ban 28, Pur 6.

Previously reported from Kolhapur of Maharashtra (Kamat 1963); U.P. (Pal 1975, Pandey 1982); W.B. (Sinha & Mukherjee 1975a).

Here trichome and spore slightly broader (trichome 3.5-4.5 μ broad instead of 3.7-4 μ and spores 10-13.5 μ broad instead of 7.5-12.5 μ).

136. *Cylindrospermum stagnale* (Kutz.) Born. et Flah. (Plate 15, figs. 2 & 3)

Revision des Nostocacees heterocystees, 250, 1888; Geitler, 819, f. 520c, 1932; Desikachary, 363, pl. 65, f. 9, 1959.

Thallus expanded, blue-green; trichomes 4-4.5 μ broad, constricted at the cross-walls, cells quadrate to cylindrical, 4-7.5 μ long; heterocyst spherical to oblong, 6-7 μ broad, 7-11 μ long; spores cylindrical 11.5-13 μ broad and 30-34 μ long, with smooth outer-wall.

From paddy field soils, Keshpur (Midnapore), Jilmili (Bankura) & Asansol (Burdwan).

Specimen nos. 119, Ban 43, A 23.

Previously this alga was reported from Maharashtra (Kamat 1962-63b, 1963); U.P. (Tiwari 1964, Chaturvedi & Pandey 1976, Chadha & Pandey 1983a); Bihar (Jha et al. 1986); Jammu (Anand 1979); Karnataka (Bongale and Bharati 1980, Somashekar 1984b); M.P. (Roy & Sen 1985); W.B. (Sinha & Mukherjee 1975b, Sengar & Pandey 1985).

137. *Cylindrospermum licheniforme* Kutz. ex Born. et Flah (Plate 14, figs. 4 & 5) Kutz., Diagnosen und Bemerkungen etc., Bot. Zeitung, 5: 197, 1847; Tab. Phycologicae, 1: 53, pl. 98, f. 6; 1859; Born. et Flah., Revision des Nostocacees heterocystees, 253, 1888; Geitler, 821, f. 520e, 1932; Prescott, 529, 1951; Desikachary, 366, pl. 65, f. 8, 1959.

Thallus mucilaginous, dark blue-green, trichomes 3.5-4 μ broad, constricted at the cross-walls; cells quadrate to cylindrical, 4-5 μ long; heterocyst oblong, 4.5-5.5 μ broad and 6.5-8.5 μ long; spores oblong to ellipsoidal, 12-15 μ broad and 19-27 μ long, epispore smooth.

From paddy field soils of Darjeeling, Jalpaiguri, Cooch Behar, Malda, West Dinajpur.

Specimen nos. WD 24, Dar 12, Cooch 3, 9, Mal 2, 8; Jal 1, 7, 9

The present form shows close resemblance to *C. licheniforme* Kutz. ex Born. et Flah. but slightly differs in size of spores.

Previously reported from Simla (Ghose 1924); North East India (Reddy et al. 1986); Kolhapur of Maharashtra (Kamat 1963); U.P. (Prasad 1952); W.B. (Prain 1905, Sinha & Mukherjee 1975a, Chatterjee & Chatterjee 1983).

138. *Cylindrospermum majus* Kutz. ex Born. et Flah. (Plate 14, figs. 16 & 17 and Plate 42D)

Kützing, Phyc. gene., 212, 1843; Bornet and Flahault, Revision des Nostocacees heterocystees, 252, 1888; Geitler, 815, f. 520b, 1932; Prescott, 530, pl. 122, f. 11,12, 1951; Desikachary, 360, pl. 80, f. 1, 1959.

Thallus expanded mucilaginous, dark blue-green; trichomes 4-5 μ broad, constricted at the cross-walls, cells as long as broad, or slightly longer, 4.5-7 μ long; heterocyst terminal, oblong, 5-6 μ diam., spores ellipsoidal, 11-13 μ broad, 22-26 μ long, epispore brownish with distinct papillae.

From rice field soils, Jhargram (Midnapore).

Specimen no. J 33.

Previously reported from Maharashtra (Kamat 1963, Kolte & Goyal 1985, 1986); Karnataka (Laloraya & Mitra 1974, Bongale & Bharat 1980); U.P. (Pandey & Pandey 1982, Chadha & Pandey 1983a) North East India (Reddy et al. 1986); Punjab (Grover & Pandhol 1975, Pandhol & Grover 1976, Sarma et al. 1979); J. & K. (Goyal et al. 1984); Bihar (Laloraya & Mitra 1974, Jha et al. 1986); W.B. (Laloraya & Mitra 1974, Sinha & Mukherjee 1975, Mukhopadhyay & Chatterjee 1981, Chatterjee & Chatterjee 1983).

139. *Cylindrospermum indicum* Rao, C.B., orth. mut. De Toni (Plate 14, figs. 11-13).

Rao, The Myxophyceae of the United Provinces India II, Proc. Indian Acad. Sci., B., 3: 169, f. 2, D, E. 1936; Desikachary, 369, pl. 64, f. 4, 11, 1959.

Thallus dark blue-green, trichomes deeply constricted at the cross-walls, cells quadrate or barrel shaped, 4-4.5 μ broad towards end regions and 2.5-3 μ broad at the median region of the trichome, cells of the terminal regions of trichomes 3-4 μ long, and cells of median region of trichomes upto 6 μ long; heterocyst spherical or subspherical, 3.5-5 μ broad and 4.5-6 μ long; spores cylindrical, with smooth outer membrane but sometimes faint lines of striations are seen; spores 8-10 μ broad and 16-20 μ long.

From paddy fields, Dharma (Midnapore) and Bankura.

Specimen nos. D 37, 58, 60, Ban 5, 10.

Spores of the present form are not so long and broad as described for the type species.

Previously reported from U.P. (Rao 1936); Maharashtra (Kolte & Goyal 1986); Karnataka (Bongale & Bharati 1980); W.B. (Sinha & Mukherjee 1975b).

140. *Cylindrospermum alatosporum* Fritsch (Plate 15, figs. 4-6)

Contributions to our knowledge of the fresh water Algae of South Africa, II, Ann. South Afr. Mus., 9: 587, f. 37, d, h, 1918; Geitler, 817, f. 521; 1932; Desikachary, 362, pl. 64, f. 9, 1959.

Thallus blue-green, trichomes closely packed, 3-4 μ broad; cells 3-5 μ long; heterocyst spherical to ellipsoidal, 5-5.5 μ broad, 6.5-8 μ long; spores yellowish brown, wall radially striated, truncated at the apex, 11-15 μ broad, 21-26 μ long.

From paddy fields of Midnapore, Howrah.

Specimen nos. 33, 39.

Previously reported from Maharashtra (Kolte & Goyal 1986); Uttar Pradesh (Chadha & Pandey 1983a); West Bengal (Sinha & Mukherjee 1975a).

141. *Cylindrospermum michailovskoense* Elenkin (Plate 14, figs. 1 & 2)

Neue selt. od. interess. Art. u. Form. d. Alg., in Mitt. Russl., II, Bull. Jardin Imp. Pierre le Gr., II: 162, 1911; Desikachary, 368, pl. 65, f. 1, 1959.

Thallus light blue-green colour, trichomes arranged in together, 4-5 µ broad, constricted at the cross-wall, cells 5.5-7 µ long; heterocysts oval, 5-6 µ broad, 6.5-8 µ long; spores ellipsoidal, 8-12 µ broad, 18-25 µ long.

From paddy fields of Bankura, Midnapore & 24 Parganas (South).

Specimen nos. Ban 24, M 15, S 35, 38.

Previously reported from U.P. (Prasad & Mehrotra 1980); North India (Randhawa 1936); W.B. (Sinha & Mukherjee 1984).

142. *Cylindrospermum sphaerica* Prasad (Plate 14, figs. 14 & 15)

Some Nostocaceae from Uttar Pradesh, J. Indian bot. Soc., 31: 358, f. 1-4, 1952; Desikachary, 363, pl. 64, f. 7, 8, 1959.

Thallus mucilaginous, yellowish brown colour, trichomes entangled 5-6 µ broad, constricted at the cross-wall; cells 4-6 µ long; heterocyst ellipsoidal, 5-6 µ broad, 7-10 µ long, spores spherical, 15- 18 µ broad, spore wall yellowish brown, thick, smooth.

From paddy fields of Howrah, Hooghly, 24- Parganas (North).

Specimen nos. How 21, 29; Hoog 7, 17; S - 4, 15, 35

Previously reported from U.P. (Prasad 1952); Karnataka (Somashekar 1983); W.B. (Chatterjee & Chatterjee 1983).

Genus: WOLLEA Born. et Flah.

143. *Wollea bharadwajae* Singh (Plate 15, figs. 7 & 8)

Wollea bharadwajae sp. nov., and its Autecology, Ann. Bot. Lod., n.a. 6: 593-606, 1942; Desikachary, 372, pl. 66, 1959.

Thallus soft, cylindrical, attached or free-floating, few cm long trichomes arranged in parallel, 4-5 µ broad, distinct constrictions at the cross-walls; cells 3.5-5.5 µ long, apical cell slighty conical; heterocysts barrel-shaped, 5.5-6.5 (-7) µ broad, 5-7.5 µ long; spores on either side of the hetercyst, spherical to subspherical, 10-15 µ diam., exopore thick, yellowish green to yellowish brown.

From water logged paddy fields of 24-Parganas (South), Bankura & Hooghly.

Specimen nos. S-39, 45; Ban - 26, 33; Hoog-21.

The present alga resembles *W. bharadwajae* Singh in most of the characters but slightly differs in sizes of cells, heterocysts and spores.

Previously reported from U.P. (Prasad 1952, Chadha & Pandey 1983); W.B. (Sinha & Mukherjee 1975, Sinha & Mukherjee 1984) also described *W. bharadwajae* f. Prasad from Bankura district.

Genus: NOSTOC Vaucher
Key to the species

1. Spores seen	..	2
1. Spores not seen	..	12
2. Trichomes clearly seen	..	3
2. Trichomes densely coiled sometimes not very clearly seen, 3.5-4 μ broad, spores subspherical, 5-6 μ broad, 5.5.-7.5 μ long	..	*N. punctiforme*
3. Thallus microscopic, trichomes 3-3.5 (-4) μ broad, spores oval to slightly oblong, 5-6.5 x 6-9 μ	..	*N. paludosum*
3. Thallus macroscopic	..	4
4. Trichomes densely coiled	..	5
4. Trichomes not densely coiled	..	9
5. Trichomes upto 3.5 μ broad	..	6
5. Trichomes broader	..	7
6. Thallus large, surface verrucose, trichomes entangled towards periphery of thallus, 3-3.5 μ broad, spores 5-6 μ broad, 5.5-7.5 μ long	..	*N. verrucosum*
6. Thallus not so large, trichomes entangled inside 2.5-3 μ broad, spores 4.5-5 μ broad, 6-7.5 μ long	..	*N. humifusum*
7. Spores oblong, 5.5-7.5 x 7.5-10 μ, trichomes 4-5 μ broad, cells 5-7 μ long	..	*N. muscorum*
7. Spores not oblong	..	8
8. Trichomes enveloped by lamellated yellowish mucilage, cells 4-5.5 μ broad, 4-5 μ long; spores spherical, 6-8 μ diam.	..	*N. linckia* var. *arvense*
8. Sheath hyaline near trichomes but thick and distinct towards periphery, cells 4.5-5.5 μ broad, 4-5.5 μ long; spores as broad as vegetative cells	..	*N. commune*
9. Spores ellipsoidal to oblong, trichomes, 3.5-4 μ broad, cells cylindrical 6-7.5 (-9) μ long, spores 6.5-8 x 12-16 μ	..	*N. ellipsosporum*
9. Spores spherical, subspherical or globose	..	10
10. Trichomes 4-5.5 μ broad, cells 4-7 μ long, spores 6-7.5 (-9) μ diam.	..	*N. piscinale*
10. Trichomes 3-4 μ broad	..	11
11. Cells moniliform, size 3-3.5 x 3-3.5(-4) μ spores 5-5.5(-6.5) μ diam.	..	*N. calcicola*

11. Cells nearly cylindrical, size 3-4 x 3.5-6 μ, spors 6-7 x 6.5-10 μ .. *N. spongiaeforme*
    12. Thallus macroscopic, trichomes irregularly curved, 4-5 μ broad .. *N. hatei*
    12. Thallus microscopic .. 13
13. Trichomes arranged more or less in parallel in a gelatinous mass, 3.5-4.5 μ broad .. *N. depressum*
13. Trichomes arranged irregularly, 4.5-5.5 μ broad .. *N. comminutum*

144. *Nostoc humifusum* Carm. ex Born. et Flah. (Plate 15, fig. 12)

Carmichael, ex Harvey, Hooker's British Flora, 2: 399, 1833; Born. et Flah. Revision des Nostocacees heterocystees, 201, 1888; Geitler, 842, 1932; Desikachary, 384, 1959.

Thallus mucilaginous, dark blue-green colour, trichomes closely packed, and twised, 2.5-3 μ broad cells barrel-shape or quadrate, 2.5-4 μ long; heterocysts broader than vegetative cells, 4-4.5 μ broad and 4-5 μ long, spores in chain, oval or slightly oblong, 4.5-5 μ broad and 6-7.5 μ long, epispore smooth and hyaline.

From paddy field soils, Contai (Midnapore).

Specimen nos. C 7.

The present form is near to *N. humifusum* Cam. ex Born. et Flah. in most of the characters but differs from the type species in having slightly larger heterocysts and spores.

Previously, it was recorded from U.P. (from paddy fields, Prasad & Mehrotra 1980, Pandey 1982); Karnataka (Bongale & Bharati 1980); Maharashtra (Gonzalves & Gangla 1949, Chapockar & Gangawane 1984); from culture of paddy field soils of Hyderabad (Laloraya & Mitra 1974); W.B. (Mukhopadhyay & Chatterjee 1981).

145. *Nostoc calcicola* Brebisson ex Born. et Flah. (Plate 16, fig. 1).

Brebisson in Meneghini, Monographia Nostochinearum italicarum, 121, 1843; Born. et Flah. Revision des Nostocacees heterocystees, 202; 1888; Geitler, 842, f. 534, 1932; Desikachary, 384, pl. 68, f. 1, 1959.

Thallus mucilaginous, sheath diffluent, trichomes losely arranged; trichomes 3-3.5 μ broad, cells moniliform, sometimes slightly longer than broad, 3-3.5 μ long (sometimes 4 μ long); heterocysts subspherical, 4-5 μ broad and 5-5.5 μ long; spores spherical to subspherical, 5-5.5 (-6.5) μ in diameter, epispore smooth.

From paddy field soils Narajole (Midnapore), Kalyani (Nadia).

Specimen nos. Na 21, N 8.

The present form differs from the type specimen in having slightly broader trichomes (3-3.5 μ broad instead of 2.5 μ), bigger spores (5-5.5(-6.5) μ instead of 1.5 μ) and heterocysts in this respect it has some similarities with the *N. calcicola* forma *variabilis* Prasadeat et al. (1977, 225, pl. 1, f. 3).

Previously reported from Maharashtra (Gonzalves & Gangla 1949, Chaporkar & Gangawana 1984, Kolte & Goyal 1985, 1986); J. & K. (Goyal et al. 1984); Karnataka (Bongale & Bharati 1980); Tamil Nadu (Anand & Revathi 1987); North East India (Reddy et al. 1986); U.P. (Prasad & Mehrotra 1980, Chadha & Pandey 1983a); from culture of soils of Shillong (Das et al. 1974), W.B. (Mukhopadhyay & Chatterjee 1981, Pal & Santra 1982, Chatterjee & Chatterjee 1983).

146. *Nostoc linckia* (Roth) Bornet ex Born. et Flah.

Bornet in Bornet and Thuret, Notes algologiques, II, 86, pl. 18, f. 1-12, 1880; Born. et Flah., Revision des Nostocacees heterocystees, 192, 1988. var *arvense* Rao C.B. (Plate 16, fig. 5).

The Myxophyceae of United Provinces, III, Proc. Indian Acad. Sci., B, 6: 358, f. 4A, 1937b; Desikachary, 377, p. 67, f. 1, 1959.

Thallus expanded, gelatinous, trichomes densely arranged, coiled, 4-5.5 μ in breadth enveloped by slightly lamellated mucilagenous yellowish brown sheath; cells spherical to barrel shaped sometimes moniliform; 4-5 μ in length; heterocyst spherical, 5.5-7 μ in diam, spores in chain, spherical 6-8 μ diam.

From paddy field soils, Dharma (Midnapore), Onda (Bankura) and Burdwan.

Specimen nos. D 74, Ban 8, Bur 16.

The present form is near to *N. linckia* var. arvense Rao, C. B. in most of the characters but differs from the type species in having slightly smaller cells (4-5 μ in length instead of 4.8-6.4 μ) and spores (6-8 μ in diam. instead of 7-8 μ).

Reported earlier from Gaya (Laloraya & Mitra 1974); Nainital (Singh 1974); Benaras (Rao 1937b); from paddy fields in Lucknow and its neighbourhood (Prasad & Mehrotra 1980); from culture of paddy field soils of Mussoorie (Das et al. 1974).

147. *Nostoc paludosum* Kützing ex Born. et Flah. (Plate 16, fig. 4)

Kützing, Tab. Phycologicae, 2; 1 pl. 1, f, 2, 1850; Bornet et Flahault, Revision des Nostocacees heterocystees, 191, 1888; Geitler, 836, f. 528a, 1932; Prescott, 524, pl. 121, f. 1-3, 1951; Desikachary, 375, pl. 69, f. 2, 1959.

Thallus microscopically small, trichomes 3-3.5(-4) μ broad, cells as long as broad or slightly longer, 3-4 μ long, heterocyst 4-5.5 μ broad and 5-6.5 μ long; spores oval to oblong, 5-6.5 μ broad and 6-9 μ long, epispore smooth.

From paddy field soils, Garhbeta (Midnapore) & Bankura.

Specimen nos. G 44, Ban 3, 17.

The present form differs from the type species in possessing slightly narrower trichomes (2.5-3 μ instead of 3-3.5 μ) and broader spores (5-6.5 μ broad instead of 4-4.5 μ).

Previously reported from Tamil Nadu (Laloraya & Mitra 1974); Maharashtra (Kamat & Harankhedkar 1976, Ashtekar & Kamat 1980, Kolte & Goyal 1986); Karnataka (Bongale & Bharati 1980); Punjab (Sarma & Kanta 1978); U.P. (Pandey & Chaturvedi 1979, Prasad & Mehrotra 1980, Pandey & Pandey 1982); Chadha & Pandey 1983a); W.B. (Sinha & Mukherjee 1975b, Chatterjee & Chatterjee 1983).

148. *Nostoc piscinale* Kutzing ex Born. et Flah. (Plate 15, fig. 14)

Kutzing, Phyc. gene. 208, 1843; Bornet and Flahult, Revision des Nostocacees, 194, 1888; Geitler, 838, f. 529, 1932; Desikachary, 377, pl. 69, f. 3, 1959.

Thallus more or less globose, gelatinous, blue-green colour, trichome, 4-5.5 μ broad, cells shorter or longer than broad, 4-7 μ long, heterocysts subspherical or oblong, 4-6 μ broad, spores globose, 6-7.5(-9) μ in diameter, arranged in chains, epispore smooth.

From paddy field soils, Diamond Harbour (South 24-Parganas) & Panagarh (Burdwan).

Specimen nos. 33, 109, Bur 33.

Previously reported from U.P. (Gupta 1956, Prasad & Mehrotra 1980, Chadha & Pandey 1983); Jalgaon of Maharashtra (Bhoge & Ragothaman 1986); Punjab (Sarma & Kanta 1978); Sagar of M.P. (Mishra & Purohit 1979); Maharashtra (Kamat 1962-63, 1963); J. & K. (Goyal et al. 1984); Mangalore (Laloraya & Mitra 1974); Karnataka (Bongale & Bharati 1980); W.B. (Sinha & Mukherjee 1975b, Maity & Santra 1985, Sengar & Pandey 1985, Santra et al. 1988).

149. *Nostoc punctiforme* (Kutz.) Hariot (Plate 15, fig. 13)

J. de Bot., 5: 31, 1891; Geitler, 834, 1932; Prescott, 525, pl. 121, f. 4, 5, 1951; Desikachary, 374, pl. 69, f. 1, 1959.

Thallus dark blue green, filaments densely packed; sheath mucilaginous, trichomes rarely straight, 3.5-4 μ broad, cells barrel-shaped, heterocysts subspherical, 5-6 μ broad; spores subspherical, 5-6 μ broad and 5.5-7.5 μ long.

From paddy field soils of Bankura, 24-Parganas, Nadia & Malda.

Specimen nos. Ban 12, S 25, N 9, Mal 12

Earlier this alga was recorded from North East India (Reddy et al. 1986); Tamil Nadu (Laloraya & Mitra 1974); Karnataka (Somashekar 1983); Punjab (Sarma & Kanta 1978); J. & K. (Goyal et al. 1984); Maharashtra (Chaporkar & Gangawane 1984, Kolte & Goyal 1985 & 1986, Bhoge & Ragothaman 1986); Bihar (Jha et al. 1986); U.P. (Pandey 1965, Chadha & Pandey 1983a, Khan 1985); W.B. (Sinha & Mukherjee 1975b, Maity & Santra 1985, Santra et al. 1988).

150. *Nostoc hatei* Dixit (Plate 16, fig. 2)

The Myxophyceae of the Bombay Presidency - I, Proc. Indian Acad. Sci., B, 3: 101, f. 30, 1936; Desikachary, 389, pl. 67, fig. 2, 1959.

Thallus attached or free floating, trich mes 4-5 μ broad, irregularly curved, densely arranged; cells ellipsoidal, heterocysts single, spherical to barrel-shaped, 4.5-5 μ broad and 4.5-5.5(-6) μ long.

From paddy field soils of Burdwan, Birbhum & Howrah.

Specimen nos. Bur 12, Bir 16, How 24

Previously recorded from Karnataka (Bongale & Bharati 1980); Maharashtra (Dixit 1936, Chaporkar & Gangawane 1984); U.P. (Prasad & Mehrotra 1980); Bihar (Jha et al. 1986); W.B. (Sengar & Pandey 1985).

151. *Nostoc muscorum* Ag. ex Born. et Flah. (Plate 17, fig. 1)

Agardh. Dispositio Algar. Sueciae, 44, 1812; Bornet et Flahault, Revision des Nostocacees heterocystees, 200, 1888; Geitler, 844, f. 535, 1932; Desikachary, 385, pl. 70, f. 2, 1959.

Thallus gelatinous, blue-green, filaments densely entangled; sheath distinct at the periphery of the thallus, trichomes 4-5 μ broad; cells subspherical, or slightly longer than broad, 5-7 μ long, heterocyst 4.5-5.5 μ broad, 5-6.5 μ long; spores oblong, many in chain, 5.5-7.5 μ broad and 7.5-9 μ long (-10) μ long.

From paddy field soils of 24-Parganas (N), Itahar (West Dinajpur).

Specimen nos. W.D. 13, PN 27, 40.

Earlier this alga was reported from Maharashtra (Kamat 1962-63b, 1963, Kolte & Goyal 1985, 1986, Patil & Satav 1986, Bhoge & Ragothaman 1986); North East India (Reddy et al. 1986); Karnataka (Bongale & Bharati 1980); J.& K. (Goyal et al. 1984); W.B. (Mukhopadhyay & Chatterjee 1981, Chatterjee & Chatterjee 1983).

152. *Nostoc ellipsosporum* (Desm.) Rabenh. ex Born. et Flah. (Plate 17, fig. 2)

Rabenhorst, Fl. Eur. Alg. 2: 169, 1865; Bornet et Flahault, Revision des Nostocacees heterocystees, 198, 1888; Geitler 841, f. 533, 1932; Prescott, 523, pl. 119, f. 14-16, 1951; Desikachary, 383, pl. 69, f. 5, 1959.

Thallus dark-blue green gelatinous; trochomes loosely arranged, 3.5-4 μ broad, constricted at the cross-walls; cells cylindrical 6-7.5(-9) μ long heterocysts subspherical to oblong, 5.5-6.5 μ broad and 6-7.5(-10) μ long; spores ellipsoidal to oblong, 6.5-8 μ broad, 12-16 μ long, epispore smooth.

From paddy fields of Goyespur, Ranaghat (Nadia); Berhampore (Murshidabad), Kaliachalk (Malda), and Bamangola (Malda).

Specimen nos. N 24, 31; Mar 14, Mul 3, 9.

In the present alga sizes of cells and spores are slightly smaller than the type specimen.

Previously reported from Karnataka (Bongale & Bharati 1980, Somashekar 1983); Maharashtra (Kamat 1963, Ashtekar & Kamat 1980, Bhoge & Ragothman 1986); U.P. (Prasad & Mehrotra 1980); from culture of paddy field soils of Balasore and Madurai (Laloraya & Mitra 1974); W.B. (Sinha & Mukherjee 1975b).

153. *Nostoc comminutum* Kutzing (plate 17, fig. 3).

Forti, A., in DC Toni, Sylloge Algarum, 5: 393, 1907; Tilden, Bibliotheca Phycologica, 165, 1910; Prasad et Mehrotra, 100, f. 5, 1978. Cells 4.5-5.5 µ in breadth and 4-6 µ in length; heterocysts 4.5-5.5 µ in breadth and 4.5-6 µ in length; spores not observed.

From paddy field soils of Midnapore, Hooghly, Howrah.

Specimen nos. 45, 63, Hoog 7, 17, How 2, 6

Previously reported from paddy fields of Northern India (Prasad & Mehrotra 1979a); Paddy fields of U.P. (Prasad & Mehrotra 1980).

154. *Nostoc depressum* Wood (Plate 17, fig. 6)

Tilden, J., The Myxophyceae of North America, Bibliotheca Phycologica, 177, 1910, Prasad et Mehrotra, p. 100, f. 9, 1978.

Trichomes many arranged in a common gelatinous mass, 3.5-4.5 µ broad; cells 2.5-5 µ long; heterocysts 4-6 µ in diameter, spores not seen.

From paddy field soils Dharma (Midnapore).

Specimen nos. 14, 25

Earlier, it was reported from paddy fields of Northern India (Prasad & Mehrotra 1979a); from paddy fields of U.P. (Prasad & Mehrotra 1980).

155. *Nostoc verrucosum* Vaucher ex Born. et Flah. (Plate 16, fig. 3)

Vaucher, Historie des Conferves d'eau douce, 225, pl. 16, f. 3, 1803; Bornet et Flahault, Revision des Nostocacees heterocystees, 216, 1888; Geitler, 854, f. 43, 1932; Desikachary, 388, pl. 70, f. 1, 1959.

Thallus very large dark blue-green spherical to sub-spherical, gelatinous, surface verrucose, trichomes densely entangled towards periphery of the colony, sheath diffluent, trichomes 3-3.5 µ broad; cells slightly shorter than broad, barrel-shaped 2.5-3.5 µ long; heterocysts 5.5-6 µ broad, 5.5-6.5 µ long, spores oval, 5-6 µ broad, 5.5-7.5 µ long, epispore smooth, yellowish.

From paddy fields of Midnapore and Murshidabad.

Specimen nos. 9, 41; Mur 6, 12.

Previously reported from U.P. (Prasad & Mehrotra 1980); J. & K. (Goyal et al. 1984); Karnataka (Bongale & Bharati 1980); W.B. (Pal & Santra 1987).

156. *Nostoc commune* Vaucher ex Born. et Flah. (Plate 17, fig. 5)

Vaucher, Histoire des Conferves d' eau douce, 222, pl. 16, f. 1, 1803; Bornet et Flahault, Revision des Nostocacees heterocystees 203, 1888; Geitler, 845, f. 536, 537, 1932; Desikachary, 387, pl. 68, f. 3, 1959.

Thallus blue-green to brownish blue-green, gelatinous, expanding, trichomes entangled expanding, trichomes entangled, 4.5-5.5 μ broad, sheath thick distinct, towards periphery, brownish blue-green; cells barrel shaped, 4-5.5 μ long; heterocyst spherical 5.5-7 μ diam., spores not very clearly seen, as broad as vegetative cell.

From paddy fields of 24 Parganas, Howrah, Nadia, Hooghly, Burdwan.

Specimen nos. S 5, 9, 17; How 22, 32; N 13, Hoog 2, 3; Bur 14, 26.

Previously reported from Bihar (Jha et al. 1986); J. & K. (Goyal et al. 1984); Tamil Nadu (Anand & Revathi 1987); Maharashtra (Kamat 1962-63b, 1963, Chaporkar & Gangawane 1984, Kolte & Goyal 1985, 1986, Patil & Satav 1986); Karnataka (Bongale & Bharati 1980); U.P. (Prasad & Mehrotra 1980); Hyderabad (Ghousuddin 1936); North East India (Reddy et al. 1986); Punjab (Sarma et al. 1979); Orissa (Mohanty 1982); W.B. (Biswas 1925, Banerjee 1935, Mukhopadhyay & Chatterjee 1981, Chatterjee & Chatterjee 1983).

157. *Nostoc spongiaeforme* Agardh ex Born. et Flah. (Plate 17, fig. 4)

C. Ag., Syst, Alg. 22, 1824; Bornet et Flahault, Revision des Nostocacees heterocystees, 197, 1888; Geitler, 839, f. 531, 1932; Desikachary, 380, 1959.

Thallus gelatinous, expanded, irregularly lobed, yellowish blue-green colour; trichomes loosely arranged, 3-4 μ broad, cells cylindrical to sub-cylindrical, 3.5-6 μ long; heterocyst barrel shaped, 4-6.5 μ broad, as long as broad, or slightly longer than broad 5-7 μ long, spores sub-spherical or oval, 6-7 μ broad and 6.5-10 μ long.

From paddy fields of Bishnupur (Bankura), Purulia and Birbhum.

Specimen nos. Ban 3, 8; Pur 6, 9, Bir 1, 12

Previously reported from Punjab (Sarma et al. 1979); U.P. (Prasad & Mehrotra 1980, Chadha & Pandey 1983a); Maharashtra (Kamat 1963, Tarar & Shewale 1984); Karnataka (Bongale & Bharati 1980); W.B. (from culture of paddy field soils of Asansole—Laloraya & Mitra 1974, Sinha & Mukherjee 1975b).

Genus: ANABAENA Bory
Key to the Species

1. Akinete contiguous to the heterocyst or otherwise .. 2
1. Akinete not contiguous to the heterocyst .. 4
    2. Akinete contiguous to the heterocyst .. 3
    2. Akinete not contiguous to the intercalary heterocyst, but contiguous to the terminal heterocyst, trichome 3-4 μ broad .. *A. oryzae*
3. Trichomes coiled .. 20
3. Trichomes straight or curved .. 9
    4. Akinete naviculoid, in series, apical cell conical .. *A. naviculoides*
    4. Akinete otherwise .. 5

5.	Akinete ellipsoidal to oblong	..	6
5.	Akinete spherical	..	7
	6. Cells 5-7.5 μ long, spores 7.5-8 x 10-12 μ	..	*A. variabilis*
	6. Cells 4-5 μ long, spores 6-8 x 7-13 μ	..	*A. variabilis* v. *ellipsospora*
7.	Trichome 3.8-4.5 μ broad, spores 7.6-8 μ diam. wall granular	..	*A. khannae*
7.	Trichome broader, spores larger	..	8
	8. Trichome 5-6 μ broad, spores 12-14 μ diam., wall smooth, terminal cell rounded	..	*A. fertilissima*
	8. Trichome 6.5-7.5 μ broad, spores 11-12 μ diam., terminal cell acute	..	*A. gelatinicola*
9.	Akinete bent or curved	..	*A. utermohlii*
9.	Akinete oval, ellipsoidal or cylindrical	..	10
	10. Akinete one or two	..	11
	10. Akinete many	..	17
11.	Akinete generally on one side of the heterocyst	..	12
11.	Akinete one on either side of the heterocyst	..	13
	12. Spore wall smooth	..	*A. volzii*
	12. Spore wall papillate	..	*A. fuellebornii*
13.	Trichomes in mucilage	..	14
13.	Trichomes not in mucilage	..	15
	14. Trichomes deeply constricted, cells barrel shaped, 4-5 x 4-4.5 μ; spores ellipsoidal to cylindrical, 9-14 x 10-15 (-18.5) μ	..	*A. ambigua*
	14. Trichomes not deeply constricted, cells cylindrical, 4-5 x 5-7.5 μ, spores oval to ellipsoidal, 9.5-10(-15) x 11-13 μ	..	*A. orientalis*
15.	Trichomes 4.5-5.5 μ broad, end cell conical with rounded ends, heterocysts 7.5-8 μ broad, spores 11.5-16 x 18-22(-30) μ spore single or on either side of the heterocyst	..	*A. iyengarii*.
15.	Trichomes 4-4.5 μ broad, heterocysts narrower	..	16
	16. End cell conical with rounded ends, spores 7.5-10 x 13-18 μ, ellipsoidal to cylindrical	..	*A. iyengarii* v. *tenuis*
	16. End cell attenuated, spores 12-16 x 18-21 μ ellipsoidal or oval	..	*A. iyengarii* v. *attenuata*

17. Trichomes 3.5-5 μ broad, spores 6-7 x 12.5-13 μ, with slight constriction at the middle .. *A torulosa*

17. Trichomes broader, generally 4-5.5 μ broad, spores broader without median constriction .. 18

18. Trichomes 4.5-5.5(-6) μ broad, heterocysts spherical, spores cylindrical, 7.5-9 x 20-33 μ .. *A. oscillarioides*

18. Trichomes narrower, spores shorter in length, heterocysts not spherical .. 19

19. Heterocysts cylindrical, 5-5.5 x 7.5-8.5 μ, cells subquadrate to cylindrical, spores 6.5-7.5 μ broad .. *A. vaginicola*

19. Heterocysts barrel shaped, 6-7 x 5.5-6 μ, cells barrel shaped, spores 7-8.5 μ broad .. *A. vaginicola* f. *fertilissima*

20. Trichomes loosely coiled .. *A. circinalis*

20. Trichomes regularly spirally coiled .. *A. spiroides*

158. *Anabaena oryzae* Fritsch (Plate 18, fig. 1)

Fritsch, 135, f. 1-16, 1949; Desikachary, 396, p. 72, f. 3, 1959.

Thallus gelatinous, trichomes straight, densely arranged, 3-4 μ broad, cells barrel shaped, 4-7 μ long, heterocysts terminal and intercalary, 4-5(-6) μ broad and 5-7(-8) μ long, spores single or two sub-spherical, 5-5.5 μ in diameter.

From paddy fields, Narajole (Midnapore), Diamond Harbour (South 24-Parganas).

Specimen nos. 51, 47.

In the present form trichomes and heterocysts are slightly broader than the type species.

Previously reported from Maharashtra (Chaporkar & Gangawane 1984, Tarar & Shewale 1984, Kolte & Goyal 1985, Patil & Satsv 1986); Bihar (Jha et al. 1986); Punjab (Grover & Pandhol 1975, Pandhol & Grover 1976, Sarma et al. 1979); J. & K. (Goyal et al. 1984); U.P. Prasad & Mehrotra 1980); W.B. (Maity & Santra 1985, Santra et al. 1988).

159. *Anabaena khannae* Skuja (Plate 18, fig. 3)

Zur Süssawasseralgenflora Burmas, 41, pl. 7, f. 1-5, 1949; Desikachary, 396, pl. 75, f. 2, 1959.

Thallus attached to the substratum, trichomes straight or curved, 3.8-4.5 μ broad, sheath colourless, cells barrel-shaped, 3.8-6 μ long; heterocysts terminal and intercalary, intercalary heterocyst barrel-shaped, but terminal heterocyst conical, 5-7.5 μ broad and 7.5-9 μ long, spores globose, in chain, 7.6-8 μ in diameter, away from the heterocyst, epispore granular.

From paddy field soils, Dharma, Contai (Midnapore)

Specimen nos. D 57, C 84.

The present form is in close agreement with *A. khannae* Skuja in most of the characters, but differ slightly in having broader trichomes and larger heterocysts.

Previously recorded from U.P. (Prasad & Mehrotra 1980); Karnataka (Bongale & Bharati 1980); J. & K. (Goyal et al. 1984); Maharashtra (Kolte & Goyal 1986); W.B. (Sinha & Mukherjee 1984).

160. *Anabaena naviculoides* Fritsch (Plate 18, fig. 4 and Plate 43A)

J. Indian bot. Soc., 28: 138, f. 17-39, 1949; De, 125, 1939; Desikachary, 410, pl. 72, f. 2, 1959.

Thallus blue-green, trichomes short, free, spices acuminate, cells barrel-shaped, as long as broad or slightly short, apical cell conical; heterocyst intercalary, single, barrel-shaped 5-6.5 μ broad, spores in series, ellispoidal, naviculoid, 10-11.5 μ long and 5.5-6.5 μ broad, epispore smooth.

From paddy fields of Bankura and 24 Parganas (North).

Specimen nos. Ban 41, S 39, 47.

Previously reported from Orissa (Laloraya and Mitra 1974); J & K. (Goyal et al. 1984); North East India (Reddy et al. 1986); U.P. (Pandey 1965, Pandey 1975); W.B. (Chatterjee & Chatterjee 1983, Sinha & Mukherjee 1984).

161. *Anabaena fertilissima* Rao, C.B. (Plate 18, fig. 6)

Proc. Indian Acad. Sci., B, 6: 363, f. 6A-C, 1937b; Desikachary, 398, pl. 74, f. 1, 1959.

Trichome single, straight or curved, 5-6 μ broad; cells barrel-shaped, 5-7 μ long, end cell of the trichome rounded; heterocyst spherical to subspherical, 8-9 μ in diam., spores in chain, spherical, wall smooth, 12-14 μ in diam.

From paddy fields of Howrah, Murshidabad and Midnapore.

Specimen nos. How 3, 7, 9; Mur 12, 16; M 29, 31.

The present algae shows slightly broader trichomes (5-6 μ instead of 5-5.6 μ) larger heterocysts (8-9 μ diam. instead of 6.4-8.4 μ) and larger spores.

Previously reported from U.P. (Prasad & Mehrotra 1980); J. & K. (Goyal et al. 1984); Maharashtra (Ashtekar & Kamat 1980, Chaporkar & Gangawane 1984, Kolte & Goyal 1985, Bhoge & Ragothman 1986); Bihar (Jha et al. 1986); Punjab (Sarma et al. 1979); North East India (Reddy et al. 1986); Karnataka (Bongale & Bharati 1980, Somashekar 1983, 1984b); W.B. (Chatterjee & Chatterjee 1983, Pal & Santra 1987).

162. *Anabaena ambigua* Rao, C.B. (Plate 21, fig. 1)

Proc. Indian Acad. Sci., B, 5: 101, f. 1, 2 1937a; Desikachary, 400, pl., 76, f. 2, 1959.

Thallus attached to submerged rice plant parts, mucilaginous, trichomes straight, or curved, 4-5 μ broad; deeply constricted at the cross-walls, cells barrel shaped 4-4.5 μ long; heterocysts spherical, 7-8 μ in diam., spores contiguous to the heterocysts, one on either side, ellipsoidal to cylindrical, with rounded ends, 9-14 μ broad, and 10-15(-18.5) μ long.

From paddy fields soils of Bankura, 24 Parganas and Haldia (Midnapore).

Specimen nos. Ban 3, 9; S 12, 18, M 7, 9

Previously reported from U.P. (Prasad & Mehrotra 1980, Pandey & Pandey 1982, Pandey 1982); Karnataka (Bongale & Bharati 1980); Maharashtra (Kamat 1963, Ashtekar & Kamat 1980, Kolte & Goyal 1986); Punjab (Grover & Pandhol 1975, Pandhol & Grover 1976); J.& K. (Goyal et al. 1984); Tamil Nadu (Anand & Revathi 1987); West Bengal (Sinha & Mukherjee, 1975a, Chatterjee et al. 1980, Chatterjee & Chatterjee 1983).

163. *Anabaena utermöhlii* Geitler (Plate 21, fig. 2)

in Pascher's Susswasserflora, 443, 1925; Kryptogamenflora, 844, 1932; Desikachary, 415, pl. 74, f. 2, 1959.

Trichomes irregularly coild, cells 4-4.5 μ broad, quadrate to slightly longer, 5.5-6.5 μ long; heterocysts 4-4.5 μ broad and 5-6 μ long; spores bent, contiguous to the heterocysts, on both sides, 6.5-7.5 μ broad and 17-20 μ long.

From paddy field soils Nabadwip (Nadia) and Murshidabad.

Specimen nos. N 14, 16, Mur 22, 36.

Previously reported from Madras, West Bengal.

164. *Anabaena oscillarioides* Bory ex Born. et Flah. (Plate 19, fig. 1 & 2)

Bory, Dic., Class. d'hist, nat., 1: 308, 1822; Bornet et Flahault, Revision des Nostocacees heterocystees, 233, 1888; Geitler, 886, f. 567e, 1932; Desikachary, 417, pl. 71, f. 7, 1959.

Thallus gelatinous, deep blue-green; trichome 4.5-5.5(-6) μ broad, cells quadrate or slightly longer, 5-6.5 μ long end cell rounded; heterocyst oval, 6-6.5 μ broad, 6.5-7.5 μ long; spores single or two on either side of the heterocysts, oval to cylindrical, sometimes upto 4 in chain, 7.5-9 μ broad and 20-33 μ long, epispore smooth.

From paddy fields of Bankura, Burdwan, Purulia & Midnapore.

Specimen no. 94, Ban 8, 16; Bur 21, Pur 5.

Previously reported from Bihar (Jha et al. 1986); U.P. (Bendre & Kumar 1975); Maharashtra (Kamat 1963, Chaporkar & Gangawane 1984); W.B. (Sinha & Mukherjee, 1975a).

165. *Anabaena Iyengarii* Bharadwaja (Plate 19, figs. 3 & 4)

Proc. Indian Acad. Sci., B, 2: 105, f, 6H-K, 1935; Desikachary, 406, pl. 78, f. 2, 1959.

Trichomes single, dark blue-green, 4.5-5.5 μ broad; cells barrel-shaped, 4-5 μ long, and cell conical with rounded ends; heterocysts 7.5-8 μ broad, 10-12 μ long, spores 11.5-16 μ broad, 18-22(-30) μ long, single or on either side of the heterocysts, ellipsoidal or cylindrical.

From water logged rice field, Midnapore Sadar.

Specimen no. 405

Previously reported from U.P. (Gupta 1956, Chaturvedi & Pandey 1976, Prasad & Mehrotra 1980, Chadha & Pandey 1983a); Bihar (Sinha and Srivastava 1980); Karnataka (Somashekar 1983, 1984, Somashekar & Ramaswami 1983); J. & K. (Gopal et al. 1984); W.B. (Santra et al. 1988).

166. *Anabaena Iyengarii* var. *tenuis* Rao (Plate 20, figs. 8 & 9)

Proc. Indian Acad. Sci., B, 6: 361, f. 5A-C, 1937b; Desikachary, 408, pl. 76, f. 1, 1959.

Thallus pale blue green colour; trichome single, straight or curved, 4-4.5 μ broad; cells as long as broad or slightly shorter than broad, 3.5-4.5 μ long; end cell conical, heterocysts, barrel-shaped, 5-7 μ broad, 6-10 μ long; spores single or on either side of the heterocyst, 7.5-10 μ broad, 13-18 μ long.

From paddy fields of Cooch Behar, Jalpaiguri and Darjeeling.

Specimen nos. Cooch 11, 19, Jal 7, 16; Dar 3, 9.

Previously reported from Maharashtra (Bhoge & Ragothaman 1986); U.P. (Rao 1937, Gupta 1956, Prasad & Mehrotra 1980, Chadha & Pandey 1983a); W.B. (Sinha & Mukherjee 1975b, Maity & Santra 1985).

167. *Anabaena Iyengarii* var. *attenuata* Rao (Plate 22, fig. 10)

Proc. Indian Acad. Sci. B, 8: 163, f. 2A-C, 1938b; Desikachary, 408, pl. 76, f. 5, 1959.

Thallus dark blue-green, trichomes single, straight or curved 4-4.5μ broad; attenuated at the apices, cells barrel shaped, 3.5-5 μ long, heterocyst 5-6 μ broad, 6-8 μ long; spores ellipsoidal or oval, single or on either side of the heterocyst; 12-16 μ broad, 18-21 μ long.

From paddy fields of Bankura, Midnapore and Purulia.

Specimen nos. Ban 6, 8; M 7, 9, Pur 3, 8

Previously reported from Orissa (Rao 1938b, Mohanty 1982, 1984); Delhi (Rao 1940); U.P. (Pal 1975); W.B. (Sinha & Mukherjee 1975a).

168. *Anabaena orientalis* Dixit (Plate 19, figs, 9-11 and Plate 43B)

Proc. Indian Acad. Sci. B, 3: 101, f. 3D, E, 1936; Desikachary, p. 405; pl. 77, f. 6, 1959.

Thallus mucilaginous, dark blue-green, trichomes straight and lying parallel, somtimes slightly curved, without individual sheath, trichomes 4-5 μ broad, constricted at the cross-walls, cells usually cylindrical, longer than broad, 5-7.5 μ long, end cells conical, sometimes with rounded apex; heterocysts single, intercalary, ellipsoidal to slightly cylindrical, 5.5-6 μ broad and 6-9 μ long; spores one on either side of the heterocyst, not in chains, ellipsoidal with rounded end-walls, 9.5-10 μ broad and 11-13 μ long, sometimes matured spores are spherical, 13-15 μ diam., epispore smooth.

From paddy fields, Salboni, Garhbeta (Midnapore) and Chakda (Nadia).

Specimen nos. G 45, G 77, G 80, S 42, N 8

The present alga is in close agreement with *Anabaena orientalis* Dixit in most of the characters but differs from the type species for the presence of slightly longer cells (5-7.5 μ instead of 3.7-4.8 μ) broader trichomes (4-5 μ instead of 2.5-4 μ) and slight differences in the measurements of heterocysts and spores. The present alga comes close to *A. orientalis* var. *ellipsospora* (Rao, C.B.), Desikachary, in many characters but differs from it in not having two spores together on either side of the heterocysts. There are many similarities between *A. orientalis* Dixit and *Anabaena iyengarii* Bharadwaja. *Anabaena iyengarii* Bharadwaja is so variable one that it is quite difficult to make clear differences between *Anabaena orientalis* Dixit and *Anabaena iyengarii* Bharadwaja specially due to the presence of one or two spores on either side of the heterocyst. Regarding the number of spores on either side of heterocysts the present alga is also near to the *Anabaena iyengarii* var. *unispora* Singh, R.N., but differs from it in other characters.

It can also be compared with *Aphanizomenon aphanizomenoides* (Forti) Horecka et Komarek (1979, 295, f. 1-6) regarding trichome breadth, shape and size of matured spore, but differs in having single, spore on either side of the heterocyst and in not having gas vacuoles within the cells.

Previously reported from Karnataka (Somashekar 1983); Maharashtra (Kamat 1963, Kolte & Goyal 1985); U.P. (Pal 1975); W.B. (Sinha & Mukherjee 1975a).

169. *Anabaena torulosa* (Carm.) Lagerh. ex Born. et Flah. (Plate 19, figs, 5 & 6) Bidrag till Sveriges Algflora, K. Vet. Akad. Forh., 40(2): 47, 1883; Bornet et Flahault, Revision des Nostocacees heterocystees, 236, 1888; Geitler, 887, f. 567d, 1932; Desikachary, 415, pl. 71, f. 6, 1959.

Thallus mucilaginous, blue-green; trichomes 3.5-5 μ broad, constricted at the cross-walls; cells quadrate to barrel shaped, 3.5-4(-5) μ long, end cell conical, 2.5 μ long and 2.5 μ broad; heterocyst barrel shaped 4-5(-7) μ broad, 4-6.5(-8) μ long, spores on both sides of the heterocysts, cylindrical, 6-7 μ broad and 12.5-13 μ long, epispore smooth.

From paddy fields, Contai (Midnapore).

Specimen nos. 17, 59.

The present form differs from the type species in having narrower trichomes (3.5-5 μ instead of 4.2-5 μ), in the measurements of heterocysts and spores; but in these characters the present alga comes close to the form described by Fritsch (1949, p. 144, f. 51-59).

Previously reported from Maharashtra (Ashtekar & Kamat 1980, Kolte & Goyal 1985, 1986, Patil & Satav 1986); Bihar (Jha et al. 1986); U.P. (Pal 1975);

Karnataka (Bongale & Bharati 1980); Punjab (Grover & Pandhol 1975, Pandhol & Grover 1976, Sarma & Kanta 1975); W.B. (Sinha & Mukherjee 1975a).

170. *Anabaena fuellebornii* Schmidle (Plate 19, figs. 7, 8 and Plate 43C) Engler's Bot., Jahrb, 32: 61, pl. 1, f. 4, 1892; Geitler, 904, f. 567e, 1932; Desikachary, 401, pl. 71, f. 11 & pl. 75, f. 1, 3, 1959.

Thallus mucilaginous, blue-green; trichomes straight or curved 4-5 μ broad, cells cylindrical, constricted at the cross-walls, 5-10 μ long; heterocysts barrel-shaped to cylindrical, 5.5-6.5 μ broad and 7.5-10 μ long; spore on one side of the heterocyst, single, rarely in pairs, ellipsoidal, 11.5-16.5 μ broad and 25-33 μ long, epispore with papillae.

From paddy fields, Panchami (Midnapore) and Bankura.

Specimen nos. 78, Ban 37.

Previously this alga was reported from Karnataka (Somashekar 1984b); Maharashtra (Kamat 1963); W.B. (Sinha & Mukherjee 1975a).

171. *Anabaena volzii* Lamm. (Plate 20, figs. 1-4; Plate 43D)

Uber die von Herrn. Dr. Walter Volz auf seiner weltreise gesammelte Susswasseralgen, Abh. Nat. Ver. Bremen, 18: 153, pl. 9, f. 4, 5, 20, 1906; Geitler, 901, 1932; Desikachary, 403, pl. 77, f. 1, 1959.

Thallus mucilaginous, blue-green; trichome 4-5 μ broad, rarely 5-5 μ broad, constricted at the cross-walls, cells cylindrical 6-8(-9) μ long, end cell rounded; heterocyst cylindrical, intercalary or terminal, 5-5.5(-7.5) μ broad, 9-13 μ long; spores on one side of the heterocyst, 11-16.5(-18) μ broad, 25-40(43-5) μ long, with smooth epispore.

From paddy fields, Jhargram, Contai, Garhbeta (Midnapore).

Specimen nos. J 95, C 82, G 91, G 113

The present form resembles the type species in all general characters but differs from the type species in having shorter cells (6-8(-9) μ long instead of 7-12 μ), rounded end cell, smaller heterocysts (5-5.5(-7.5) μ x 9-13 μ instead of 5.5-7 μ x 12-15 μ) and larger spores (11-16.5(-18) μ x 25-40(43.5) μ instead of 15-21 μ x 32-33 μ).

It has some similarities with *A. volzi* v. *crassa* (Rao) Fritsch.

Earlier this alga was reported from Kolhapur of Maharashtra (Kamat 1963); Bihar (Jha et al. 1986); U.P. (Prasad & Mehrotra 1980 a forma of *A. volzii*); W.B. (Sinha & Mukherjee, 1975a, reported presence of *A. volzii* v. *crassa* from Bankura district).

172. *Anabaena spiroides* Klebahn (Plate 21, figs. 3 & 4).

Flora, 80: 125, pl. 4, f. 11-13, 1895; Geitler, 881, 1932; Desikachary, 395, pl. 71, f. 9, 1959.

Thallus blue-green, mucilaginous, trichomes spirally coiled, spirals 48-50 μ broad, and distance between two spirals 38-45 μ cells spherical, with vacuoles, 6-7.5 μ in diameter, heterocysts subspherical, 7-7.5 μ in diam.; spores spherical to slightly elongate.

From paddy fields of Kanchrapara, Krishnagar (Nadia), and Berhampore (Murshidabad).

Specimen nos. N 18, 22, Mur 4, 12.

Previously reported from Orissa (Mohanty 1982); J. & K. (Goyal et al. 1984); Karnataka (Bongale & Bharati 1980); Maharashtra (Chaporkar & Gangowane 1984); Bhagalpur (Nasir & Munshi 1976, Saha & Pandit 1987) and West Bengal (Biswas 1942 from river Hooghly).

173. *Anabaena gelatinicola* Ghose (Plate 20, figs. 5 & 6)

J. Linn. Soc., Bot., Lond., 46: 341, pl. 31, f. 8, 1924; Desikachary, 398, pl. 76, f. 6, 1959.

Thallus gelatinous; trichomes densely arranged; cells spherical, 6.5-7.5 μ in diam.; apex acute; heterocysts spherical, 7.5-9 μ diam.; spores spherical away from heterocysts, 11-12 μ diam.

From paddy fields of 24 Parganas (N), Howrah and Hooghly.

Specimen nos. DN 46, 34; How 10, 13; Hoog 7, 9.

In the present form spores slightly smaller in sizes than that of type speices.

Previously reported W.B. (Chatterjee & Chatterjee 1983, Maity & Santra 1985, Santra et al. 1988).

174. *Anabaena vaginicola* Fritsch et Rich (Plate 20, fig. 7).

Contributions to our knowledge of the fresh water algae of Africa, 7. Fresh water Algae from Girgualand West. Trans. Roy. Soc. S. Afr., 18(1): 87, 1929, Desikachary, 401, pl. 73, f. 5, 1959.

Thallus mucilaginous; many trichomes, more or less parallel, each trichome 4-4.5 μ broad, cells quadrate to elongated, 4-5.5 μ long; end cell conical, heterocyst cylindrical, 5-5.5 μ broad and 7.5-8.5 μ long, spores oblong, contiguous with the heterocysts, sometimes in chains, 6.5-7.5 μ broad and 12-15 μ long.

From paddy fields of Bankura, Purulia and Birbhum.

Specimen nos. Ban 3, 7, Pur 13, 19, Bir 27.

Previously reported from Karnataka (Bongale & Bharati 1980); U.P. (Prasad & Mehrotra 1980, Chadha & Pandey 1983a), W.B. (Sinha & Mukherjee 1975b).

175. *Anabaena vaginicola* f. *fertilissima* Prasad (Plate 18, fig. 5)

J. Indian bot. Soc., 31: 361, f. 14-17, 1952; Desikachary, 401, pl. 73, f. 3, 1959.

Thallus light-blue-green, soft, trichomes lying in the mucilaginous sheath, sheath diffluent; trichomes 4-5.5 μ broad, cells 3.5-5.5 μ long; heterocyst 6-7 μ broad; 5.5-6 μ long; spores oval to ellipsoidal, arranged in chain, 7-8.5 μ broad, 11-14.5 μ long.

From water-logged paddy fields of Bankura & Midnapore.

Specimen nos. Ban 6, 9; Mid 33, 47.

The present form is similar to *A. vaginicola* f. *fertilissima* Prasad in all characters, but differs from it in possessing slightly narrower trichomes (4-5.5 μ instead of 4.8-5.6 μ) slightly narrower heterocysts (6-7 μ broad instead of 6.6-7.2 μ) and slightly different spore size (7-8.5 μ x 11-14.5 μ instead of 6.4-8 μ x 11.2-13.4 μ).

Previously reported from Maharashtra (Kolte & Goyal 1986); Sagar of M.P. (Mishra & Purohit 1979); W.B. (Sinha & Mukherjee 1975a).

176. *Anabaena variabilis* Kutz. ex Born. et Flah. (Plate 20, figs 10 & 11) Kützing, Phyc. gene., 210, 1843; Bornet et Flahult, Revision des Nostocacees heterocystees, 226, 1888; Geitler, 876, f. 558, 1932; Desikachary, 410, pl. 71, f. 5, 1959.

Trichomes long, free, flexuous, 4-5 μ broad, constricted at the cross-walls, cells barrel shaped 5-7.5 μ long, end cell rounded or slightly conical; heterocyst terminal or intercalary, barrel shaped, 5.5-6 μ broad and 6.5-7.5 μ long; spores in series, not contiguous with the heterocyst, oblong, 7.5-8 μ broad, 10-12 μ long, epispore smooth.

From paddy fields, Midnapore, Bankura, Howrah and West Dinajpore.

Specimen nos. M 74, WD 179, Ban 41, How 5, 7.

Previously reported from Maharashtra (Kolte & Goyal 1985); Karnataka (Bongale & Bharati 1980, Somashekar 1984b); Bihar (Sinha & Srivastava 1980, Saha & Pandit 1987); Tamil Nadu (Anand & Revathi 1987); Orissa (Mohanty 1982); U.P. (Chadha & Pandey 1983a); W.B. (Pal et al. 1986).

177. *Anabaena variabilis* var. *ellipsospora* Fritsch (Plate 18, fig. 2) J. Indian bot. Soc., 28, 142, f. 40-50, 1949; Desikachary 411, pl. 72, f. 1, 1959.

Thallus blue-green, trichomes free, flexuous, 4-5.5 μ broad, slightly constricted at the cross-wall; cells barrel shaped, 4-5 μ long, end cell slightly conical with rounded apex; heterocysts spherical, 5-6 μ broad, 6-7 μ long, spores in chain, not contiguous to the heterocyst, ellipsoidal to oblong, 6-8 μ broad, 7-13 μ long.

From paddy fields of Bankura, Hooghly, Howrah, Nadia and Murshidabad.

Specimen nos. Ban 21, 36; Hoog 20, 23; How 10, 15; N 23, 29; Mur 13.

Previously reported from Bihar (Jha et al. 1986); North East India (Reddy et al. 1986); Maharashtra (Dixit 1936, Chaporkar & Gangawane 1984); Karnataka (Bongale & Bharati 1980); U.P. (Prasad & Mehrotra 1980, Pandey 1982, Chadha & Pandey 1983a); Tamil Nadu (Anand & Revathi 1987); Simla (Ghosh 1924); W.B. (Banerjee 1938, Biswas 1942, Sinha & Mukherjee 1984).

178. *Anabaena circinalis* Rabenhorst ex Born. et Flah.

var. *crassa* Ghose (Plate 30, fig. 1)

On some Myxophyceae from Rangoon, J. Burma Res. Soc., 15: 249, pl. 6, f. 14, 1926; Desikachary, 414, pl. 77, f. 5, 1959.

Trichome single loosely coiled; cells barrel-shaped or spherical, 5.5-7 μ broad, 5-6 μ long; heterocysts spherical 7-8 μ broad, spores not seen.

From paddy fields of Bankura & Midnapore.

Specimen nos. Ban 8, 16; M 3, 19

Previously reported from U.P. (Chaturvedi & Pandey 1976); Karnataka (Bongale & Bharati 1980); W.B. (Sinha & Mukherjee 1984).

Genus: PSEUDANABAENA Lauterborn

179. *Pseudanabaena schmidlei* Jaag. O. (Plate 30, fig. 2)

Die Kryptogamenflora des Rheinfalls und des Hochrheins von Stein bis Englisau, Mitt. Naturf. Ges. Schauffhausen, 14: 1938; Desikachary, 419, pl. 75, f. 4, 1959.

Trichomes single, sheath absent, 6-7 μ broad; cells 6-6.5 μ long, deeply constricted at the cross-wall, end cell rounded.

From paddy fields of Bankura & Midnapore.

Specimen nos. Ban 25, 30; M 7, 21.

Previously reported from U.P. (Pal 1875); W.B. (Sinha & Mukherjee 1984).

Genus: NODULARIA Mertens

180. *Nodularia spumigena* Mertens ex Born. et Flah. (Plate 21, fig. 11) in Jurgens, Alg. Dec., 15, nr. 4, 1822; Bornet et Flahault, Revision des Nostocaoacees heterocystees, 245, 1888; Geitler, 866, f. 554b,c, 1932; Desikachary, 423, pl. 80, f. 13, 14, 1959.

Filaments single, straight or bent, 10-12 μ broad; sheath thin, hyaline; cells 8.5-9.5 μ broad, 3-4.5 μ long, heterocyst spherical to subspherical, 9-10 μ broad, 5-6 μ long, spores single or in chain, spherical to sub-spherical, 11-12 μ broad and 7-9 μ long, epispore smooth and yellowish brown.

From paddy fields of Bankura & Purulia.

Specimen nos. Ban 11, Pur 14.

Previously reported from U.P. (Prasad & Mehrotra 1980); North East India (Reddy et al. 1986); Karnataka (Bongale & Bharati 1980); Punjab (Sarma et al. 1979); Sagar of M.P. (Mishra & Purohit 1979); Bihar (Sinha & Srivastava 1980); W.B. (Sinha & Mukherjee 1975a).

Genus: AULOSIRA Kirchner

1. Spore not seen .. 2
1. Spores seen .. 3

2. Trichome 4.5-5(-6) μ broad, sheath thin, cells cylindrical	..	*A. prolifica*
2. Trichome 6-10 μ broad, sheath thick, cells sub-quadrate to slightly longer, false branching on hormogone germination	..	*A. pseudoramosa*
3. Spores generally alternating with dead cells	..	4
3. Spores not alternating with dead cells	..	5
4. Trichomes 4.5-7.5 μ broad, sheath thick	..	*A. fertilissima*
4. Trichomes 4-5 μ broad, sheath thin	..	*A. fertilissima* v. *tenuis*
5. Trichomes 4.5-6 μ broad, spores 7 μ broad, 4.6-10, 5 μ long	..	*A. aenigmatica*
5. Trichomes 9-10 μ broad	..	6
6. Sheath thin, spores 9-10 μ broad, 15-17.5 μ long	..	*A. implexa*
6. Sheath thick, upto 3 μ	..	*A. fritschii*

181. *Aulosira prolifica* Bharadwaja (Plate 22, fig. 1; and Plate 44A)

Contributions to our knowledge of Myxophyceae of India, Ann. Bot., Lond., 47: 131, f. 5, 6, 1933a; Desikachary, 426, pl. 81, f. 1-6, 1959.

Thallus blue-green forming a mucilagenous scum; filaments 6-8(-9) μ broad; sheath firm, unlamellated, rearly with pseudobranches; trichomes with slightly constricted or unconstricted cross-walls, 4.5-5(-6) μ broad; cells 5-10 μ long but in some specimens upto 14 μ long; end cells narrowly elongated with rounded apex; heterocyst intercalary, 5.5-6(-7.5) μ broad and 14-23 μ long, ellipsoidal, or slightly cylindrical, rarely hormogones seen; spores not seen.

From paddy fields, Contai, Jhargram, Silda, Dharma (Midnapore), Bankura, Burdwan, Birbhum and Malda.

Specimen nos. 13, 15, 51, 89; Ban 6, Bur 4, Bir 6, 8, Mal 4, 14.

The present alga differs from the type species in having broader filaments (6-8(-9) μ instead of 4.8-5.6 μ), broader trichomes (4.5-5(-6) μ broad instead of 3.1-5.2 μ) and in not having markedly constricted cross-wall.

Previously this alga was reported from U.P. (Bharadwaja 1933a, Pandey 1982, Pandey & Pandey 1982); Maharashtra (Ashtekar & Kamat 1980, Chaporkar & Gangawane 1984, Kolte & Goyal 1985); Punjab (Grover & Pandhol 1975, Pandhol & Grover 1976, Sarma & Kanta 1978); A.P. (Rao 1938a), W.B. (Sinha & Mukherjee 1975a, Sengar & Pandey 1985).

182. *Aulosira aenigmatica* Fremy (Plate 22, figs. 2 & 3)

Cyanophycees de l'Inde meridionale, Blumea, Suppl. 2: 37, f. 11, 1942; Desikachary, 428, pl. 81, f. 15, 17, 1959.

Thallus dark blue-green, floating, filaments 5-7 μ broad, sheath thin; trichomes 4.5-6 μ broad, slightly constricted at the cross-walls; cells 5.5-7 μ long, quadrate to slightly longer than broad; contents granular; end cells obtuse, heterocystes as broad as trichomes, 7-10.5 μ long, spore in series, 7 μ broad, 4.6-10.5 μ long.

From paddy fields, Digha (Midnapore), Diamond Harbour (South 24-Parganas).

Specimen nos. 32, 432.

Here the filaments and trichomes slightly narrower than the type species.

Filaments 5-7 μ broad instead of 6.5-8 μ and trichome 4.5-6 μ broad instead of 5.5-6 μ.

Previously reported from Maharashtra (Tarar & Shewale 1984); Patiala (Sarma & Kanta 1978); Karnataka (Bongale & Bharati 1980); W.B. (Maity & Santra 1985, Pal & Santra 1987, Santra et al. 1988).

183. *Aulosira pseudoramosa* Bharadwaja (Plate 22, figs. 6 & 9)

Contribution to our knowledge of Myxophycae of India, Ann. Bot., Lond., 47: 137, f. 7, 9, 1933a; Desikachary, 430, pl. 81, f. 7-14, 1959.

Thallus blue-green; filaments densely entangled 12-15.5 μ broad; sheath thick hyaline sometimes outer surface straified trichomes, 6-10 μ broad, constricted at the cross-walls; cells sub-quadrate to slightly longer than broad; 6-10 μ broad, 6-12 μ long, heterocysts intercalary, as broad as trichome, and 12-20 μ long; spores not seen, hormogones and false branching present.

From paddy fields, Thakurchak (Midnapore), Burdwan and Hooghly.

Specimen nos. 104, T 125, Bur 3, 6, Hoog 7, 10.

The present alga differs from the type species in having slightly broader filaments (12-15.5 μ instead of 9.5-14.7 μ) and narrower trichomes 6-10 μ broad, instead of 6.3-10.5 μ).

Previously this alga was reported from U.P. (Bharadwaja 1933a, Pal 1975, Prasad & Mehrotra 1980); Punjab (Sarma et al. 1979); W.B. (Sengar & Pandey 1985).

184. *Aulosira implexa* Bornet et Flahault (Plate 22, figs. 4 & 5)

Note sur le genre *Aulosira*, Bull. Soc. bot. Fr., 32: 120, pl. 4, f. 4, 1885; Revision des Nostocacees heterocystees, 257, 1888; Geitler, 675, f. 434b, c, 1932; Desikachary, 430, 1959.

Filaments slightly curved or straight, 15-16 μ broad, sheath hyaline, thin; trichomes 9-10 μ broad; cells subquadrate to quadrate, sometimes cylindrical, 8-15 μ long, cross-wall constricted; heterocysts quadrate, or slightly cylindrical, 9-10 μ broad and 9-12 μ long, spores 9-10 μ broad, and 15-17.5 μ long.

From water logged rice field of Burdwan, Murshidabad & Midnapore.

Specimen nos. Bur 4, 12, Mur 5, M 13.

Previously reported from Karnataka (Bongale & Bharati 1980); U.P. (Pandey 1965); Maharashtra (Dixit 1936, Gonzalves & Kamat 1954); Punjab (Sarma et al. 1979); W.B. (Mukhopadhyay & Chatterjee 1981, Sengar & Pandey 1985, Pal & Santra 1987).

185. *Aulosira fritschii* Bhardawaja (Plate 22, figs. 7 & 8)

Contributions to our knowledge of the Myxophyceae of India, Ann. Bot, Lond., 47: 123, f. 3, 4, 1933a; Desikachary, 432, pl. 80, f. 7-12, 1959.

Thallus dark-green; filaments unbranched, 12-15 μ broad; sheath hyaline, upto 3 μ thick, trichome 9-10 μ diam. constricted at the cross-walls, cells 15-20 μ long, heterocysts cylindrical, 9.5-13 μ broad and 13.5-21 μ long; spores in series, 11-13 μ broad and 12-22 μ long.

From paddy fields of Bankura, Purulia, Birbhum, Burdwan.

Specimen nos. Ban 12, 16, Pur 4, 7, Bir 3, Bur 4.

Previously reported from Punjab (Sarma et al. 1979); U.P. (Pal 1975); W.B. (Sinha & Mukherjee 1975b).

186. *Aulosira fertilissima* Ghose (Plate 21, figs. 5 & 6)

A systematic and ecological account of a collection of blue green algae from Lahore and Simla, J. Linn. Soc. Bot., 46: 342, pl. 31, f. 9, 1924; Geitler, 675, f. 435, 1932; Desikachary, 431, pl. 80, f. 6, 1959.

Thallus dark blue-green; trichomes straight, parallel, 4.5-7.5 μ broad; cells 7-10 μ long, quadrate to slightly cylindrical; sheath thick; heterocysts intercalary, elliptical 8-8.5 μ broad, and 9-14 μ long; spores alternating with dead cells, 8-11 μ broad and 13-18.5 μ long, in long chain.

From paddy fields, Diamond Harbour (South 24-Parganas) and Digha (Midnapore), Santipore (Nadia), Raigunj (West Dinajpur).

Specimen nos. DH 68, 416, N 17, WD 19.

Previously reported from U.P. (Pandey 1965, Pandey 1982); J & K (Anand 1979, Goyal et al. 1984); Karnataka (Bongale & Bharati 1980); Punjab (Grover & Pandhol (1975, Pandhol & Grover 1976); North East India (Reddy et al. 1986); Tamil Nadu (Anand & Revathi 1987); W.B. (Sinha & Mukherjee 1975a, Maity & Santra 1985, Sengar & Pandey 1985, Santra et al. 1988).

187. *Aulosira fertilissima* Ghose var. *tenuis* Rao, C.B. (Plate 21, figs. 7-10, Plate 44B).

The Myxophyceae of the United Provinces - III, Proc. Indian Acad. Sci., B, 6: 353, f. 3F-I, 1937b; Desikachary, 431, pl. 80, f. 2-5, 1959.

Thallus pale blue-green colour; filaments 4.5-6.5 μ broad, trichomes 4-5 μ broad, generally not constricted at the cross-wall; but sometimes constricted (specimen from Contai); sheath thin; cell cylindrical 7-13 μ long; end cell slightly tapering with rounded apex; heterocyst oblong-cylindrical with rounded ends, 5.5-7 μ broad, slightly buldged, 16-23 μ long; spores in chain, ellipsoidal to cylindrical, sometimes slightly angular, 6-9 μ broad, 10.5-18.5 μ long, alternating with moribund cells.

From paddy fields Midnapore, Bankura, Purulia, Malda, West Dinajpore, Cooch Behar, Jalpaiguri.

Specimen nos. C 12, 17; Ban 6, 8; Pur 3, 8, Mal 4, 8, WD 3, 8, Cooch 2, 4; Jal 3, 5.

Previously reported from U.P. (Chaturvedi & Pandey 1976, Pandey & Pandey 1982) Maharashtra (Kamat 1963, Ashtekar & Kamat 1980); W.B. (Sinha & Mukherjee 1975b).

Family: SCYTONEMATACEAE Raben.
Key to the Genera

1. Heterocyst absent, false branching seen, single or geminate .. *Plectonema*
1. Heterocyst present .. 2
    2. False branches single or gemiante .. *Scytonema*
    2. False branches usually single and arise next to the heterocyst .. *Tolypothrix*

Genus: PLECTONEMA Thuret
Key to the species

1. Trichomes 10-11 μ broad .. *P. tornasinianum*
1. Trichomes narrow .. 2
    2. Trichomes 2.5-3 μ broad, not constricted, cells 5-7 μ long .. *P. notatum*
    2. Trichomes 1-2 μ broad, not constricted or slightly constricted, cells 2-3.5 μ long .. *P. nostocorum*

188. *Plectonema notatum* Schmidle (Plate 27, fig. 3)

in Simmer's IV, Ber. uber d. Kryptogamenflora v. Kreuzecker, Allg. bot. Zeitschr., n. 3/4; 84, f. 8 & 9, 1901; Geitler, 693, 1932; Prescott, 540, pl. 126, f. 6, 7, 1951; Desikachary, 440, pl. 83, f. 5, 1959.

Filaments blue-green, bent; sheath thin, rarely branched; trichomes 2.5-3 μ broad, not constricted at the cross-walls, cells 5-7 μ long, end cell rounded.

From paddy field soils, Thakurchak (Midnapore).

Specimen no. T 115

Here the trichome is slightly broader than type species (2.5-3 μ instead of 1.75-2 μ).

Previously reported from Karnataka (Bongale & Bharati 1980); Maharashtra (Kolte & Goyal 1985); from culture soil of Mussoorie of U.P. (Das et. al. 1974).

189. *Plectonema tomasinianum* (Kutz.) Born. ex Gomont (Plate 26, fig. 4).

Bornet, Les Nostocacees heterocystees du Systema Algarum de C. Agardh et leur synonymie, Bull. Soc. bot. Fr., 36: 155, 1889, Gomont, Monogr. Oscillariees, 99, 1892; Geitler, 688, f. 442, 1932; Desikachary, 438, pl. 83, f. 3, 1959.

Filaments densely entangled forming brownish green thallus, showing false branches, sheath colourless or yellowish brown, trichomes slightly constricted at the cross-walls, 10-11 μ broad, cells 4-6 μ long, end cell rounded.

From paddy fields of Burdwan and Hooghly.

Specimen nos. Bur 8, 12; Hoog 6, 8.

Previously reported from Maharashtra (Kamat & Harankhedkar 1976, Kolte & Goyal 1985); W.B. (Sengar & Pandey 1985).

190. *Plectonema nostocorum* Bornet ex Gomont (Plate 27, fig. 2)

Bornet, in Bornet and Thuret, Notes algologiques, 2: 137, 1887; Gomont, Monogr. Oscillariees, 102, pl. 1, f. 11, 1892; Geitler, 692, f. 445, 1932; Prescott, 539, pl. 126, f. 4, 5, 1951; Desikachary, 439, pl. 83, f. 7, 1959.

Thallus pale blue-green; filaments straight or flexuous, 2-3 μ broad, false branches rare, single or geminate; trichomes 1-2 μ broad, cross-wall not constricted or slightly constricted, septa not granulated; cells cylindrical, 2-3.5 μ long, end cell rounded.

From paddy field soil of Howrah & 24-Parganas (N).

Specimen nos. How 4, 6; PN 5, 8.

Previously reported from Karnataka (Bongale & Bharati 1980); Maharashtra (Kolte & Goyal 1985); J & K (Goyal et al. 1984); U.P. (Pandey 1965 described new forma *major*) and W.B. (Mukhopadhyay & Chatterjee 1981).

Genus: SCYTONEMA Ag.

Key to the species

1. Sheath lamellated .. 2
1. Sheath not clearly lamellated .. 9
    2. Sheath with distinctly divergent lamellation .. *S. myochrous*
    2. Sheath with parallel lamellation .. 3
3. Thallus in floating condition .. 4
3. Thallus not in floating, but on wet soils .. 6

	4. Cells quadrate to shorter than broad trichome 10-11.5 μ broad, cells 6.5-10 μ long	..	*S. javanicum*
	4. Cells quadrate to longer than broad	..	5
5.	Trichome 5.5-6 μ broad	..	*S. leptobasis*
5.	Trichome 10-12 μ broad	..	*S. quyanense*
	6. Sheath with slightly divergent lamellation	..	7
	6. Sheath not so, only parallel lamellation	..	8
7.	Filaments 17-20 μ broad, trichomes 7.5-11 μ broad	..	*S. mirabile*
7.	Filaments 10-14 μ broad trichomes 9-11 μ broad	..	*S. tolypothrichoides*
	8. Trichomes 5-6 μ broad	..	*S. caldarium*
	8. Trichomes 10-12 μ broad	..	*S. burmanicum*
9.	Thallus in floating condition	..	10
9.	Thallus on wet soils	..	13
	10. Trichome 16-25 μ broad	..	*S. cincinatum*
	10. Trichome narrower	..	11
11.	Filaments 17-20 μ broad, trichomes 9.5-13 μ broad, cells subquadrate to cylindrical, 7.5-20(-27) μ long	..	*S. coactile*
11.	Filaments narrower, cells quadrate to cylindrical	..	12
	12. Trichomes 9-10(-11.5) μ broad	..	*S. simplex*
	12. Trichomes 6.3-7 μ broad	..	*S. fritschii*
13.	Sheath sometimes lamellated, filaments 18-23 μ broad	..	*S. stuposum*
13.	Sheath not lamellated, filaments narrower	..	14
	14. Trichomes 10-11.5 μ broad, not constricted at cross-walls	..	*S. hofmanii*
	14. Trichomes constricted	..	15
15.	Trichomes 6.5-9 μ broad, cells subquadrate, 4-5 μ long	..	*S. schmidtii*
15.	Trichomes 4-5 μ broad cells longer than broad 4.5-7.5 μ long	..	*S. fremyii*

191. *Scytonema simplex* Bharadwaja (Plate 23, figs. 1-4; Plate 44C)

The taxonomy of Scytonema and Tolypothrix, Rev. algol., Paris, 7: 157, f. 1A, B, 1934; Desikachary, 455, pl. 89, f. 1. 1959.

Thallus dark blue green, filaments 12.5-15 μ broad, loosely entangled false branches long; sheath firm, hyaline, unstratified; trichomes 9-10(-11.5) μ broad; cells generally cylindrical, sometimes quadrate, 12.5-22(-30) μ long; heterocyst single, as broad as trichome 10-11 μ broad and 27-40 μ long.

From paddy fields, in floating condition, Jhargram (Midnapore), Bankura, Burdwan, Howrah, Hooghly, Birbhum, Murshidabad.

Specimen nos. 415, J 49, J 122, Ban 3, 4; Bur 6, 8; How 2, 6; Hoog 4, 6; Bir 5, 6; Mur 3, 4.

Previously reported from Bihar (Sinha & Srivastava 1980); Gujarat (Vasishta 1968); Maharashtra (Vasishta 1968, Kolte & Goyal 1985, Patil & Satav 1986); U.P. (Khan & Kumari 1972; Khan 1985); Andaman & Nicobar Islands (Prasad & Srivastava 1984); Karnataka (Somashekar 1983, 1984b); W.B. (Sinha & Mukherjee 1975a, Mukhopadhyay & Chatterjee 1981, Pal & Santra 1982, Sengar & Pandey 1985).

192. *Scytonema coactile* Montagne ex Born. et Flah. (Plate 24, figs. 3-5; Plate 44D).

Montagne in Kützing Species Algarum, 305, 1849; Bornet et Flahault, Revision des Nostocacees heterocystees, 90, 1887; Geitler, 753, f. 479, 1932; Desikachary, 455, pl. 90, f. 2, 1959.

Thallus wooly, green to blue-green; filament 17-20 μ broad, false branches long; sheath firm, 2-2.5 μ thick, hyaline to yellowish; trichome 9.5-13 μ broad; cells subquadrate to cylindrical, 7.5-20(-27) μ, at the tip region of the trichome cells are 6-7.5 μ long; heterocysts longer than broad, (9-) 11-13 μ broad and 12-25 (-32.5) μ long.

From paddy fields, Kolsanda (Midnapore).

Specimen nos. 104, 171.

Previously reported from Allahabad (Chadha & Pandey 1983a).

193. *Scytonema fritschii* Ghose (Plate 23, figs. 5-7).

J. Linn. Soc. Bot. Lond., 46: 342, pl. 31, f. 11; 1924; Geitler, 755, f. 480, 1932; Desikachary, 458, pl. 95, f. 1, 1959.

Thallus pale blue-green to brownish-green; filaments 15-17.8 μ broad; sheath thick, hyaline; trichome 6.3-7 μ broad; cells 6.3-10 μ long, cross-walls distinct; false branches geminate; heterocysts 7.5-8.5 μ broad and 14-21 μ long.

From paddy fields of Burdwan and Hooghly.

Specimen nos. Bur 15; Hoog 6, 8.

Previously reported from Allahabad (Chadha & Pandey 1983a); Punjab (Sarma et al. 1979); A.P. (Rao 1938a); W.B. (Sengar & Pandey 1985).

194. *Scytonema hofmanni* Ag. ex Born. et Flah. (Plate 25, fig. 3).

Agardh, Synopsis Algar. Sueciee, p. 117, 1817; Bornet et Flahault, Revision des Nostocacees heterocystees, 97, 1887; Geitler, 772, f. 495, 1932; Desikachary, 476, pl. 91, f. 2, 1959.

Thallus blue-green, terrestrial; filaments 12-15 μ broad, false branches common, sheath firm, membranceous; trichomes 10-11.5 μ broad, cells variable in length 7.5-12(-20) μ long, heterocysts oblong, 11.5-12.7 μ broad and 20-30(-35) μ long.

From paddy field soils of Contai (Midnapore), 24-Parganas (N), Nadia, Howrah, Malda, Jalpaiguri, Darjeeling, Cooch Behar.

Specimen nos. C 104, PN 22; N 6, 8; How 7, 9; Mal 12, 16, Jal 7, 14; Darj 5, 7; Cooch 10, 12.

The present alga resembles *S. hofmanni* Ag. ex Born. et Flah. in many characters, but differs in the measurements of heterocyst.

Previously reported from U.P. (Pandey 1965, Pandey & Chaturvedi 1979), Prasad & Mehrotra 1980, Pandey & Pandey 1982, Chadha & Pandey 1983a); J & K (Goyal et al. 1984); North East India (Reddy et al. 1986); W.B. (Chatterjee & Chatterjee 1983, Sengar & Pandey 1985).

195. *Scytonema fremyii* Desikachary (Plate 24, figs. 1 & 2)

Desikachary, 474, pl. 100, f. 3, 1959.

Thallus greenish black, prostrate; filaments 10-11.5 μ broad, sheath thick, hyaline, not lamellated; trichomes 4-5 μ broad, and 5.5 μ broad at the apices, deeply constricted at the apical regions, but slightly constricted at remaining region of the trichome; cells 4.5-7.5 μ long; heterocysts rectangular 5-6 μ broad and 7.5-11.5 μ long.

From paddy field soils, Panchkuri (Midnapore) and Burdwan.

Specimen nos. Panch 128, Bur 14, 23.

Here filament slightly broader than the type speices.

Previously reported from Madras (Fremy, 1942); Punjab (Sarma et al., 1979); W.B. (Sengar & Pandey, 1985).

196. *Scytonema mirabile* (Dillw.) Born. (Plate 26, fig. 1)

Nostoc. heteroc. du Syst. Alg. de C.A. Agargh, Bull, Soc. bot. Fr., 36: 155, 1889; Geitler, 775, f. 497, 498, 1932; Desikachary, 483, pl. 91, f. 3, 1959.

Thallus tomentose, balckish green; filaments 17-20 μ broad, sheath lamellated, slightly divergent, yellowish brown; trichome 7.5-11 μ broad, cells cylindrical; heterocyst quadrate to longer than broad.

From paddy fields, Bankura, West Dinajpur, Purulia.

Specimen nos. W 28, Pur 7, Ban 12.

Previously reported from Tamil Nadu (Suxena 1983); U.P. (Prasad & Mehrotra 1980); W.B. (Sinha & Mukherjee 1975b).

197. *Scytonema schmidtii* Gom. (Plate 25, fig. 1)

In Johs. Schmidt, Flora of Koh - Chang, Myxo. hormogon, 124, 1901; Bot. Tidsskr., 24, 210, pl. 5, f. 1-4, 1901; Geitler, 759, f. 483b, 1932; Desikachary, 459, pl. 92, f. 1959.

Thallus brownish black, crustaceous; filaments irregularly undulated, prostrate, repeatedly false branched 10-14 μ broad with yellowish brown sheath, trichomes 6.5-9 μ broad, cells subquadrate, 4-5 μ long constricted, heterocysts quadrate.

From paddy fields soil of Burdwan.

Specimen nos. Bur 24, 32.

Previously reported from Punjab (Sarma et al. 1979); U.P. (Chadha & Pandey 1983a); Karnataka (Bongale & Bharati 1980); Maharashtra (Chaporkar & Gangawane 1984; Kolte & Goyal 1986); W.B. (Sengar & Pandey 1985).

198. *Scytonema caldarium* Setchell (Plate 25, fig. 2)

Notes on Cyanophyceae III, Erythea, 7: 48, pl. 3, f. 3, 1899; Geitler, 754, 1932; Desikachary, 458, 1959.

Thallus expanded; filaments 14-15.5 μ broad, with false branches, branches geminate; sheath showing parallel lamellations, colourless to yellowish brown, trichomes 5-6 μ broad and cells subquadrate to cylindrical, about 5-12.5(-15) μ long; heterocyst 5-6 μ broad and 5-7.5(-8) μ long.

From paddy field soils of Burdwan.

Specimen nos. Bur 14, 16.

Previously reported from W.B. (Sengar & Pandey 1985).

199. *Scytonema javanicum* (Kutz.) Bornet ex Born. et Flah. (Plate 28, figs, 4 & 5; Plate 45A).

Bornet, E. and Flahault, C., Revision des Nostocacees heterocystees, III. 7(6): 95, 1887; Tilden, 218, pl. 12, f. 6, 1910; Geitler, 765, f. 490, 1932; Desikachary, 461, pl. 100, f. 4, 1959.

Thallus brownish blue-green; filaments densely aggregated; filaments 16-18 μ broad; sheath lamellated, colourless to yellowish brown, upto 3.5 μ thick, trichomes 10-11.5 μ broad, not constricted at the cross-wall; cells quadrate or shorter than broad, 6.5-10 μ long; heterocysts quadrate to cylindrical 10-12 μ broad and 17-25(-35) μ long.

From paddy fields in floating condition, Keshpur (Midnapore), Burdwan, Hooghly.

Specimen nos. 120, Bur 13, 18, Hoog 6, 8.

The present form is close to *Scytonema javanicum* (Kütz.) Bornet ex Born. et Flah. in many characters but differs from the type species in having broader filaments and quadrate to cylindrical heterocysts.

Previously reported from Maharashtra (Marathe 1970, Chaporkar & Gangawane 1984, Kolte & Goyal 1985, 1986; Patil & Satav 1986); North India (Prasad & Mehrotra 1979); Banaras (Rao 1937); Nainital (Singh 1959); Dehra Dun (Khan & Kumari 1972); Meerut (Bentre and Kumar 1975); from paddy fields in Lucknow and its neighbourhood (Prasad & Mehrotra 1980); Ghaziabad (Pal 1975); Andaman and Nicobar Islands (Prasad & Srivastava 1984); W.B. (Sengar & Pandey 1985).

200. *Scytonema leptobasis* Ghose (Plate 25, figs. 4-6; Plate 45B)

Five more myxophyceae from Burma. J. Indian bot. Soc., 10(1): 35-37, f. 3-5, 1931; Desikachary, 457, 1959.

Thallus brownish blue-green, filaments 13-15 μ in diameter, false branches geminate; sheath thick, colourless to brownish blue green, lamellated, lamellae more or less parallel, sheath upto 7.5 μ thick; trichomes 5.5-6 μ broad at base and 6-9 μ broad at tips, trichomes constricted at the cross-walls towards the tip regions but not constricted along the basal regions; cells 6-10 μ long, barrel-shaped to cylindrical, apical cells swollen with rounded ends; heterocysts 5-6 μ broad and 12-13 μ long.

From water logged rice field, Pachra (Midnapore), Burdwan and Howrah.

Specimen nos. 498, Bur 22, How 15, 21.

Previously reported from Andaman and Nicobar Islands (Prasad & Srivastava 1984); West Bengal (Sengar & Pandey 1985).

201. *Scytonema burmanicum* Skuja (Plate 28, figs. 1-3)

Zur Susswasseralgenflora Burmas, 34, pl. 4, f. 1-10, 1948; Desikachary, 460, pl. 97, 1959.

Thallus prostrate, bluish-brown in colour, sometimes blackish; filaments densely aggregated 14.5-16 μ broad; pseudo-branches solitary or geminate; sheath showing parallel lamellations, colourless to brown; trichomes 10-12 μ broad; cells quadrate to cylindrical, slightly constricted or unconstricted at the cross-wall, 10-16 μ long; heterocysts 10-13 μ broad, 12-17.5 μ long.

From paddy field soils of Burdwan, Hooghly and Howrah.

Specimen nos. Bur 6, 12, Hoog 4, 8, How 3, 5.

Previously reported from Maharashtra (Kamat & Harankhedkar 1976) U.P. (Pal 1975); West Bengal (Sengar & Pandey 1985).

202. *Scytonema guyanense* (Mont.) Born. et Flah. (Plate 27, fig. 1)

Revision des Nostocacees heterocystees, 94, 1887; Geitler, 770, f. 493, 1932; Desikachary, 469, 1959.

Thallus brownish-green colour, sometimes blackish green. Filaments 7.5-20 μ in diameter; sheath yellowish brown, lamellated, upto 5 μ thick; trichomes 10-12 μ broad; cells quadrate to longer than broad, 10-17 μ long; heterocysts cylindrical 10-12 μ broad and 10-20 μ long.

From water-logged rice field of Burdwan and Purulia.

Specimen nos. Bur 26, 28; Pur 13, 17.

Previously reported from U.P. (Rao 1937, Singh 1959, Pandey & Chaturvedi 1979, Chadha & Pandey 1982); A.P. (Rao 1938); Tamil Nadu (Fremy 1942); Andaman and Nicobar Islands (Prasad & Srivastava 1984); W.B. (Sengar & Pandey 1985).

203. *Scytonema stuposum* (Kutz.) Bornet ex Born. et Flah. (Plate 30, fig. 5).

Bornet in Bornet and Thuret, Notes algoligigues, 2: 146, 1880; Bornet and Flahault, Revision des Nostocacees heterocystees, 92, 1887; Geitler, 756, f. 482, 1932; Desikachary, 459, pl. 93, f. 4, 1959.

Thallus felt-like, blackish green colour; filaments 18-23 μ broad, sheath thick, generally gelatinous, sometimes lamellated, false branches single and geminate; trichomes 14-16 μ broad, constricted at the cross-wall, cells shorter than broad, sometimes quadrate; heterocysts 15-16 μ broad, 14-18 μ long.

From paddy fields of Bankura, Purulia and Birbhum.

Specimen nos. Ban 8, 12; Pur 10, 12; Bir 3, 12.

Previously reported from Allahabad (Pandey & Pandey 1982, Chadha & Pandey 1983a); Rohilkhand of U.P. (Pandey 1982); from culture soils, Panchmari of M.P. (Das et al. 1974); Panhala of Maharashtra (Kamat 1961-62); W.B. (Sinha & Mukherjee 1984).

204. *Scytonema cincinnatum* Thuret ex Born. et Flah. (Plate 29, fig. 1)

Thuret, Essai de classification des Nostochinees, Ann. Sci. nat., 6 ser., Bot. 1: 380, 1875; Bornet et Flahault, Revision des Nostocacees heterocystees, 89, 1887; Geitler, 748, f. 477, 1932; Desikachary, 453, pl. 93, f. 1, 1959.

Thallus woolly, blue-green to brownish green colour, filaments 22-27 μ broad, false branches geminate, trichomes 16-25 μ broad, constricted at the cross-wall, sheath membranaceous; cells 5-10 μ long, heterocysts quadrate or depressed quadrate 20-23 μ broad, 11-14 μ long.

From paddy fields of Bankura, Midnapore & Purulia.

Specimen nos. Ban 13, 17; M 2, 4, Pur 6, 8

Previously reported from U.P. (Chadha & Pandey 1983a; Khan, 1985); North East India (Reddy et al. 1986); Punjab (Sarma et al. 1979); W.B. (Sinha & Mukherjee 1975a).

205. *Scytonema myochrous* (Dillw.) Ag. ex Born. et Flah. (Plate 29, figs. 2 & 3)

Agardh, Disp. Alg. Succiae, 38, 1912; Bornet et Flahault, Revision des Nostocacees heterocystees, 104, 1887; Geitler, 780, f. 49, 501, 502, 1932; Desikachary, 487, pl. 90, f. 3, pl. 99, f. 2, 1959.

Thallus blackish green, forming woolly stratum; filaments 20-30 μ broad, trichomes 6.5-11 μ broad, generally not constricted at the cross-wall, sometimes very slightly constricted, sheath with divergent lamellae; cells cylindrical below and quadrate to discoid above, 5-15 μ long, heterocysts 10-12 μ broad, 6-13 μ long.

From paddy fields of Bankura, Midnapore and Birbhum.

Specimen nos. Ban 18, 22; M 7, 31; Bir 6, 8.

Previously reported from Maharashtra (Kamat 1962-63, Chaporkar & Gangawane, 1984); W.B. (Sinha & Mukherjee 1975a).

206. *Scytonema tolypothrichoides* Kutzing ex Born. et Flah. (Plate 26, figs. 2 & 3)

Kutzing, Species Algarum, 307, 1849; Tab. Phycologicae, 2: 6, pl. 22, f. 4, 18; Bornet et Flahault, Revision des Nostocacees heterocystees, 100, 1887; Geitler, 779, f. 500; 1932; Desikachary, 479, pl. 88, f. 3, pl. 91, f. 1, 1959.

Thallus blue-green to brownish-green; filaments densely entangled, 10-14 μ broad, pseudobranches geminate, similar to main filaments, trichomes 9-11 μ broad, not constricted at the cross-wall, sheath thick, lamellated sometimes with slight divergent lamellation; cells quadrate to subquadrate, 6-9.5 μ long, cell contents granulated, end cell rounded; heterocysts elliptical or cylindrical, 11-12 μ broad, 16-20 μ long.

From paddy fields of Bankura, 24-Parganas (S), Nadia and Howrah.

Specimen nos. Ban 35, 40; S - 17, 28; N - 3, 13, How 1, 19.

Previously reported from North India (Prasad & Mehrotra 1979a); U.P. (Darbal et al. 1978, Prasad & Mehrotra 1979b, 1980); Maharashtra (Gonzalves 1947, Vasishta 1968); Andaman & Nicobar Islands (Prasad & Srivastava 1984); W.B. (Bruhl & Biswas 1923; Banerji 1938, Sinha & Mukherjee 1984a).

Genus: TOLYPOTHRIX Kutzing

Key to the species

1. Filaments 5-6 μ broad, sheath thin, trichomes constricted, cells cylindrical, heterocysts ellipsoidal to cylindrical .. *T. nodosa*

1. Filaments 16-18 μ broad, sheath thick, sometimes lamellated, trichomes unconstricted or slightly constricted, cells quadrate to slightly cylindrical, heterocysts discoid .. *T. tenuis*

207. *Tolypothrix nodosa* Bharadwaja (Plate 27, fig. 4)

The Taxonomy of *Scytonema* and *Tolypothrix*, Rev. algol, Paris, 7: 175, f. 7c, 1934; Desikachary, 494, pl. 98, f. 5, 1959.

Thallus mucilaginous, blue-green; filaments densely entangled, 5-6 μ broad, with false branches, sometimes with dead cell adjoining to the heterocyst; sheath firm, thin, hyaline, bulged out at the point of branching; trichomes constricted at the cross-walls, 4-4.5 μ broad; cells 6-8 μ long, heterocysts intercalary or terminal, ellipsoidal to slightly cylindrical in shape, 6-8 μ broad and 8-18 μ long.

From water-logged paddy fields, Kolaghat (Midnapore).

Specimen no. 252.

Previously reported from U.P. (Pandey, D.C. 1965, Pandey, U.C. 1982); W.B. (Pal & Santra, 1985).

208. *Tolypothrix tenuis* (Kütz.) Johs. Schmidt. em. (Plate 32, fig. 1)

Bot. Tidsskr., 22: 383, 1899; Lemmermann, Kryptogamenflora der Mark Brandenburg, 3: 217, 1910; Desikachary, 494, pl. 102, f. 4, 1959.

Thallus blue-green to yellowish blue-green; filaments branched, 16-18 μ broad, sheath colourless to yellowish brown, sometimes lamellated, trichomes 9-12 μ broad, unconstricted or slightly constricted at the cross-wall, cells quadrate to slightly cylindrical, 6-13 μ long; heterocysts discoid, 9-14 μ diam., single.

From paddy fields of Bankura, Birbhum, Howrah, Nadia, Malda, Jalpaiguri, Darjeeling.

Specimen nos. Ban 10, Bir 7, How 3, 6, N 7, 14, Mal 3, 9, Jal 12, 16, Dar 18.

Previously reported from U.P. (Rao 1937b, Prasad & Mehrotra 1980, Pandey 1982); Delhi (Rao 1940); North-East India (Reddy et al. 1986); Maharashtra (Dixit 1936, Kamat & Harankhedkar 1976, Chaporkar & Gangawane 1984, Kolte & Goyal 1985, Patil & Satav 1986); W.B. (Sinha & Mukherjee 1975b, Chatterjee & Chatterjee 1983).

Family : MICROCHAETACEAE Lemmermann

Genus : MICROCHAETE Thuret

Key to the Species

1. Filaments 14-15 μ broad, sheath firm, cells generally subquadrate, sometimes cylindrical, ge heterocysts cylindrical .. *M. uberrima*
1. Filaments 6-8 (-10) μ broad, sheath thin, cells cylindrical at the base, and quadrate at the apices, heterocysts spherical to ellipsoidal. .. *M. tenera*

209. *Microchaete uberrima* Carter, N. (Plate 31, figs 1-3; Plate 45C)

Fresh water algae from India, Rec. bot. surv. of India, 9: 268, pl. f. 1 - 3, 1926; Geitler, 669, f. 430, 1932; Desikachary, 511, pl. 104, f. 5 - 7, 10, 13, 16, 18, 1959.

Filaments straight, cylindrical, stellately arranged from bases, filaments 14-15 μ broad; sheath firm, trichome 9.5-13 μ broad, not attenuated; cells generally subquadrate 8-12.5 μ long, sometimes cylindrical and upto 20 μ long, constricted at the cross-walls; heterocysts basal and intercalary, 10-12 μ broad and 15-23 μ long; spore single or in chain, 9.5-11 μ broad and 12-18 μ long.

Epiphytic on submerged portion of rice plant, Narajole (Midnapore)

Specimen no. 51.

Previously reported from—Maharashtra (Kamat 1963, Kamat & Harankhedkar 1976); Karnataka (Bongale & Bharati 1980); U.P. (Prasad & Mehrotra 1980); Bihar (Jha et al. 1986).

210. *Microchaete tenera* Thuret ex Born. et Flah. (Plate 31, fig. 6)

Thuret, Essai class, nost., Ann. Sci. nat. Bot., Ser., 6, 1: 378, 1875; Bornet et Flahault, Revision des Nostocacees heterocystees, 84, 1887; Geitler, 668, f. 429, 1932; Desikachary, 513, pl. 105, f. 1, 2, 1959.

Filament single, sometimes in clusters, slightly bent at the base, 6-8 μ broad upto 10 μ broad at the base; trichome 4-5 μ broad and upto 7 μ broad at the base, sheath thin, hyaline, unlamellated, unconstricted or slightly constricted at the cross walls; cells cylindrical at the base, upto 10 μ long and quadrate at the apices; heterocysts basis and intercalary.

Spherical to ellipsoidal 6-7.5 μ broad and 9-11 μ long; spores in chain, long ellipsoidal, 6-8 μ broad, and 13.5-16 μ long.

From paddy fields of Falta (24 Parganas), Bankura, Malda & Jalpaiguri.

Specimen nos. F 21, Bank - 31, Mal - 17, Jal - 21, 23.

Previously reported from U.P. (Rao 1937b, Prasad & Mehrotra 1980) Karnataka (Bongale & Bharati 1980); W.B. (Sinha & Mukherjee, 1984, Maity & Santra, 1985, Santra et al. 1988)

Family: RIVULARIACEAE Rabenhorst

Key to the Genera

1. Trichomes with heterocysts .. 2
1. Trichomes without heterocysts .. *Homoeothrix*
    2. Filaments single or in bundles, trichome with a firm sheath, spore present or absent .. *Calothrix*
    2. Filaments many in a spherical or hemispherical thallus, sheath gelatinising; single large spore at the base of the trichome .. *Gloeotrichia*

Genus: HOMOEOTHRIX (Thuret) Kirchner

211. *Homoeothrix juliana* (Menegh.) Kirchn. (Plate 31, fig. 4)

Engler-Prandl. Naturlichen Pflanzenfam. I, 1a, 348, 1898; Geitler, 575, f. 359, 1932; Desikachary, 519 pl. 107, f. 7, 1959.

Filaments unbranched, 11-12 μ broad; sheath thin; colourless not lamellated; trichomes 8-9 μ broad produced into a long heir, cells discoid 4-5 μ long, but 6 μ long in the attenuated region.

Epiphytic on submerged portion of rice plants of Tamluk (Midnapore) & West Dinajpur.

Specimen nos. 292, WD 12, 16.

The present alga shows slightly narrower trichome (9-12 μ broad instead of 9-12.5 μ) than the type species.

Previously reported from Punjab (Sarma et al. 1979); W.B. (Pal & Santra 1985).

Genus: CALOTHRIX Ag.

1.	Spores present, about 12 μ broad, 16-25 μ long	..	*C. ghosei*
1.	Spores absent	..	2
	2. Trichomes upto 5 μ broad	..	3
	2. Trichomes broader	..	6
3.	Trichomes 4.5-5 μ broad at the base, terminating into a hair	..	*C. marchica*
3.	Trichomes narrower not ending into hairs	..	4
	4. Heterocysts spherical or globose, end cell rounded	..	5
	4. Heterocysts subspherical, end cell narrow, trichomes 3.5-4.5 μ broad, thallus membranous	..	*C. membranacea*
5.	Trichomes 50-70 μ long, 3-4 μ broad, heterocysts 3 μ diam.	..	*C. brevissima*
5.	Trichomes 32-45 μ long, 4-5 μ broad, heterocysts 4-5 μ diam.	..	*C. brevissima* v. *moniliforme*
	6. Thallus surface hairy	..	7
	6. Thallus surface not hairy	..	8
7.	Trichomes 8.5-9 μ broad at base, sheath thin	..	*C. castellii*
7.	Trichomes 11-13 μ broad at base, sheath thick	..	*C. castellii* v. *somastipurense*
	8. Filaments generally bent or curved at base, sheath thin	..	9
	8. Filaments not so	..	10
9.	Trichomes slightly bent at base, sometimes straight, 6.5-7.5(-8) μ broad at base, ends terminating into hairs	..	*C. brauni*
9.	Trichomes in tufts, bent at base, 5.5-7 μ broad at base, apex not ending into hair	..	*C. elenkinii*
	10. Filaments generally single	..	12
	10. Filaments very closely arranged	..	11
11.	Trichomes 8-10 μ broad at the base, sheath distinct	..	*C. parietina*
11.	Trichomes narrower, sheath distinct only towards base	..	*C. scytonemicola*

12. Trichomes 7-8 μ broad at the base .. *C. fusca*

12. Trichomes 10-11 μ broad at the base .. *C. fusca* var. *crassa*

212. *Calothrix fusca* (Kutz.) Bornet et Flahault (Plate 32, fig. 2)

Revision des Nostocacees heterocystees, 364, 1886; Geitler, 610, f. 384; 1932; Desikachary, 527, pl. 107, f. 10, 1959.

Filaments single, rarely aggregated in a gelatinous thallus, filament 11-15 μ broad at the base; sheath colourless, trichome 7-7.5 μ broad at the middle region and 8 μ broad at the base, terminating in a long hair, heterocysts basal, hemispherical, smaller than the basal cell of the trichome.

From paddy fields, on dead plant part, with *Oedogonium* sp.

Dharma (Midnapore).

Specimen no. 377.

Previously reported from Orissa (Rao 1938b); Punjab (Sarma et al. 1979); Bihar (Jha et al. 1986, Saha & Pandit 1987); U.P. (Chaturvedi & Pandey 1976), Maharashtra (Kamat 1962, 1963; Ashtekar & Kamat 1980); Tamil Nadu (Fremy 1942); W.B. (Pal & Santra 1987).

213. *Calothrix fusca* var. *crassa* Rao, C.S. (Plate 30, fig. 4)

The Myxophyceae of the Bihar Provinces, I, Proc. Indian Acad. Sci., B, 9: 145, f. 1 B-E, 1939; Desikachary, 529, pl. 113, f. 7, 8, 10, 1959.

Filaments single, straight or curved, 16-17 μ broad at the base, trichome 10-11 μ broad at the base, 4.5-5.5 μ broad in the middle and terminating into a hair; not constricted at the cross-wall, sheath yellowish brown lamellated, sheath absent on terminal hair; cells 4-6(-7.5) μ long; heterocysts hemispherical, 8-9 μ broad and 6-7.5 μ long.

From paddy fields of Joyrambati (Bankura), Purulia.

Specimen nos. Ban 41, Pur 14, 22.

Previously reported from Bihar (Rao 1939); Karnataka (Gonzalves and Kamat 1960); U.P. (Pal 1975); W.B. (Sinha & Mukherjee 1975b).

214. *Calothrix parietina* Thuret ex Born. et Flah. (Plate 30, fig. 3)

Thuret, Essai, class. Nostochinees, Ann. Sci. nat. Bot., ser. 6, 1: 381, 1875; Bornet et Flahault, Revision des Nostocacees heterocystees, 366, 1886; Geitler, 604, f. 380, 1932; Desikachary, 538, pl. 108, f. 6-8 & pl. 115, f. 1, 1959.

Thallus dark blue-green, filaments closely arranged, 10-14 μ broad at the base; trichome 8-10 μ broad at the base, 4-5 μ in the middle and ending into a hair; cross-wall not constricted or slightly constricted cells 5-6.5 μ long at base, and 6-8 μ long in the middle, heterocysts generally basal, subspherical, 7.5-8.5 μ broad, 6-8 μ long.

From paddy fields of Hooghly, Burdwan, Howrah.

Specimen nos. Hoog 8, 13; Bur 6, 8; How 5, 7.

Previously reported from Tamil Nadu (Fremy 1942); U.P. (Prasad & Mehrotra 1980); W.B. (Prain 1905, Chatterjee & Chatterjee 1983).

215. *Calothrix ghosei* Bharadwaja (Plate 31, fig. 7)

The Myxophyceae of the United Province, I, Proc. Indian Acad. Sc. B, 2: 99, f. 3 C-F, 1935; Desikachary, 526, pl. 112, f. 2, 1959.

Filaments in groups, straight or curved, 8-8.5 μ broad; sheath hyaline; trichomes 8-8.5 μ broad near heterocysts, and 4-5 μ broad towards hairs, slightly constricted at the cross-walls; apical cell rounded, cells of the middle region of the trichome quadrate to barrel shaped, cells towards hair region longer than broad; heterocyst single, basal, nearly spherical, 6-7 μ in diam., enclosed within a sheath; spore single, in between vegetative cell and heterocyst, 12 μ broad, 16-25 μ long.

From paddy fields of Ghatal (Midnapore).

Specimen no. 292.

In the present form the heterocysts are larger (6-7 μ in diam. instead of 4.2-6.3 μ) but spores are smaller (12 μ x 16-25 μ instead of 10.5-12.5 x 31.5-52.5 μ) than the type species.

Previously reported from U.P. (Bharadwaja 1935); W.B. (Pal and Santra 1985).

216. *Calothrix brevissima* West G.S. (Plate 30, fig. 6)

Freshwater Alg. Tanganyika, J. Linn. Soc. Bot., Lond., 38: 180, pl. 10, f. 8, 1907; Geitler, 624, f. 396, 1932; Desikachary, 533, pl. 114, f. 1, 1959.

Filaments densely arranged, 50-70 μ long, 5-6 μ broad, very slightly attenuated; trichomes 3-4 μ broad, end cell rounded or slightly conical, cross-walls not constricted or slightly constricted; cells 2-3 μ long, heterocysts single, basal, spherical, 3 μ diam.

From paddy fields of Burdwan, Howrah, 24-Parganas (S).

Specimen nos. Bur 12, 16; Howrah 3, 8; S 7, 22.

Previously reported from Karnataka (Bongale & Bharati 1980); U.P. (Pal 1975); Maharashtra (Kolte & Goyal 1985, 1986); J & K (Goyal et al., 1984); W.B. (Chatterjee & Chatterjee 1983).

217. *Calothrix brevissima* var. *moniliforme* Ghose orth. mut. Geitler (Plate 31, fig. 5).

Myxophyceae of Rangoon, III, J. Burma Res. Soc., 17: 242, pl. 2, f. 13, 1927; Geitler, 624, 1932; Desikachary, 533, pl. 112, f. 8, 1959.

Filaments densely arranged, 5.5-7 μ broad, ends not attenuated; trichomes 32-45 μ long, 4-5 μ broad; cross-wall constricted, cells (2.5)-3-4 μ long, heterocyst globose, 4-5 μ diam.

From paddy fields of Bankura and Purulia.

Specimen nos. Ban 9, Pur 15.

Previously reported from Karnataka (Bongale & Bharati 1980); W.B. (Sinha & Mukherjee 1975a).

218. *Calothrix braunii* (A. Br.) Bornet et Flahault (Plate 32, fig. 3)

Revision des Nostocacees heterocystees, 368, 1886; Geitler, 606, f. 381; 1932; Desikachary, 535, pl. 114, f. 3, 1959.

Filaments straight or slightly curved, base swollen, slightly bent, 9.5-10.5 μ broad at base; sheath thin close to trichome; trichome 6.5-7.5 (-8) μ broad at base, and gradually tapering towards upper end and terminating in a hair, cross-wall constricted; cells 3-4.5 μ long at base, 5-8 μ long in the middle, heterocysts basal, hemispherical, 5-6 μ broad.

From paddy fields of Bankura, Hooghly, Nadia, Malda.

Specimen nos. Ban 13, 18 Hoog 5, 9; N 7, 8; Mal 6, 9.

Previously reported from Punjab (Grover & Pandhol 1975, Pandhol & Grover 1976); Bihar (Saha & Pandit 1987); U.P. (Prasad & Mehrotra 1980); Maharashtra (Kolte & Goyal 1985); W.B. (Sinha & Mukherjee 1975b).

219. *Calothrix scytonemicola* Tilden (Plate 34, fig. 2)

Minnesota algae, 1: 265, pl. 17, f. 7, 1910; Geitler, 627, f. 398, 1932; Desikachary, 537, pl. 112, f. 14, 1959.

Filaments 6.5-7.5 μ diam, a few in a group, basal portion of the filaments attached with other algae (*Lyngbya* sp.) and the remaining portion free and ends into a hair; sheath distinct only towards base; heterocyst globose, 6-7.5 μ diam.

From paddy fields of Bankura & Purulia.

Specimen nos. Ban 21, 25; Pur 16, 22.

Previously reported from W.B. (Sinha & Mukherjee 1984).

220. *Calothrix castellii* (Massal.) Born. et Flah. (Plate 31, fig. 10)

Revision des Nostocacees heterocystees, 369, 1886; Geitler, 611, f. 385, 1932; Desikachary, 529, pl. 114, f. 7, 1959.

Thallus blue-green, ends of filaments projected outside; filaments aggregated; sheath thin, hyaline; trichome broad at the base 8.5-9 μ broad near heterocyst, ending into a long hair; cells 3-4 μ long near heterocyst; heterocyst basal 4-6 μ long and 5-7.5 μ broad; spore absent.

From paddy fields, Diamond Harbour (South 24-Parganas), Howrah, Midnapore.

Specimen nos. S 7, 8; How 6, 8; M 17, 23.

Previously reported from U.P. (Prasad & Mehrotra 1980); Tamil Nadu (Anand & Revathi 1987); W.B. (Maity & Santra 1985; Santra et al. 1988).

221. *Calothrix castellii* var. *somastipurense* Rao, C.S. (Plate 32, fig. 4)

The Myxophyceae of the Bihar Province I, Proc. Indian Acad. Sci., B, 9: 145, f. 1 F-1, 1939; Desikachary, 531, pl. 113, f. 11-14, 1959.

Filaments densely arranged, 15-19 μ broad towards base; base swollen, sheath unstratified, upto 3.5 μ thick, trichome swollen at the base and 11-13 μ broad, 7-8.5 μ broad in the middle, 2-2.5 μ broad at the apex, cells 5.5-8 μ long at the base, 4-5.5 μ long in the middle and 7-11 μ long in the hair, heterocysts basal, single or in pairs, hemispherical, 8-12 μ broad, 8-10 μ long.

From paddy fields of Bankura, Hooghly, Purulia.

Specimen nos. Ban 19, 34; Hoogh 5, 8; Pur 13.

Previously reported from Punjab (Sarma et al. 1979); U.P. (Prasad & Mehrotra 1980); Maharashtra (Ashtekar & Kamat 1980); W.B. (Sinha & Mukherjee 1975b).

222. *Calothrix elenkinii* Kossinskaja (Plate 34, figs. 6-8)

Not. Syst. Crypt. Inst. Horti Bot. Petropol., 3, 11, 1924; Geitler, 609, f. 383 (5-6), 1932; Desikachary, 531, pl. 114, f. 5, 6, 11, 1959.

Filaments arranged in tufts, bent and swollen at the base, 6.5-9.5 μ broad, at the base, in the middle 5.5-6 μ broad; sometimes falsely branched, trichomes 5.5-7 μ broad at the base, 4-4.5 μ broad in the middle, constricted at the cross-wall, apex of the trichome not ending into hair, sheath thin, hyaline, open at the terminal ends, cells quadrate to sub-quadrate heterocysts basal, spherical to hemispherical 5.5-7 μ broad, 3.5-4 μ long.

From paddy fields of 24-Parganas (N), Howrah and Nadia.

Specimen nos. S 43, 45; How 23, N 25.

Previously reported from A.P. (Rao 1938a); Karnataka (Bongale & Bharati 1980); Maharashtra (Ashtekar & Kamat 1980; Kolte & Goyal 1985); North East India (Reddy et al. 1986); U.P. (Prasad & Mehrotra 1980); Orissa (Laloraya & Mitra 1974); W.B. (Chatterjee & Chatterjee 1983).

223. *Calothrix membranacea* Schmidle (Plate 31, fig. 11)

Beitr. zur Algenfl. Afrikas, Engler's Bot. Jahrb., 30: 61, pl. 2, f. 12-14, 1901; Geitler, 629, 1932; Desikachary, 542, pl. 106, f. 10, 1959.

Thallus blue-green, membranous; filaments long, curved, attenuated gradually but not ending into a hair, sheath thin, hyaline, trichome 3.5-4.5 μ broad constricted at the cross-walls; cells 2.5 μ long; end cell narrow, heterocyst basal, subspherical, 3.5-4.5 μ broad and 3.5-5 μ long, spore absent.

From paddy fields of Bankura, Nadia and Murshidabad.

Specimen nos. N 13, Ban 24, Mur 13, 18.

Previously reported from J & K (Goyal et al. 1984); Maharashtra (Kolte & Goyal 1985, 1986); Karnataka (Bongale & Bharati 1980); U.P. (Mitra 1951, Pandey D.C. 1965, Prasad & Mehrotra 1980, Pandey, U.C. 1982, Chadha & Pandey 1983a); W.B. (Sinha & Mukherjee 1984).

224. *Calothrix marchica* Lemmermann (Plate 31, figs. 8 & 9)

Abh. Naturh. Ver. Bremen, 23(1): 248, f. 1, 2, 1914; Geitler, 607, f. 382a, 1932; Desikachary, 543, pl. 114, f. 4, 1959.

Filaments straight or slightly curved, blue-green with a thin sheath; trichome gradually attenuated into a hair, constricted at the cross-wall, 4.5-5 μ broad at the base; cells 4-5 μ long; heterocyst hemispherical, 4.5-5 μ broad.

From paddy fields of Bankura & West Dinajpur.

Specimen nos. WD 20, Ban 23.

Previously reported from Karnataka (Gonzalves & Kamat 1960, Bongale & Bharati 1980); Maharashtra (Chaporkar & Gangawane 1984, Tarar & Shewale 1984, Kolte & Goyal 1986); U.P. (Prasad & Mehrotra 1980); J & K (Goyal et al. 1984), Sagar of M.P. (Mishra & Purohit 1979); Tamil Nadu (Laloraya & Mitra 1974; Anand & Revathi 1987); W.B. (Sinha & Mukherjee 1975b).

Genus: GLOEOTRICHIA Ag.
Key to the Species

1.	Sheath of the spore saccate	.. 2
1.	Sheath close to the spore, not saccate	.. 5
	2. Spore straight	.. 3
	2. Spores straight or bent	.. 4
3.	Spores 45-52(-63.5) μ long, heterocysts 7.5-10 μ diam.	.. *G. raciborskii*
3.	Spores 44-62 μ long, heterocysts 10-12.5 μ diam., thallus large	.. *G. raciborskii* v. *kashiense*
	4. Spores straight, 43-54 μ long, sheath of spores constricted at intervals	.. *G. ghosei*
	4. Spores straight or bent 55-82 (-115) μ long	.. *G. natans*
5.	Heterocysts generally two, sometimes three in number, one basal and other intercalary between two spores, spores oblong with rounded ends	.. *G. pilgeri*
5.	Heterocyst one, spores cylindrical	.. 6
	6. Heterocysts spherical, 8-10 μ diam., trichomes 6-7.5 μ broad at base	.. *G. pisum*
	6. Heterocysts spherical to subspherical broader	.. 7
7.	Heterocysts 9.5-12.5 μ broad, spores 60-85 μ long	.. *G. intermedia*
7.	Heterocysts 10-13 μ broad, spores 48-65 μ long	.. *G. intermedia* v. *kanariensis*

225. *Gloeotrichia natans* Rabenhorst ex Born. et Flah. (Plate 33, fig. 3; Plate 45D)

Rabenh., Kryptogamenflora, 90, 1847; Bornet et Flahault, Rev. des Nostocacees heterocystees, 369, 1886; Geitler, 639, f. 406, 407, 1932; Desikachary, 561, pl. 118, f. 7, 15. 1959.

Thallus spherical, 1-4 cm broad, soft, blue-green to brown; trichomes 7.5-9 μ broad towards base and 4.5-5 μ broad towards hair, gradually attenuated into a long hair, cells 4-5 μ long towards base, 7.5-10 μ long towards hair, and higher up upto 4 times as long as broad; heterocysts basal, spherical, 12-14 μ in diam., spore cylindrical, straight or bent, without sheath 10-15 μ broad, and 55-82 μ (-115) μ long, with sheath 35-40 μ broad, sheath saccate.

From paddy fields of Midnapore, Bankura, Murshidabad, Malda.

Specimen nos. Ban 23, Mur 8, 10, Mal 3, 12.

Previously reported from North East India (Reddy et al. 1986); U.P. (Gupta 1956, Chaturvedi & Pandey 1976, Pandey & Pandey 1982); W.B. (Sinha & Mukherjee 1975a, Pal et al. 1986).

226. *Gloeotrichia raciborskii* Woloszynska (Plate 33, fig. 1)

Bull. Acad. Sc. Cracovic, Sect. B, 6: 687, 12, 1912; Geitler, 637, f. 405a, 1932; Desikachary, 562, pl. 118, f. 14. 1959.

Thallus spherical; trichome ending in a long hair about 700-800 μ long, sheath at the base lamellated; cells at the base of the trichome shorter than broad, 10 μ (approx.) broad and 3-5 μ long, higher up as long as broader, or longer, 5-5.2 μ broad and 12.5-13 μ long; heterocysts spherical, 7.5-10 μ in diameter; spore long, ellipsoidal, sheath more or less saccate, without sheath 14-18 μ broad and 45-52(-63.5) μ long; spore with yellowish epispore.

From water logged paddy fields, Dharma (Midnapore Sadar), 24-Parganas (South).

Specimen nos. D 54, S 28.

Previously reported from Maharashtra (Kamat 1963); Karnataka (Somashekar 1983); Bihar (Jha et al 1986), U.P. (Pal 1975); Tamil Nadu (Anand & Revathi 1987); W.B. (Maity & Santra 1985, Santra et al. 1988).

227. *Gloeotrichia raciborskii* var. *kashiense* Rao, C.B. (Plate 33, fig. 2).

The Myxophyceae of the United Provinces, III, Proc. Indian Acad. Sci., B, 6: 351, f. 3, A-E, 1937, Desikachary, 563, pl. 117, f. 2-6, 1959.

Thallus large, 5-6 cm diam.; trichomes 8-10 μ broad at base, 5-7 μ broad in the middle and 3.5-5 μ broad at apex; heterocyst globose, 10-12.5 μ diam., spores 13-16 μ broad without sheath and with sheath 20-25 μ broad and 44-62 μ long.

From paddy fields of 24-Parganas (N), Midnapore and Hooghly.

Specimen nos. S - 12, 18; M - 25, 26; Hoog - 14, 16.

Previously reported from U.P. (Pandey 1965, Bendre & Kumar 1975, Pandey 1982); Orissa (Mohanty 1982); W.B. (Sinha & Mukherjee 1975b, Maity & Santra 1985, Santra et al. 1988).

228. *Gloeotrichia intermedia* (Lemm.) Geitler (Plate 33, fig. 6)

in Pascher's Susswasserflora, 12: 233, 1925; Kryptogamenflora, 635, 1932; Desikachary, 560, pl. 116, f. 8, 1959.

Thallus globose, soft 5-9 mm diam., trichomes curved 5-7 μ broad in the middle, towards base cells broader than length 7.5-9 μ broad, towards tip cells longer than broad, 3-4.5 μ broad, 9-11 μ long and at the tip terminating into hair; heterocyst spherical, 9.5-12.5 μ broad, spores cylindrical without sheath 10-12 μ broad and with sheath 13.5-15 μ broad, 60-85 μ long.

From paddy fields of Murshidabad & Midnapore.

Specimen nos. Mur 6, 8; M 20, 25.

Previously reported from Jammu (Anand 1979); U.P. (Pal 1975, Khan 1985); W.B. (Pal & Santra 1982, Pal et al. 1986).

229. *Gloeotrichis intermedia* var. *kanwaensis* Rao, C.B. (Plate 33, figs. 4 & 5).

The Myxophyceae of the United Provinces II, Proc. Indian, Acad. S, B, 3. 168, f. 2A-C, 1936; Desikachary, 560, pl. 118, f. 9-11, 1959.

Thallus spherical, soft, trichomes with constrictions at the joints at the base and ending in a hair, cells 7.5-10 μ broad and 3-5 μ long at the base, 5 μ broad and 7.5-9 μ long in the middle region and 4-4.5 μ broad and 15 μ long at the tail region; heterocyst spherical or subspherical, 9 μ long, 10-13 μ broad; spores cylindrical, sheath more or less close to the spores, not saccate, with sheath 18-20 μ broad and without sheath 13-15 μ broad and 48-65 μ long.

From paddy fields of Keshpur (Midnapore).

Specimen no. K 39

The present form is similar to *G. intermedia* var. *kanwaensis* Rao, C.B. in many characters, but differs from it in having narrower trichomes, smaller heterocysts and spores.

Previously reported from Punjab (Sarma et al. 1979).

230. *Gloeotrichia pisum* Thuret ex Born. et Flah. (Plate 32, fig. 6)

Thuret, Essai de Class. des Nostocacees, Ann. Sci., nat. Sci. ser. 6, 1: 382, 1875; Bornet et Flahault, Revision des Nostocacees heterocystees, 366, 1886; Geitler, 632, f. 401, 1932; Desikachary, 555, pl. 116, f. 4 & 5, 1959.

Thallus globose, 3-4 mm diam., dark blue-green; filaments densely arranged, trichomes 6-7.5 μ broad towards base and terminating into a hair, cells 3-4 μ long towards base, and 6-9 μ long towards apex; heterocysts spherical, 8-10 μ diam., spores cylindrical 10-12.5 μ broad and 55-80 μ long, wall smooth.

From paddy fields of Onda (Bankura), Howrah, 24-Parganas (S).

Specimen nos. Ban 32, How 17, S 30.

Previously reported from Bihar (Jha et al. 1986); U.P. (Pandey 1982); Karnataka (Gonzalves & Kamat 1960); Kolhapur of Maharashtra (Kamat 1963); W.B. (Biswas 1932, Banerjee 1938, Sinha & Mukherjee 1975a).

231. *Gloeotrichia pilgeri* Schimidle (Plate 32, fig. 5)

Bras. Alg., Hedwigia, 40: 54, pl. 4, f. 14-15, 1901; Geitler, 641, f. 408, 1932; Desikachary, 558, pl. 116, f. 6, 7 & pl. 118, f. 8, 1959.

Thallus spherical or hemispherical, filaments radially arranged in the thallus, trichomes 6-8 μ broad at base, terminating into a short hair; cells torulose, 4.5-7 μ long, heterocysts generally two, sometimes three in number, one basal and other intercalary between two spores; heterocysts 5.5-8 μ diam.; spherical; spores ellipsoidal or oblong with rounded ends; arranged in a curve, 11-13 μ broad, 26-35 μ long.

From paddy fields of Purulia and Bankura.

Specimen nos. Pur 16, Ban 24.

Here trichomes slightly narrower, spores slightly smaller.

Previously reported from Karnataka (Gonzalves & Kamat 1960, Somashekar 1983, 1984b); Kolhapure of Maharashtra (Kamat 1963); W.B. (Sinha & Mukherjee 1975a).

232. *Gloeotrichia ghosei* Singh, R.N. (Plate 33, figs. 7 & 8)

The Myxophyceae of the United Provinces, IV-Proc. Indian Acad. Sci., B, 9: 64, f. A-C, 1939; Desikachary, 561, pl. 118, f. 1-3, 1959.

Thallus free floating, upto 2 cm diam.; generally spherical in shape; trichome at base 8-10 μ broad, higher up 3.5-6 μ broad, constricted at the cross-wall; at base cells 3-4 μ long, higher up 5.5-9 (-12) μ long, ending in along cylindrical hair; heterocysts spherical, 10.5-12.5 μ diam., spores with sheath 21-25 μ broad, sheath yellowish brown, and constricted at intervals; spore without sheath 13-18 μ broad, 43-54 μ long.

From paddy fields of Bankura and Birbhum.

Specimen nos. Ban 20, Bir 11.

Previously reported from Punjab (Grover & Pandhol 1975, Pandhol & Grover 76, Sarma et al. 1979); Karnataka (Somashekar 1983); U.P. (Prasad & Mehrotra 1980, Pandey 1982, Chadha & Pandey 1983a); W.B. (Sinha & Mukherjee 1975a).

Order : STIGONEMATALES Geitler

Family : MASTIGOCLADACEAE Geitler

Genus : MASTIGOCLADUS Cohn

233. *Mastigocladus laminosus* Cohn (Plate 34, fig. 1)

Uber die Algen des Karlsbader Sprudels, Abh. Schles. Ges. Vaterl. Cultur, 2: 391, 1893; Geitler, 558, f. 350-352, 1932; Desikachary, 581, pl. 127, f. 1-9, 1959.

Thallus membranous, dark blue green, hard, with calcium carbonate; filaments entangled, 5-6.5 μ broad, curved, sheath distinct, sometimes sheath indistinct in lateral branches; lateral branches 2-3 μ broad; cells in a single row; cells 4.5-5 μ broad and 5-6.5 μ long in the main filament and 3-4.5 μ broad and 6-12 μ long in the lateral branches; heterocysts spherical, intercalary, 5-6.5 μ broad and 8-12 μ long.

From paddy fields of Bankura, Burdwan, 24-Parganas (S) and Howrah, Hooghly.

Specimen nos. Bur 27, Ban 16, S 23, How 18, Hoog 15.

Previously reported from Sagar of M.P. (Mishra & Purohit 1979); North East India (Reddy et al. 1986); W.B. (Chatterjee & Chatterjee 1983, Sinha & Mukherjee 1984).

Family: STIGONEMATACEAE Kirchner

Key to the Genera

1. Filaments crescent shaped, hormocysts formed at the apex of the branches, sheath lamellated .. *Camptylonema*
1. Filaments not crescent shaped, hormocysts absent .. 2
    2. Terminal portions of lateral branches form chains of pseudohormocysts (endospores) .. *Westiellopsis*
    2. Pseudohormocysts absent, hormogones formed .. 3
3. Filaments 18-40 μ broad, sheath thick, lamellated, cells polyhedral, arranged 1-2 rows of cells .. *Stigonema*
3. Filaments narrower, sheath not so thick, not lamellated, cells subquadrate, quadrate or cylindrical, arranged generally in a single, sometimes in two rows of cells .. *Hapalosiphon*

Genus: HAPALOSIPHON Näg.

Key to the species

1. Cells of main filaments arranged in single or two rows of cells, cells of main filaments spherical to subspherical 6-11 μ diam., sheath thick, heterocysts spherical to subspherical, 8-12 μ broad .. *H. fontinalis*
1. Cells of main filaments arranged only in single row of cells, cells of main filaments subquadrate, quadrate or cylindrical, sheath thin .. 2
    2. Thallus blue-green, filaments 5-7 μ broad, sheath thin hyaline sometimes not visible, cells 3.5-4.5 μ broad, heterocysts 4-5.5 μ broad .. *H. intricatus*
    2. Thallus yellowish brown, sometimes blue-green when young, filaments broader, sheath distinct; cells and heterocysts broader .. 3

3. Thallus blue-green sometimes yellowish blue-green when old, filaments 6-8(-9) μ broad, sheath thin lateral branches short, as broad as the main filaments but slightly attenuated at the apex, cells 4-7 μ broad, heterocysts 5-7 μ broad .. *H. welwitschii*

3. Thallus yellowish brown, filaments 9-10.5 μ broad, sheath firm yellowish, lateral branches as broad as the main filament not attenuated at apex, cells of the main filaments 4.5-6 μ broad, heterocysts 5-9 μ broad .. *H. luteolus*

234. *Hapalosiphon welwitschii* W. et G.S. West (Plate 35, fig. 1)

J. Bot., Lond., 35: 242, 1897; Ann. Roy. Bot. adn. 6, 240, 1907; Geitler, 531, f. 328a, 1932; Desikachary, 588, 137, f. 5, 1959.

Filaments entangled, flexuous, blue-green; when old becomes yellowish blue-green, 6-8(-9) μ broad; sheath thin, close to the trichome; lateral branches short, as broad as the main filament, slightly attenuated at the apex; trichomes slightly constricted, cells 4-7 μ broad, quadrate to cylindrical, 6-12(-16) μ long; heterocysts rare, intercalary, 5-7 μ broad and 7.5-10.5 μ long; spores subspherical 7.5-10 μ broad and 10-12 μ long.

From paddy field soils of Panchkuri (Midnapore Sadar).

Specimen no. Panch 102.

Previously reported from Maharashtra (Gonzalves and Gangla 1949, Marathe 1966-67, 1969, 1970, Kamat 1968, Tiwari 1972, 1979; Chaporkar & Gangawane 1984, Kolte & Goyal 1985, 1986); Orissa (Laloraya & Mitra 1974), U.P. (Rao 1937, Prasad & Srivastava 1968, Khan 1970, Das et al. 1974, Prasad & Mehrotra 1980), M.P. (Bendre & Agarker 1965, Das et al. 1974); Himachal Pradesh (Vasishta 1968), North India (Prasad & Mehrotra 1979a); Karnataka (Bharati & Bongale 1975, 1976a, b, c, d, Bharati et al. 1978, Bongale 1978, 1982, Bongale & Bharati 1980); Tamil Nadu (Tiwari 1972, 1979); J & K (Goyal et al. 1984); W.B. (Kachroo 1959).

235. *Hapalosiphon fontinalis* (Ag.) Bornet (Plate 36, fig. 4)

Bull. Soc. bot. Fr., 36: 156, 1889; Geitler, 535, f. 332, 1932; Desikachary, 592, pl. 130, f. 3, 1959.

Thallus blue-green to brownish blue-green, filaments interwoven, branched, branches lateral, straight or curved; main filament 18-22 μ broad; in the main filaments cells in a single or sometimes in two rows of cells; sheath thick colourless or yellowish brown, cells spherical to subspherical, 6-11 μ diam., lateral branches thinner than main filament, 8-10 μ broad, here sheath thin, cells 4.5-6.5 μ broad, cells quadrate to cylindrical, upto 11 μ long, but towards apex quadrate to subquadrate in length; heterocysts spherical to subspherical. 8-12 μ diam.

From paddy fields of Joyrambati (Bankura).

Specimen no. Ban 41

Previously reported from Maharashtra (Kolte & Goyal 1986); Karnataka (Gonzalves & Kamat 1960); U.P. (Pandey 1965), W.B. (Sinha & Mukherjee 1984).

236. *Hapalosiphon luteolus* W. & G.S. West (Plate 35, figs. 2 & 3) Welwitsch. Afr. Alg., J. Bot. Lond., 35: 241, 1897; Geitler, 537, f. 333, 1932; Desikachary, 593, pl. 130, f. 1, 1959.

Thallus yellowish brown, filaments in tangled, main filaments 9-10.5 μ broad, lateral branches flexuous, more or less as broad as the main filaments, 7-10 μ broad, sheath firm, yellowish, cells arranged in a single row cells subquadrate, quadrate or cylindrical 4.5-6 μ broad and 4.5-7.5 μ long in the main filament; 3-4.5 μ broad and 5-9.5 μ long in the lateral branches; heterocysts spherical to cylindrical 5-9 μ broad, 7-9.5 μ long, hormogones 8-9 μ broad, 40-60 μ long with 8-15 cells at the tip of lateral branches.

From paddy fields of Bankura, Jhalda (Purulia).

Specimen nos. Ban 13, Pur 27.

Previously reported from Karnataka (Bongale & Bharati 1980); W.B. (Sinha & Mukherjee 1975a).

237. *Hapalosiphon intricatus* W. et G.S. West (Plate 36, fig. 3)

Alg. West Indies, J. Linn. Soc. Bot., Lond., 30: 271, pl. 15, f. 16-28, 1894; Geitler, 533, f. 330, 1932; Prescott, 546, pl. 129, f. 1, 1951; Desikachary, 591, pl. 129, f. 1-3, 1959.

Thallus blue-green, densely entangled, branched profusely, 5-7 μ broad, lateral branches are as broad as the main filament; sheath hyaline, thin close to the trichome, sheath sometimes not visible; cells cylindrical, 3.5-4.5 μ broad, 6-10 μ long, heterocysts intercalary, subquadrate to cylindrical, 4-5.5 μ broad, 5-12 μ long.

From paddy fields of Moynapur (Bankura), Birbhum, Burdwan

Specimen nos. Ban 31, Bur 10, Bir 8.

Previously reported from J & K (Goyal et al. 1984); Karnataka (Bongale & Bharati 1980); Tamilnadu (Fremy 1942); Punjab (Vasishta 1965); North India (Prasad & Mehrotra 1979); Himachal Pradesh (Vasishta 1968); U.P. (Prasad & Srivastava 1968, Chadha & Pandey 1983a); Andaman & Nicobar (Prasad & Srivastava 1986); Maharashtra (Kolte & Goyal 1985, 1986, Patil & Satav 1986); W.B. (Sinha & Mukherjee 1975b).

Genus: WESTIELLOPSIS Janet

238. *Westiellopsis prolifica* Janet (Plate 34, figs. 3-5).

Ann. Bot., Lond., ns., 5: 167-170, 1942; Desikachary, 596, pl. 131, f. 1-12, 1959.

Filaments with branchings, without sheath, with single row of cells, two kinds of filaments, main filaments prostrate, with quadrate, barrel shaped or slightly longer cells, constriction at the cross-walls, cells 8-10 μ broad, and 7-11 μ long; lateral branches erect, narrower, not constricted or slightly constricted at the cross-walls, cells 4.5-6 μ broad, μ long, heterocysts quadrate to oblong, 6-7 μ broad and 12-16 μ long, the dialated terminal portions of lateral branches forming chains of pseudohormocysts, gonidia 8-9 μ diam., formed singly in each cell of the pseudohormocyst.

From paddy fields of Purulia, Onda (Bankura).

Specimen nos. Pur 14, Ban 22.

Previously reported from Tamil Nadu (Janet 1942); Sagar of M.P. (Mishra & Purohit 1979); J & K (Goyal et al. 1984); U.P. (Prasad & Mehrotra 1980); Bihar (Jha et al. 1986); Maharashtra (Chaporkar & Gangawane 1984, Kolte & Goyal 1985, 1986, Patil & Satav 1986); W.B. (Sinha & Mukherjee 1984);

Genus: CAMPTYLONEMA Schmidle

239. *Camptylonema indicum* Schmidle (Plate 36, figs. 1 & 2)

Ueber einige von Prof. R. Hansgirg in Ostindien gesammelte Susswasseralgen, Hedwigia, 39: 161, 1900; Desikachary, Proc. Indian Acad. Sci., B. 28: 35, f. 1-15, 1948; Cyanophyta, 598, pl. 132, f. 1-15, 1959.

Thallus brownish, filaments entangled, sparsely branched, both false and true branching present; main filaments 14-16 μ broad, and erect branched 9.5-11 μ broad; sheath lamellated; trichome of the main filaments 7-10 μ broad, torulose and trichome of the erect branches 4-5.5 μ broad, slightly constricted or unconstricted; cells 4.5-8 μ long in the main filament and 6-10 μ long in the erect branches; heterocysts cylindrical, 6.5-10 μ long and as broad as the trichome; hormocysts formed at the apex of the erect branches, with 4-8 cells, with lamellated sheath.

From paddy fields of Bishnupur, Joyrambati (Bankura), Purulia and Birbhum.

Specimen nos. Ban 14, Ban 38, Pur 7, Bir 9.

Previously reported from Sagar of M.P. (Mishra & Purohit 1979); W.B. (Sinha & Mukherjee 1975a).

Genus: STIGONEMA Ag.

240. *Stigonema ocellatum* (Dillw.) Thuret ex Born. et Flah. (Plate 37, figs. 1-4).

Thuret, Essai class. nostoch., Ann. Aci. nat. Bot. ser. 6, 1: 380, 1875; Bornet et Flahault, Revision des Nostocacees heterocystees, 69, 1887; Geitler, 504, f. 305-307, 1932; Desikachary, 607, pl. 138, f. 2, 1959.

Thallus wooly, brownish, filaments with prostrate and erect portions; 18-40 μ broad; branches irregular, branched filaments more or less as broad as the main filament; sheath thick, lamellated, colourless to yellowish brown; cells in a single or in two rows, cells 5-11 μ diam.; heterocyst 6-8 μ broad, 9-12 μ long; hormogones formed at the end of branches without sheath 8-10 μ broad and 52-63 μ long.

From paddy field soils of Bankura, Birbhum, Malda, Purulia, Cooch Behar.

Specimen nos. Pur 7, 11, Ban 5, Bir 12, Mal 6, 8, Cooch 14.

Previously reported from W.B. (Sinha & Mukherjee, 1975b).

4

IMPORTANCE OF BLUE-GREEN ALGAE

Introduction

Blue green algae are considered to be one of the major algal forms which are useful to mankinds in various ways. The major uses may be categories under the following heads:

* Utilisation of blue green algae as food.
* Utilisation of blue green algae in agriculture particularly with reference to nitrogen economy of soil.
* Utilisation of blue greens for conversion of solar energy through biophotolysis.
* Biogas and biomass production by the blue green algae.
* Reclamation of saline and alkaline soil by the blue green algae.

In addition, the abundant growth of blue greens may be a source of considerable nuisance to man in some situations, as in the management of water reservoirs. Some blue greens may also produce toxin that causes gastric disorders of farm animals and also dermal and other allergenic disorders of man.

Utilisation of Blue Green Algae as Food

Blue-greens have recently been considered to be potentially excellent sources of food for animals and even for man. The introduction of blue greens into the human diet would by no means constitute a genuine innovation. The Chinese consider *Nostoc commune* as a delicacy and *Aphanothece sacrum, Nostoc rucosum, N. commune* and *Brachytrichia* sp. have been used as side dishes since ancient time in Japan. In the northern parts of the African republic of Chad there are stretches of shallow alkaline lakes with high salt content which produce abundant growth of the planktonic *Spirulina platensis*.The planktonic material has been collected for food by the local people and dried in sun. The fried algal cake, called 'die' is being sold at the market throughout the country. It is used for the preparation of soup and sauce. It is likely that one of the basic food stuffs of the Aztecs was of a similar nature, which they collected from shallow Mexican lakes. More recently, for the production of nutritions 'health food' *Nostoc* is utilized as food in Peru and in the Far East. *Microcystis* has also been grown on domestic wastes and used to feed fish in lagoons and fish ponds in India and Pakistan.

Chemical analysis of *Spirulina* showed that it contains 55-60% proteinaceous material, 16-18% carbohydrate, 2-3%, lipid as well as mineral salts. It also contains a number of essential amino acids, but the contents of sulphur-containing amino acids (cysteine and methionine) is quite low. Feeding experiments with animals indicate that the *Spirulina* diet is acceptable and well digested by farm animals.

Utilisation of Blue Green Algae in Agriculture

Due to enormous rise of population on earth surface, the demand for food in the world is really unprecedented and thus malnutrition and starvation are already desperate problems. In view of the World's food crisis, greater attention has been turned to traditional farming practices. But the prospects of increased foodgrain production are aggregated by the ever-increasing cost of fuel and chemical fertilizers.

In this context, the recent inventions which enable the more extensive and efficient utilization of biological nitrogen fertilizers in the field agriculture have greater significance. During past few decades, blue-greens are considered to be potent biological nitrogen fixer and thus recognised as 'nitrogenous biofertilizers'.

Many tropical paddy fields receive no chemical fertilizer nor natural manure, yet they remain productive and capable of supporting large populations with basic food. Here the fertility of paddy soil is maintained by the activities of blue greens which grow spontaneously and often luxuriantly in the water logged field. They provide fixed nitrogen to rice plants through both secretion of nitrogenous substances and on their decay and subsequent mineralisation of organic substances in the soil.

Since De (1939) attributed the inherent fertility of the tropical rice field soils to the activity of nitrogen-fixing blue-green algae, considerable interest has been generated in exploiting the potential of these biological systems. In the south and south-east Asian regions, only about 4% of the 851 localities examined were found to harbour N_2-fixing blue-greens (Watanabe, 1959). The dominant forms in Japanese rice soils largely belong to the genera *Tolypothrix*, *Anabanea* and *Nostoc* (Okuda and Yamaguchi, 1952, 1956). In India, Singh was the first to observe a characteristic succession of blue-greens in the rice fields of Uttar Pradesh. However, it is interesting that only about 33% of the 2213 rice-field soil samples examined in India were found to harbour useful N_2-fixing blue-greens and their relative abundance showed a wide variation, with localisation of specific forms in certain regions (Venkataraman, 1975). Although blue-greens normally prefer warmer habitats they have been shown to play a major role as nitrogen fixers in temperate, polar and subpolar regions. The distribution and role of nitrogen-fixing blue-green in temperate regions have been investigated in Russia (Gollerbakh and Shtina, 1969; Shtina, 1969), northern and western Europe (Granhall, 1975; Henriksson et al., 1972) and Northern America (Jurgensen and Davey 1968; Shields and Dumell, 1964).

There is now positive evidence for the nitrogen-fixing ability of more than 100 species of blue greens. A nitrogen accretion of 14 to 40Kg/ha has been attributed to the activities of the N_2-fixing blue greens in different localities (De and Mandal, 1958; Singh, 1961; Watanabe, 1962; Henriksoon, 1971; Sankaram, 1971; Yoshida and Ancaja, 1973; Alexander, 1975).

In India alone, the distribution of rice field blue-green algae in various states were carried for past two decades by various workers as stated in Table 4.1.

Table 4.1: Bibliography of Rice-Fields Blue-Green Algal Survey in India

States	*References*
Andhra Pradesh, Andaman & Nicobar, Bihar	Laloraya & Mitra, 1973; Prasad et al.; 1987; Laloraya & Mitra, 1973; Jha et al., 1986.
Goa	Bongale, 1981.
Gujrat	Mahajan and Patel, 1983
Himachal Pradesh	Kamat, 1970.
Jammu and Kashmir	Goyal et al., 1984; Khan, 1957.
Karnataka	Tiwari, 1972,1975; Tiwari & Pandey, 1976; Kamat, 1979, Bongale & Bharati, 1980; Bongale, 1981, 1984; Shivaprakash & Shivappa Shetty, 1983; Bongale, 1987.
Kerala	Parukutty, 1940; Aier, 1965; Verma, 1965; Anand & Shanthakumar Hopper, 1987.
Madhya Pradesh	Tiwari, 1972, 1975; Mishra & Purohit, 1979.
Maharashtra	Gonzalves & Gangla, 1949; Marathe, 1963; Marathe & Navalkar, 1963; Sardeshpande & Goyal, 1981; Tarar and Shewale, 1984; Kolte and Goyal, 1985; Patil & Satav, 1986.
Meghalaya & other North-eastern states	Reddy et al., 1986.
Orissa	Rao, 1938; Laloraya & Mitra, 1973.
Punjab	Sarma & Swarnakanta, 1978; Grover & Pandhol, 1975; Sarma et al; 1979.
Rajasthan	Gupta & Kumar, 1968; Gupta, 1972.
Sikkim	Santra, 1987.
Tamil Nadu	Khurana & Venkataraman, 1968; Tiwari, 1972, 1975; Laloraya & Mitra, 1973; Tiwari & Pandey, 1976; Anand & Revathy, 1987, Anand, 1989.
Uttar Pradesh	Mitra, 1951; Gupta, 1957; Pandey 1965; Gupta, 1966; Singh & Chaturvedi, 1970; Prasad and Mehrotra, 1980, Pal & Yadav, 1974; Khan & Mathur, 1976; Prasad and Srivastava, 1968.
West Bengal	Biswas, 1926; De, 1939; Laloraya & Mitra, 1973; Sinha & Mukherjee, 1975, 1984; Saha & Mandal, 1980; Mukhopadhyay & Chatterjee, 1981; Pal & Santra, 1982; Chatterjee & Chatterjee, 1983; Sinha & Mukherjee, 1984; Maity & Santra, 1985; Pal & Santra, 1985, Santra, 1991.

Recent advances in methodology have helped a great deal in our understanding of blue-green algal N_2-fixation. Nitrogenase is oxygen-sensitive and blue-green algae have evolved special devices such as heterocysts for the protection of their nitrogenone against oxygen damage (Fay et al., 1968; Stewart et al., 1969). On the whole, the basic mode of nitrogenase action appears to be similar with that of other N_2-fixing organisms. The mapping of nif^+ gene was also made in several N_2-fixing blue-green algae.

The role of N_2-fixing blue green algae in the maintenance of the fertility of rice fields has been well substantiated and documented all over the world. In India alone, the beneficial effect of blue green algae on yield of many rice varieties has been demonstrated in a number of localities (Singh, 1961; Sankaram, 1971; Subrahmanyam, 1972; Venkataraman, 1972, 1975, 1979, 1980, 1981; Saha and Mandal, 1981; Santra, 1991). In many areas, 10 to 20% increase in the grain yield has been observed as a result of algal application in the absence of any added chemical N-fertiliser (Table 4.2).

Table 4.2: Response of BGA-Application in Rice Cultivation

(Santra 1991)

Variety	*Trial season*	*N+P+K*	*Gross yield (kg/ha)* BGA+ P+K	P+K	BGA+ P+K	*Additional yield over schedule doses of NPK (kg/ha)*
High yield varieties	Aman	3620	3329	3015	3925	305 (10.5%)
(HYV) RATNA	Boro	3919	3050	2720	4520	601 (15.3%)

Beneficial effects of algal inoculation have also been reported on a number of other crops such as barley (Fuller et al., 1960), oats (Shtina, 1957), tomato (Kaushik and Venkataraman, 1979; Roger et al., 1979), radish (Roger et al., 1979), cotton (Kuchkarova and Maksudov, 1964), sugarcane and maize (Singh, 1961), chilli & lettuce (Dadhich *et al.*, 1969).

Considering these above stated facts, blue-green algal bio-fertiliser application in rice field has earned immense potentialities. Unlike *Rhizobium* and *Azospirillum* bacteria which are mainly produced in sterile conditions, blue green algae, such as *Nostoc, Anabaena* and *Aulosira,* are grown in open air troughs, pits and tanks on the farm to raise a biomass. Upon sun drying, the algal flakes are collected and broadcast on the rice fields, one week after transplantation of rice seedlings at a rate of 10-15 kg/ha. The cost-benefit ratio of algal production in farmer's field indicates a net income of Rs.2000 from half an acre of land (Venkataraman, 1980).

The beneficial effect of N_2-fixation for crop production are even more apparent in paddy fields which are populated by the water fern *Azolla* during the water-logged period. The importance of *Azolla* as an organic fertilizer in rice cultivation was first demonstrated in North Vietnam in 1957. Afterwards its potentiality was recognised in the USA, Indonesia, Japan, Philippines, China, India. *Azolla* is a tiny floating freshwater fern and is ubiquitous in distribution. Each leaf has a dorsal and a ventral lobe. The dorsal lobe is green and has a blue green algal symbiont (*Anabaena azollae*) within a central cavity. The alga fixes atmospheric nitrogen, hence the importance of *Azolla* in rice cultivation. Estimates of the annual quantities of N_2-fixed by *Azolla* range between 120 and 312 kg nitrogen per hectare. The plant supplies substantial amounts of green manure between 150-300 ton per hectare per year), which supports the growth of soil micro-organisms including heterotrophic N_2-fixers (Kannaiyan, 1979, 1985).

Solar Energy Conversion Through Biophotolysis

Heterocystous blue-greens possess the unique ability to simultaneously evolve O_2 in photosynthesis (in vegetative cells) and H_2 by nitrogenase catalyzed electron transfer to H^+- ions (in heterocysts) in the absence of N_2 or other substrates of nitrogenase. This is the basis for the attempts of several workers to exploit this potential through the development of a 'biophotolytic system' for solar energy conversion. Following earlier laboratory studies successful outdoor systems were established in which *Anabaena cylindrica* was used to produce H_2 and O_2 in light when sporged continuously with a gas mixture containing argon and CO_2.

Among blue green algae *Anabaena flos-aquae, A. cylindrica, Mastigocladus laminosus*, species of *Lyngbya, Plectonema, Nostoc, Tolypothrix, Scytonema, Gloeothece* etc. are known to photoproduce hydrogen.

Hydrogen has many advantages as a fuel energy source for future. It is the most abundant element in the universe and has maximum energy per unit weight (122 KJ/g). It has a greater energy conversion efficiency over petroleum. Hydrogen, when used as a fuel does not pollute the environment because its only conversion by-product is water. Besides as fuel, hydrogen is useful in industrial applications like iron ore reduction and ammonia (fertilizer) production.

Biogas and Biomass Production by Blue-Green Algae

It is possible to make sewage productive by optimising the conditions of waste treatment by algal forms, particularly blue-green algae. The system developed long by Oswald et al. (1957) using a stabilization pond for waste water treatment coupled with algal production. This system is efficient, as it utilizes atmospheric carbon dioxide and sun light to obtain an yield of algal biomass. The oxygen released during photosynthesis favours the aerobic bacteria which degrade the organic material. In India, the work of BGA production from sewage was limited. At National Biological Research Institute, Lucknow, India extensive attempt has been made for large scale production of *Spirulina* from sewage with or without bicarbonate supplement. Then the harvested biomass either used for biogas production or as food substrate particularly for livestocks.

Reclamation of Saline and Alkaline Soil

Blue green algae is one of the most well adopted algal forms in saline habitats. It has tremendous capacity to transform saline soil into productive neutral soil with greater amount of available nitrogen and carbon. Incubation of saline or alkaline soils with native blue green algae *in vitro* or *in vivo* resulted an improvement of physico-chemical properties of soils (Singh 1961, Subhashini and Kaushik, 1981, 1982, 1984). Blue green algae are known to secrete substances like polysaccharides and lipids which have the capacity to transform the structure of saline and alkaline soil (Lewin, 1972; Hellebust, 1974). Again majority of the blue green algae are endowed with capacity to fix atmospheric gaseous nitrogen and this fixed nitrogen is released during the growth and decay of algae along with organic carbon compounds into the soil (Stewart, 1968). Thus, these beneficial effects of blue green algae in improving the physical, chemical and nutritional properties of saline and alkaline soils, suggest the possibility of exploiting this biological system to ameliorate the salt affected soils.

It was revealed from the past study that the species of *Oscillatoria, Lyngbya, Phormidium, Microcoleus, Hydrocoleum* and *Calothrix* etc. were most dominant in saline and alkaline habitats (Singh, 1950, Marathe 1965, Prasad and Srivastava, 1968; Santra *et al.*, 1988, Kaushik, 1989). An example of soil quality improvement upto BGA application is depicted in the Table 4.3.

The dominant blue-green algal forms in salt affected soils of different states of India, were presented separately in the Table 4.4. It was also further interesting to note the fact that limestone and phosphate rocks were biologically weathered by blue- green algae *in vitro* and *in vivo* studies. In this connection, it was reported earlier that some blue green forms are capable of secreting acidic substances which in turn was able to dissolve the rock components. That is why they are also called as pioneer of soil formation on rock surface.

The ultimate success of bioameliorant cyanobacteria is reflected in the successful establishment of a crop with moderately high yields. The various package of practices recommended for amelioration of such soils have advocated the use of paddy as the first crop. The reclamation trials conducted for a period of 3 years in sodic soils of Delhi during summer season are given in Table 4.5.

In contrary to the above stated utility of blue green forms, there are a number of nuisance activities of blue greens viz., nuisance in water supplies, water toxicity and as allergy causative agents. Extensive growth of algae can create severe problems in the maintenance of water supplies and in meeting the ever increasing demand for portable water. Numerous episodes of poisoning of farm animals (e.g. cattle, horse, pig) and of waterfowl and wildlife were described throughout the world and attributed to the ingestion by drinking animals of lethal doses of toxic blue-greens accumulating at the surface of water or along the shores of temperate lakes, ponds and reservoirs. Several species have been indicated but toxic strains of three common bloom-forming species—*Microcystis aeruginosa, Aphanizomenon flos-aquae* and *Anabaena flos-aquae,* appear to be responsible for most incidents of serious consequences. Several freshwater and marine blue-greens were also implicated in outbreaks of human gastro-enteritis and in various allergic responses like dermal itching. The frequently occurring 'swimmers itch' is attributed to contact with *Lyngbya majuscula, Schizothrix calcicola* and *Oscillatoria nigroviridis,* which are commonly found in tropical and subtropical seawaters. The toxin responsible are lipid-soluble phenolic compounds.

Table 4.3

Changes in the Physico-Chemical Properties of Soil after Algalization and/or Gypsum Treatment for 7 days *in vitro* Experiment (Santra et al., 1987b)

Treatment	*Soil properties*			
	pH	*Organic carbon (%)*	*Exchangeable cations (mg/gms soil)*	
			Ca^{++}	Mg^{++}
Soil (R)	6.3	0.17	0.20	0.20
Soil (R) + Algae (100:1)	6.5	0.77	0.14	0.20
Soil (R) + Gypsum (100:0.05)	6.4	0.31	0.20	0.20
Soil + Algae + Gypsum (100: 1: 0.05)	7.0	0.43	0.14	0.19
Soil(S)	6.5	1.27	0.21	0.16
Soil (S) + Algae (100: 1)	7.0	3.33	0.15	0.16
Soil (S) + Gypsum (100: 0.05)	6.8	2.67	0.19	0.16
Soil (S) + Algae + Gypsum (100: 1: 0.05)	7.2	3.24	0.15	0.15

R = River bank soil; S = Soil of cultivated land.

Future Prospect

(A) Cyanobacterial Genetical Studies

Recent development in gene transfer techniques and their use in obtaining mutants of unicellular non-diazotrophic cyanobacteria have opened up new approaches to the study of Cyanobacterial genetics at molecular level. However, a well defined reproducible gene transfer system is yet to be developed for the heterocystons diazotrophic cyanobacteria. Singh (1977) and Stewart and Singh (1975) reported genetic recombination in mixed cultures of the mutants of *Anabaena doliolum* and *Nostoc muscorum*. Similarly rice field herbicide resistance genes of aerobic diazotrophic *Gloeocapsa* strain have also been found to be efficiently transformable to *Nostoc muscorum* (Singh et al., 1987).

Table 4.4

Dominant Blue-Green Algal Forms in Salt Affected Areas of Different States of India

States	*Soil type*	*Dominant Algal Reference genera*
Andhra Pradesh	Sodic	*Nostoc, Plectonema*
Delhi	Sodic	*Anabaena, Oscillatoria, Plectonema*
Haryana	Sodic	*Anabaena, Calothrix, Nostoc, Plectonema, Westiellopsis*
Maharashtra	Saline	*Nostoc*
Punjab	Saline	*Calothrix, Hapalosiphon, Plectonema*
Rajasthan	Saline	*Nostoc, Scytonema, Plectonema*
Tamil Nadu	Saline-Sodic	*Nostoc, Tolypothrix, Plectonema*
Uttar Pradesh	Saline	*Calothrix, Plectonema, Hapalosiphon*
West Bengal	Saline	*Nostoc, Oscillatoria, Lyngbya, Hydrocoleus*

Table 4.5

Amelioration of Sodic Soils and the Yield (Q/ha) of Paddy Variety Pusa-33 due to Algalization for Three Consecutive Seasons (Kaushik, 1989)

Treatment	*1985*		*1986*		*1987*	
	Paddy	*Straw*	*Paddy*	*Straw*	*Paddy*	*Straw*
Control	0	0	0	0	6.15	5.15
BGA	1.46	3.43	8.50	22.66	41.00	30.60
Gypsum	5.41	10.08	24.83	43.50	42.00	30.40
BGA + Gypsum	7.96	13.00	32.33	49.50	45.40	33.20

Once the mechanism of gene transfer in heterocystous diazotrophic cyanobacteria is well developed through transformation or transposon mutagenesis, as has been successfully developed in unicellular cyanobacteria, it would be possible to understand the molecular basis of nitrogen fixation, heterocyst differentiation and other related physiological processes. Finally acquisition of such knowledge would finally become the most powerful tool for manipulation of desired cyanobacteria genes or genetic systems.

(B) Investigation on Stress Compatibility of Cyanobacteria

It is well known that blue-green algae occur in a very wide range of habitats. In many fresh water bodies like streams, ponds, rivers and lakes, and surface layers of soil both upland and lowland they constitute one of the major life form along with other photosynthetic organism. They were also found grow in healthy state in thermal habitat, ice zone, or even in saline soils and marine water bodies. On their occurrence and growth in various habitat were dependant on the following components like, concentration of O_2 and CO_2, availability of nitrogen, pH of soil and water and temperature regime.

Furthermore it was interesting to note the fact that various species of blue green forms were also tolerance to toxic heavy metals, herbicides and pesticides. As such inspite of higher application of pesticides like BHC, carbofuran, endosulphan, parathion, captan etc. and herbicides like 2, 4-D, Ordram, butachlor etc. in rice field, there are a number of blue green species which can tolerance to such stressed polluted condition. Not only that the effect of certain toxicant significantly accelerate both nitrogen fixation and growth of tolerant blue green algal species.

Wide variations in the cyanobacterial response to different ecological stresses, perceptible even at intraspecies level, point to the possibility of developing region specific and stress compatible strains through selection from natural populations. A systemic survey, isolation and screening of the autochthonous algal population will not only help in preparing an algal map of the country and identification of algal deficient areas but will also help in raising a bank of germplasm collection of cyanobacteria with known attributes.

BIBLIOGRAPHY

Aiyer, R.S. 1965. Comparative algological studies in rice fields in Kerala State. Agric. Res. J. Kerala, 3: 100-104.

Aiyer, R.S., S. Salahuddin and G.S. Venkataraman, 1972. Long-term algalisation, field trial with high yielding rice varieties. Indian J. Agric. Sci., 42: 382.

Alexander, V. 1975. Nitrogen fixation by Free-living microorganisms (WDP Stewart ed.) Cambridge University Press, Cambridge, p. 175.

Allen, M.B. and D.I. Arnon. 1955. Studies on nitrogen fixing bluee green algae. I. Growth and nitrogen fixation by *Anabaena cylindrica* Lemms. Pl. Physiol. 30: 366-372.

Anand, N. 1980. Studies on blue green algal populations of Rice fields. In Proc. Nat. Workshop on Algal Systems (Ed. C.V. Sheshadri *et al.*), Indian Society of Biotechnology, IIT/New Delhi, pp. 51-54.

Anand, N., L. Radha, R.S. Shanthakumar Hopper, G. Revathi and T.D. Subramanian, 1987. Blue green algal as Biofertilizers certain view points on the choice of suitable isolates. Proc. International Symposium on Phycology, January 8-13, 1987, Madras, India (In press).

Anand, N. and G. Revathi, 1987. Blue-green algal from rice-fields of Tamil Nadu. Phykos, 26: 17-21.

Anand, N. and R.S. Shanthakumar Hopper, 1987. Blue-green algae from rice fields in Kerala State. India. Hydrobiologia, 144: 223- 232.

Anand, N. 1989. Handbook of Blue-green algae. Bishen Singh Mahendra Pal Singh, Dehra Dun, p. 80.

Banerjee, J.C. 1935. On algal found on soil samples from alluvial paddy field of Faridpur, Bengal, Sci. Cult. I: 298.

Biswas, K. 1926. Flora of Salt Lakes, Calcutta. J. Dept. Sci. Cal. Unit, 8: 1-47.

Bisoyi, R.N. and Singh, P.K. 1988. Effect of Phosphorus fertilization on BGA inoculum production and nitrogen yields under field condition. Biol. fertil. Soils. 5, 338-343.

Bongale, U.D. 1981. On the soil algae from paddy fields of Panjim (Goa) and Chikkamanchali (Raichur District, Karnataka), India. J. Indian bot. Soc. 60: 326, 329.

Bongale, U.D. 1984. Soil algae in relation to crop soil groups in certain soils of Karnataka State, India. Comp. Physiol. Ecol. 9 (Suppl.): 384-393.

Bongale, U.D. and S.G. Bharati 1980. Algal flora of cultivated soils of Karnataka State, India. Phykos, 19: 95-109.

Carr, N.G. and I.W. Craig 1970. The relationship between bacteria, blue green algae and chloroplast. In Phytochemical Phylogamy (J.B. Harbau ed.) p. 119-143, Academic Press, London & New York.

Carr, N.G. and B.A. Whitton 1973. The biology of Blue green algae. Univ. California Press, Berkeley & Los-Angeles, p.676.

Carr, N.G. and B.A. Whitton eds. 1982. The Biology of Cyanobacteria. Blackwell Scientific Publ., Oxford, London, p. 688.

Castenzolz, R. W. 1970. Laboratory culture of thrmophilic cyanophytes, Schweiz. Z. Hydrocol. 32: 538-551.

Chatterjee, M. and S.P. Chatterjee, 1983. Nitrogen fixing cyanobacteria from the paddy field soils of Burdwan district. Phykos. 22: 64-66.

Chu, S. 1942. The influence of the mineral composition of the medium on the growth of planktonic alga—I Methods & culture media. J.Ecol. 30, 284-325.

Dadhich, K.S; A.K. Verma and G.S. Veenkataraman, 1969. Plant and Soil,3: 377.

De, P.K. 1939. The role of blue-green algae in nitrogen fixation in rice fields. Proc. Roy. Soc. B. 127: 121-139.

De, P.K. and L.N. Mandal, 1956. Fixation of nitrogen by algae in soils. Soil Sci., 81: 453.

Desikachary, T.V. 1959. Cyanophyta, Indian Council of Agric. Res., New Delhi, P. 686.

Dutta, N. and G.S. Venkataraman, 1958. An exploratory study of the algae of some cultivated soil. Indian J. Agron., 3: 109-115.

Eedelman, M. D. Swinton, J.A. Schiff., H.T. Epsteeein and B. Zeld 1967. Deoxyribonucleic acid of the blue green algae (cyanophyta). Bact. Rev. 31, 315-331.

Ellis, R.J. and M.R. Hartley 1971. Sites of synthesis of chloroplast protein. Nature, London, 233, 193-196.

Fay, P., W.D.P. Steward; E. Walsby and G.E. Fogg. 1968. Nature, 220, 810.

Fogg, G.E.; W.D.P. Stewart; P. Fay; A.E. Walsby, 1973. The blue green algae. Academic Press, London & New York, p. 459.

Fogg, G.E. 1949. Culture media for algae. Appendix III. In G.S. Venkataraman (Ed.). The cultivation of algae 1969, ICAR, New Delhi: 237-246.

Fuller, W.H; R.E. Cameron and N. Rajee 1960. Trans. 7th Int. Cong: Soil Sci.

Gerloff, G.C; G.P. Fizgeald and F. Skoog 1950. The isolation, purification and culture of blue green algae. Amer. J. Bot. 36, 216-218.

Ghosh, T.K. 1989. Studies on nitrogen fixing blue-green algae in rice soils of West Bengal. Ph.D. Thesis submitted to BCKV: W.B.

Gollerbakh, M.M and E.A. Shtina, 1969. Pochvennyle Vodorosli (Soil Algae) idz., 'Nauka', Leningrad, p.228.

Gonzalves, E.A. and K.S. Gangla, 1949. Observations on the algae of paddy field soils. J. Univ. Bombay, 18: 51.

Gonzalves, E.A. and V.S. Yalavigi, 1959. Algae in the rhizospheres of some crop plants. Proc. Sym. Alg.: 335-342.

Goyal, S.K., B.M. Sharma and R.S. Gupta, 1984. Algal flora of rice field soils of Jammu & Kashmir State. Phykos, 23: 59-64.

Goyal, S.K. and G.S. Venkataraman, 1971. Response of high yielding rice varieties to algalization II. Interaction of soil types with algal inoculation. Phykos, 10: 32-33.

Granhall, U. 1975. Nitrogen fixation by free living Microorganisms (WDP Stewart, ed.), Cambridge University Press, Cambridge, p.189.

Grover, I.S. and R.K. Pandhol, 1975. Algal flora of paddy fields of Ludhiana and its adjacent areas. Phykos, 14: 89-97.

Gupta, A.B. 1957. The algal flora of some paddy fields and its importance in soil economy, Part I. J. Res., 4: 1-24.

Gupta, A.B. 1964. Algal flora and its importance in the economy of rice fields. Hydrobiologia, 38: 213-222.

Gupta, A.B. 1966. Algal flora and its importance in the economy of rice fields. Hydrobiologia, 28: 213-222.

Gupta, R.S. 1972. Blue green algal flora of Rajasthan. Nova Hedwigia, 23: 481-489.

Gupta, R.S. and H.D. Kumar, 1968. Blue green algal flora of Udaipur and its neighbourhood. Revue algol, 9: 91-103.

Hellebust, J.A. 1974. Algal physiology and biochemistry (Stewart, WDP ed.) Blackwell Sci. Pub. Oxford, p.838.

Henriksson, E. 1971. Plant and Soil (Spl.vol.40).

Henriksson, E; B. Englund; M.B. Henden and I. Uas. 1972. Taxonomy and biology of algae (T.V. Desikachery ed.) Madras Univ. Press, Madras, p.265.

Hopwood, D.A. and A.M. Glauert 1960. The fine structure of the nuclear material of a blue green alga *Anaoaena cylindrica* Lemm. J. Biophys. Biochem. Cytol. 8, 813-823.

Ichimura, S. 1954. Jap. J. Bot. 14: 269.

Jha, M.N., U.N. Jha, N. Ahmad and M.K. Mallik, 1986. Cyanobacterial flora of rice field soils of Pusa and its adjoining areas. Phykos, 25: 97-101.

Jurgenoen, M. and C.B. Davey, 1968. Can J. Microbiol., 14: 1179.

Kamat, N.D. 1972. Oscillatoriaceae of the Mysore State. Phykos, 11: 59-63.

Kamat, N.D. 1974. The Nostocales of the Mysore State. Phykos, 13: 33-37.

Kannaiyan, S. 1979. Blue green algal biofertilizer for rice crop. Farmers and Parliament, 14(II): 19-25.

Kannaiyan, S. 1985. Algal Biofertilizers for low land rice. Tamil Nadu Agriculture University, Coimbatore, India, P. 14.

Kannaiyan, S., S.K. Govindarajan and H.D. Lewin, 1980. Blue-green algal biofertilizer for rice crop. Proc. National Workshop on Algal Systems (Ed. C.V. Seshadri *et al.*) Indian Society of Biotechnology, IIT/New Delhi, pp. 21-24.

Kaushik, B.D. and G.S. Venkataram, 1979. Plant and Soil, 52: 135.

Kaushik, B.D. and G.S.R. Krishnamurti, 1981. Effect of blue green algal and gypsum application on the physico-chemical properties of saline alkaline rice soils. Sci. & Cult, 47: 169-170.

Kaushik, B.D. and G.S.R. Krishnamurti, 1981. Effect of blue green algae and gypsum application on the physico-chemical properties of alkaline soils. Phykos, 20: 91-94.

Kaushik, N.D. 1989. Reclamative potential of Cyanobacteria in salt-affected soils. Phykos, 28(1 & 2): 101-109.

Khan, M. 1985. Algal flora of Sultanpur (U.P.) India, Phykos, 24: 52-57.

Khan, M. and A. Mathur, 1976. Algal flora of the rice fields around Dehra Dun, India. Rev. Algol, N.S. XI: 333-337.

Khan S.I. 1957. Cultures of the algae of rice fields of Kashmir. Biologia, 3: 29-43.

Khurana, A.S. and G.S. Venkataraman, 1968. Algal and fungal flora of paddy soils. Indian J. Microbiol., 8: 91-94.

Klein, R.M. 1970. Relationships between blue green algae and red algae. In Phylogenesis and Morphogenesis in the Algae: (J.F. Fredrick & R.M. Klein eds.) pp.623-633, Annals of New York Academy of Sciences, *175*.

Kratz, W.A. and J. Myers, 1955. Nutrition and growth of several blue-green algae. Amer. J. Bot., 42: 282-287.

Koltze, S.O. and S.K. Goyal, 1985. Distributional pattern of blue-green algae in rice field soils of Vidarbha region of Maharashtra State. Phykos, 24: 156-162.

Kuchkarova, E. and T.M. Maksudov, 1964. Microbiologiya, 33: 472.

Kurasawa, H. 1965. Bull. Res. Sci. Japan. Part 1 & 2, 41-42: 86,45: 73.

Kumar, H. 1970. Cyanophyceae of Sardhana. Phykos, 9: 79-85.

Kusum, Lata, 1965. List of algae from a few paddy fields of Kanpur. Agra Univ. J. Res. 14: 29-35.

Laloraya, V.K. and A.K. Mitra, 1973. Studies on the blue-green algal of the paddy fields of India. Part I - Cultural studies, general considerations and distribution pattern of the blue-green algae in paddy fields of India. Nova Hedwigia, Beiheft, 47.

Lang, N.J. 1968. The fine structure of blue-green algae. Ann. Rev. Microbiol., 22: 15-46.

Lang, N.J. 1972. Cellular differentiation in blue green algae. In Taxonomy and biology of Blue-green algae (Ed. T.V. Desikachary), Madras Univ., pp. 6-12.

Lang, N.J. and P. Fay, 1971. The heterocysts of blue-green algae II—Details of ultrastructure. Proc. Roy. Soc. B. 178: 193- 203.

Leak, L.V. 1965. Electron microscopic autoradiography incorporation of 3_H thymidine in a blue green algae *Anabaena* sp. J. Ultrastructure Research. 12, 135-146.

Leak, L.V. 1967. Studies on the preservation and organisation of DNA-containing regions in a blue green algae: A cytological and ultrastructural studies. J. Ultra. Res. 20, 190-205.

Lewin, I.M. 1972. Studies on free and combined amino acid composition in some blue green algae Phykos, 2: 50-58.

Mahajan, A.D. and R.J. Patel, 1983. Some nitrogen fixing blue green algae from paddy fields of Kaira district, Gujarat. In: Advances in Applied Phycology (Ed. Shukla, A.C.), Kanpur: 302- 308.

Maity, H. and S.C. Santra, 1985. Blue green algal flora of 24- Parganas, W.B. (India). Phykos, 24: 46-51.

Maity, H., G. Bandyopadhyaya and S.C. Santra, 1987. Algal flora of saline habitats of Sundarban, W.B. and its possible role in reclamation of soil. J. Indian Soc. Coastal Agric. Res. 5(1): 325-331.

Marathe, K.V. 1963. A study of the effect of fertilizers on the subterranean algal flora of paddy field soils from Karjat. J. Univ., Bombay, 31: 1.

Marathe, K.V. and B.S. Navalkar, 1963. A study of the effect of fertilizers and manures on the subterranean algal flora of paddy field soils, J. Univ., Bombay, 31: 11-19.

Marathe, K.V. 1964. A study of the effects of green manures on the subterranean algal flora of paddy fields soils. J. biol. Sci. 7: 1.

Marathe, K.V. 1964. J. Biol. Sci. 7:1.

Marathe, K.V. 1967. Studies on soil algae of India. I. Soil algae from the cultivated fields of Dhulia (Maharashtra). J. Univ. Bombay, 35: 100-105.

Mishra, G.P. and Jaishree Purohit, 1979. A preliminary list of soil algae from Sagar. Phykos, 18: 63-67.

Mitra, A.K. 1951. The algal flora of certain Indian soils. Indian J. Agric. Sci., 21: 357-373.

Mukhopadhyay, A. and P. Chatterjee, 1981. A check list of blue-green algae from the paddy fields of 24-Parganas and Howrah districts of West Bengal - I. Phykos, 20: 81-84.

Nichols, H.W. and H.C. Bold, 1965. *Trichosarcina polymorpha* gen et sp. nov. J. Phycol. I, 34-38.

Okuda, A. and M. Yamaguchi, 1952. Mem. Res. Int. Food Sci. Kyoto, Univ. 2:1.

Okuda, A. and M.Yamaguchi, 1956. Nitrogen fixing microorganisms in paddy soils. II. Distribution of BGA in paddy soils and the relationship between the growth of them at soil & soil propeties. Soil & Plant, 2, 4-7.

Pal, S. and A.K. Yadav, 1974. Some cyanophyceae from Sardhanpur district, a taxonomic enumeration. Phykos, 13: 38-47.

Pal, T.K. and S.C. Santra, 1982. Contribution to the Cyanophyceae of Murshidabad. Phykos, 21: 150-152.

Pal. U.C. and S.C. Santra, 1985. Algal flora of Midnapore, West Bengal (India) I. Cyanophyceae. Phykos, 24: 12-17.

Pandey, D.C. 1965a. A study of the algae from paddy soils of Ballia and Ghazipur districts of Uttar Pradesh: I. Cultural and ecological conditions. Nova Hedwigia, 9: 299-334.

Pandey, D.C. 1965b. A. study on the algae from paddy field soils of Ballia and Ghazipur districts of Uttar Pradesh, India. Part II (A) - Taxonomic considerations. Cyanophyceae. Nova Hedwigia 10: 177-209.

Pandey, D.C. and A.K. Mitra 1965. Certain new Myxophyceae from rice field soils of India. Nova Hedwigia, 10(1 & 2): 85-96.

Pankratz, H.S. and C.C. Brown, 1963. Cytology of blue green algae I. The cells of *Symploca muscorum.* Am. J. Bot. 50, 387- 399.

Parukratz, P.R. 1940. The Myxophyceae of the Travancore State, India. Proc. Indian Acad. Sci. B.11: 117-124.

Patil, P.L. and S.D. Satav, 1985. A study on nitrogen fixing blue-green algae from rice fields of Western Maharashtra. Phykos, 25: 113-116.

Prasad, B.N., R.K. Mehrotra and Yashpal Singh, 1976. Some new taxa of the genus *Lyngbya* Ag. from crop field soils. New Botanica, III. 1 & 2: 61-65.

Prasad, B.N. and R.K. Mehrotra, 1976. *Calothrix lucknowense* - a new species from paddy field soil. Phykos, 15 (1 & 2): 69-73.

Prasad, B.N. and R.K. Mehrotra, 1976. *Nostoc citrisporum* sp. nov. from paddy field soil. Phykos, 15 (1 & 2): 65-67.

Prasad, B.N. and R.K. Mehrotra, 1977. *Aphanocapsa gigantea* - a new species from a crop-field soil. New Botanist. IV: 1-4: 99-100.

Prasad, B.N. and Mehrotra 1977. Two new taxa of Genus *Phormidium* Kutz. from paddy fields. Proc. Nat. Acad. Sci. 47(B), I: 29-32.

Prasad, B.N., R.K. Mehrotra and Yashpal Singh, 1977. Some new taxa of *Nostoc* Vaucher from North Indian crop fields. Geophytology, 7(2): 222-228.

Prasad, B.N., R.K. Mehrotra and Yashpal Singh, 1978. Four new taxa of the genus *Lyngbya* Ag. from paddy fields of Uttar Pradesh. Acta Bot. Indica, 6: 75-80.

Prasad, B.N. and R.K. Mehrotra, 1979. Cyanophyceae flora of some North Indian crop fields. Geophytology, 8(2): 147-157.

Prasad, B.N. and R.K. Mehrotra, 1980. Blue green algae of paddy fields of U.P. Phykos, 19: 121-128.

Prasad, B.N. and P.V. Srivastava, 1968. Systemic and ecological studies on algae of alkaline (Usar) soils. Phykos, 7: 102-111.

Prekhodkova, L.P. 1968. Urk. Bot. Zh. 25: 590.

Rao, C.B. 1938. The Myxophyceae of the Madras Presidency, India I. J. Indian bot. Soc., 17: 81-96.

Rama Kant and D.C. Pandey, 1984. A preliminary study on blue-green algae of the paddy field of Assam, India. Univ. Allahabad Studies, 16(N.S.) 6: 221-224.

Reddy, P.M., D.D. Yamnam and T.Y. Imchen, 1985. Investigations on the Blue-green algae of North East India. Distribution and habitat preferences. Phykos, 25: 148-158.

Reddy, P.M. and Roger, P.A. 1988. Dynamics of algal population and acetylene reducing activity of fine rice soils inoculated with BGA. Biol. ferti. Soils. 6: 14-21.

Relwani, L.L. and R. Subrahmanyan, 1953. Role of blue-green algae-chemical nutrients and partial soil sterilisation on paddy yield. Curr. Sci., 32: 441.

Roger, P. and P. Raynaud, 1976. Rev. Ecol. Biol. Sol., 13: 545.

Roger, P. and P. Raynaud, 1979. Nitrogen & Rice, IRRI, Los Banos. Laguna, Philippines, p. 289.

Ris, H. and R.N. Singh, 1961. Electron microscopic studies on blue green algae. J. Biophys. Biochem. Cytol., 9, 63-80.

Saha, K.C. and L.N. Mandal, 1980. A green house study on the effect of inoculation of N_2-fixing blue green algae in an alluvial soil treated with P and M_o on the yield of rice and changes in the N content of soil. Plant & Soil, 57: 23-30.

Santra, S.C. 1984. Algal flora of Sikkim himalayas I: A contribution to our knowledge of the blue-green algae of North- east Sikkim (India). J. Econ. Tax. Bot., 5(5): 1205-1219.

Santra, S.C. 1987a. Studies on algal flora of West Bengal - An ecofloristic survey. PRS Thesis, Univ. of Calcutta.

Santra, S.C. 1987b. Freshwater algae of West Bengal. Proc. Int. Symp. on Phycology, Madras, India (In press).

Santra, S.C., U.C. Pal, H. Maity and G. Bandyopadhyaya, 1988. Blue-green algae in saline habitats of West Bengal: A systematic accounts. Biol. Mem. 14(1): 81-108.

Sankaram, A. 1971. Work done on blue green algae in relation to agriculture. ICAR, New Delhi.

Santra, S.C. 1991. Rice field blue green algae (Cyanobacteriá) and its utilisation prospect as biofertiliser in West Bengal, India. Proc. Nat. Symp. on Cyanobacterial Nitrogen fixation. New Delhi: 385-389.

Sardeshpande, J.S. and S.K. Goyal, 1981. Distributional pattern of blue-green algae in rice field soils of Konkan region of Maharashtra State. Phykos, 20: 102-106.

Sarma, T.A., Swarn Kanta, 1978. Algal flora of Patiala and its environs. I. Phykos, 17: 105-111.

Sarma, T.A., Swarn Kanta and Sunita, 1979. Algal flora of Patiala and its environs - Cyanophyceae. II. Phykos, 18: 13-19.

Schopf, J.W. 1974. Paleobiology of Precambrian: The algae of blue green algae In. Dobzhanoky, T. Hecht, M.K. and Steere W.C. (eds.). Evolutionary biology, 7, p.43 Plenum Press, New York.

Sengar, M. and D.C. Pandey, 1985. Contributions to the study of rice field algae I - A preliminary survey of Cyanophytes from paddy fields of W.B. (India). Univ. of Allahabad Studies 17, 5: 555-567.

Shestakov, S.V. and N.T. Khyen, 1980. Evidence for genetic transformation in blue green algae, *Anacystis* nidulans. Molec. Gen. Genet., 107, 372-375.

Shields, L.I. and L.W. Dumell, 1964. Bot. Rev. 30, 92.

Shtina, E.A. 1969. Biologija Sina-Zelenikh Vodoroslaj. 2: 21.

Shivaprakosha, M.K. and K. Shivappa Shetty 1983. Distribution of N_2-fixing blue green algae in paddy soils of Shimoga district of Karnataka and their growth rate and N_2-fixing capacity. J. Soil Biol. Ecol., 3(2): 77-82.

Shivaram, S. and K. Shivappa Shetty, 1988. A check list of heterocystous blue-green algae from the paddy soils of Cauvery Command area of Karnataka. Phykos, 27: 44-47.

Shukla, A.C. 1971. Systematic description of algae from Rice fields, Kanpur, India. Rev. Algal. 10(3): 257-270.

Singh, P.K. 1977. Growth and nitrogen fixation of algae from unicellular blue green alga *Apanothece castagnei*. Biologia Plantanum,19: 156-157.

Singh, D.T., K. Nirmala, D. Modi, S. Katiyar and H.N. Singh, 1987. Genetic transfer of herbicide resistant gene(s) from *Gloeocapsa* sp. to *Nostoc muscorum*. Mol. Gen. Genet., 208: 436-438.

Singh, K.P. and U.K. Chaturvedi, 1970. Myxophyceae of the Rohilkhand Division, U.P. India, III. Phykos. 9: 36-40.

Singh, P.K. 1978. Nitrogen economy of rice soils in relation to nitrogen fixation by blue-green algae and Azolla. In. Proc. National, Symp. on increasing rice yield in kharif. (CFRI, India. p. 12-23).

Singh, P.K. and Bisoyi, R.K. 1989. Blue green algae in rice fields. Phykos 28 (1 & 2): 181-195.

Singh, H.N. and Sonie, K.C. 1977. Isolation and characterization of chlorate resistant mutants of the blue-green algal, *Nostoc muscorum*. Mutation Res., 43: 205-212.

Singh, R.N. 1939. An investigation into the algal flora of paddy field soils of the United Provinces - I. Indian J. Agri. Sci., 9(1): 55-77.

Singh, R.N. 1942. The fixation of elementary nitrogen by some of the commonest blue-green algae from paddy field soils of United Provinces and Bihar. Indian J. Agric. Sci., 12: 743-756.

Singh, R.N. 1950. Reclamation of Usar lands in India through blue-green algae. Nature, Lond. 165: 325-326.

Singh, R.N. 1961. Role of blue-green algae in Nitrogen economy of Indian agriculture. Indian Council of Agricultural Research, New Delhi.

Singh, R.N., D.N. Tiwari and V.P. Singh, 1972. Genetic basis of cellular differentiation in Blue-green algae. In Taxonomy & Biology of Blue-green algae (Ed. T.V. Desikachary) Madras University, P. 27-37.

Sinha, J.P. and D. Mukherjee, 1975a. Blue-green algae from the paddy fields of Bankura district of West Bengal — I. Phykos, 14: 117-118.

Sinha, J.P. and D. Mukherjee, 1975b. Blue-green algae from the paddy fields of Bankura district of West Bengal — II. Phykos. 14: 119-120.

Sinha, J.P. and D. Mukherjee, 1984. Blue-green algae from the paddy fields of Bankura district of West Bengal — III. Phykos, 23: 142-143.

Stanier, R.Y.; Kunisawa, R; M. Mandel and G. Cohen-Bazire 1971. Purification and properties of unicellular blue-green algae (order-chroococcales). Bact. Rev. 35, 171-205.

Stewart, W.D.P. 1968. Nature, 200, 1020.

Stewart, W.D.P. 1969. Proc. Roy Soc. London, 172B: 367.

Stewart, W.D.P. and H.N. Singh, 1975. Transfer of nitrogen fixing (nif) genes in the blue green algae *Nostoc muscorum*. Biochem. Biophys. Res. Commun., 62: 62-69.

Stewart, W.D.P., P.P. Rowell, J.K. Ladha, and J.A.M. Sampaio, 1979. Blue green algae (Cyanobacteria): some aspects related to their role as sources of fixed nitrogen in paddy soils. In. Nitrogen and Rice, I.R.R.I, Los Banos, Phillippins. pp. 263-386.

Stewart, W.D.P. 1980. Some aspects of structure and function in nitrogen-fixing Cyanobacteria. Ann. Rev. Microbiol., 34: 497-536.

Subba Raju, N. 1972. The Blue-green algae from the soils of India. In: Taxonomy & Biology of Blue-green algae (Ed. T.V. Desikachary), Univ. of Madras, pp. 332-352.

Subhashini, D. and B.D. Kaushik, 1981. Amelioration of sodic soils with blue green algae. Aust. J. Soil Res., 19: 361-367.

Subhashini, D. and B.D. Kaushik, 1982. Nitrogen fixing potential of blue-green algae from saline and alkaline soils. Acta. Bot. Indica, 10: 321-322.

Subhashini, D. and B.D. Kaushik, 1984. Amelioration of salt- affected soils with blue green algae I. Influence of algalization on the properties of saline-alkali soils. Phykos, 23, 273-277.

Subrahmanyan, R. 1972. Taxonomy and biology of Blue-green algae (T.V. Desikachary ed.) Madras University Press, Madras, India, p. 281.

Subrahamanyan, R. and M.N. Sahai, 1964. Some observations on the utilisation of blue-green algae and remarks on its potentialities in rice culture. Proc. Indian Acad. Sci. B, 60: 145-154.

Subrahmanyan, R. 1972. Some observations on the utilisation of blue-green algal mixtures in rice cultivation in India. In Taxonomy and Biology of Blue-green algae (Ed. T.V. Desikachary), Univ. of Madras, pp. 281-293.

Subrahamanyan, R., L.L. Rolwani and G.B. Manna, 1965. Nitrogen enrichment of rice soils by blue-green algae and its effect on the yield of paddy. Proc. Nat. Acad. Sci. India, 35: 382-386.

Tarar, J.L. and T.H. Shewale, 1984. Studies on the effects of some fungicides on soil algae of paddy field. Phykos, 23: 191- 201.

Tiwari, G.L. 1972. Study of blue green algae from paddy field soils of India. Hydrobiologia, 29: 335-350.

Tiwari, G.L. 1975. A study of the blue green algae from paddy field soils of India II. Taxonomic considerations of non-heterocystous, BGA. Nova, Hedwigia, 26: 765-797.

Tiwari, G.L. and R.S. Pandey 1976. A study of the BGA from paddy field soils of India. III. Nostocaceae, Nova Hedwigia, 27: 701- 730.

Tiwari, D.N. 1987. Advances in Cyanophyte Research, Banaras Hindu University, Varanasi, India, P. 77.

Vaara, T; M. Varra; S. Niemela, 1982. Two improved methods for obtaining auxenic cultures of Cyanobacteria. Appl. Environ. Microbiol. 38: 1011-1014.

Venkatarkishnan, J. and K. Narayanan, 1982. Algal in rice fields of Chingleput district, Tamil Nadu, India. IRRN News letter, 7(4): 18.

Venkataraman, G.S. and S. Neelakantan, 1967. Effect of the cellular constituents of the nitrogen fixing blue-green algae - *Cylindrospermum muscicola* on the root growth of seedlings. J. Gen. Appl. Microbiol., Tokyo, 13: 53.

Venkataraman, G.S. 1972. Algal biofertilizers and rice cultivation. Today & Tomorrow's Pub., New Delhi, P. 75.

Venkataraman, G.S. 1975. The role of blue green algae in tropical rice cultivation. In. Nitrogen fixation by free living Micro organismis (Ed. W.D.P. Stewart) Cambridge Univ. Press, Camb. 207-218.

Venkataraman, G.S. 1980. Algal biofertilizers: Potentials and problems. Proc. Nat. Workshop on Algal System, Indian Society of Biotechnology, III, Madras, India.

Venkataraman, G.S. 1981. Blue-green algae for rice production — a manual for its promotion. FAO Soil Bulletin. 46: 102.

Watanabe, A., S. Nishigaki and C. Konishi, 1951. Effect of nitrogen fixing blue-green algae on the growth of rice plants. Nature, 168: 748-749.

Watanabe, A. 1956. On the effect of the atmospheric nitrogen fixing blue-green algae on the field of rice. The Bot. Mag. Tokyo, 69: 530-536.

Watanabe, A. 1959. Distribution of N_2-fixing blue-green algae in various areas of South and West Asia. J. Gen. Appl. Microbiol. Tokyo, 5: 21-29.

Watanabe, A. 1960. Collection and cultivation of nitrogen fixing blue-green algae and their effect on the growth and crop yield of rice plants. Proc. Symp. Algology, ICAR, New Delhi, 162-166.

Watanabe, A. 1962. Effect on N_2-fixing blue-green algae: *Tolypothrix tenuis* on the nitrogenous fertility of paddy soil and on the crop yield of rice plant. Jour. Gen. Appl. Micro., 8(2): 85-91.

Watanabe, I., K.K. Lee., B.V. Alimagno., M, Sato., D.C. deal Rosario and M.R. De guzman. 1977. IRRI Res. Paper Ser.3.

Watanabe I., K.K. Lee and B.V. Alimagno. 1978. Soil Sci. Plant Nutri. 24:1.

Watanabe, I. and Furusaka C. 1980. Microbial Ecology of flooded rice soils. Advances in Microbial Ecology. 4: 125-168.

Watanabe, A. and Y. Yamamoda. 1970. Proc. 2nd Symp. Nitrogen fixation and nitrogen cycle send Japan, p.22.

Yamaguchi, M. 1976. Nitrogen fixation by micro organisms in paddy soils in relation to their fertility. In: The fertility of paddy soils and fertilizer applications for Rice. Food and Fertilizer Technology Centre for Asian Pacific Region, Taipei, Tiwan, pp. 60- 75.

Yoshida, T. and Ancajas, R.R. 1973. Nitrogen fixing activity in upland and flooded rice fields. Soil Sci. Soc. Amer. Proc. 37: 42-46.

Zarrouk, C. 1966. Contribution of A 1'Etude d'Une cyanophyceae influence de Divers facteure physiouses et chemiques sur la croissance et al photosynthese de *Spirulina maxima* Stetch at Gardner Geitler, Thesis, Paris.

GLOSSARY

Aerobic	:	needing oxygen.
Akinete	:	thick-walled resting spore.
Alphagranule	:	protoplasmic structure containing myxophycean starch in the cyanophyceae.
Autotroph	:	not needing an external source of an organic compound as an energy sources. Energy is obtained from light or inorganic chemical reactions.
Bactriocin	:	antibiotic secreted by blue-green alage that kills related strains of blue-green algae.
Baeocyte	:	endospore produced in the chamaesiphonales of the cyanophytes.
Biomass	:	the amount at any one time of living matter in a habitat.
Bloom	:	heavy growth of planktonic algae in a body of water.
Capsule	:	extracellular mucilage.
Carotene	:	Oxygenfree, unsaturated, hydrocarbon carotenoid.
Carotenoid	:	Yellow, orange or red hydrocarbon fat-soluble pigment.
Chlorophyll	:	far-soluble, green, porphyrin-type pigment.
Chloroplast	:	plastid with chlorophyll.
Chromatic adaption	:	Change in the proportions of different photosynthetic pigments enabling optimum absorption of available wavelengths of light.
Compensation depth	:	depth of water where there is sufficient light so that photosynthesis equal respiration over a 24 hour period.
Compensation point	:	light intensity at which respiration equals photosynthesis over a 24 hour period.
Cyanelle	:	endosymbiotic blue-green alga.
Cyanobacterocin	:	an antibiotic produced by a blue-green alga that inhibits growth of related blue-green algae.
Cyanome	:	host cell containing a cyanelle.
Cyanophage	:	Virus in cells of the cyanophyceae.
Cyanophycin granule	:	polypeptide storage granule in the cyanophyceae.
Cyst	:	Cell resistant to poor conditions with a thick or silicified wall, usually not including the original wall, if one was present.

Electron-dense or electron-opaque : term used to describe a material that absorbs electrons and appears dark in electron micrographs.

Electron-transparent : term used to describe a material that does not absorb electrons and appears light in electron micrographs.

Endospore : a sexual spore in the cyanophyta formed by the internal division of a cell.

Endosymbiotic : term that describe an organism living inside a host in a symbiosis.

Endozoic : living inside an animal.

Eucaryotic : a cell having membrane-enclosed organelles such as the nucleus and mitochondria.

Extant : living today.

Extinct : not living today.

False branching : in the cyanophyta, breakage of a trichome through a seath, giving the appearance of a branch.

Filament : in the cyanophyceae, one or more trichomes enclosed in a seath.

Fragmentation : type of a sexual reproduction where a thallus breaks into two or more parts, each of which forms a new thallus.

Gas vacuole : a collection of gas vesicles.

Gas vesicle : hollow cylindrical gas-filled structures in the cyanophyta.

Glycopeptide or Glycoportein : polysaccharide composed of sugars and amino acids or peptides.

Hermaphroditic or homothallic : producing both male and female gametangia or the same thallus.

Intercalary : in between two cells or tissues.

Metachromic granule : protoplasmic body containing stored polyphosphate in the cyanophyta.

Mucopeptide : polysaccharide of the walls of the cyanophyta composed of sugars and amino acids.

Myxophycean starch : Storage polysaccharide of cyanophyta, similar to glycogen.

Necridium or seeparation disc : a cell that dies in a trichome of the cyanophyta resulting in the formation of a hormogonium from part of the trichome.

Nitrogen fixation : the intracellular fixation of nitrogen gas from the atmosphere to ammonia in cyanophyta or bacteria.

Organelle : a membrane-bounded part of a cell.

Peptidoglycan : Polysaccharide composed of sugars and amino acids in the walls of the cyanophyta.

Photoautotropic	:	term describing autotrophic plant that obtains energy from photosynthesis.
Photoreceptor	:	the part of the cell that receives the stimulus in phototaxis, usually a dense area in a flagellar swelling.
Phycobiliprotein or phycobilin	:	Water soluble blue-green or pink pigment in the cyanophyta, Rhodophyta, and Cryptophyta.
Phycobilisome	:	an aggregation of phycobiliproteins on the surface of a thylakoid.
Phycobiont	:	algal partner in a lichen.
Phycocolloid	:	polysaccharide colloid formed by an alga.
Phycocyanin	:	blue-green-colored phycobiliprotein.
Phycoerythrin	:	pink-colored phycobili-protein.
Polyglucan granule	:	protoplasmic structure containing the storage product in the cyanophyta.
Polyhedral body	:	protoplasmic structure in the cyanophyta associated with DNA microfibrils, it may contain the carbondioxide-fixing enzyme ribulose-1,5-biphosphate carboxylase.
Polyol	:	sugar alcohol.
Polyphosphate body	:	protoplasmic structure containing stored phosphate in the cyanophyta.
Procaryotic cell	:	a type of cell lacking membrane-bounded organelles.
Red tide	:	Water with a large number of dinoflagellates or other organisms that colour the water red.
Symbiosis or reciprocal parasitism	:	two organisms living together to the mutual benefit of each.
Zygote	:	product of the fusion of two gametes.

INDEX

Akinetes 12
Allophycocyanin 7, 8
Allen and Arnon's medium 22
Anabaena 108
A. oryzae 110
A. khannae 110
A. naviculoides 111
A. fertilissima 111
A. ambigua 111
A. utermohlii 112
A. oscillaroides 112
O. iyengarii 112
A. iyengarii var. *tenuis* 113
A. iyengarii var. *attenuata* 113
A. orientalis 113
A. torulosa 114
A. fuellebornii 115
A. volzii 115
A.spiroides 115
A. gelatinicola 116
A. vaginicola 116
A. vaginicola f. *fertilissima* 116
A. variabilis 117
A. variabilis var. *ellipsospora* 117
A. circiralis 117
Anacystis 12
Aphanizomenor 1
Aphanocapsa 47
A. roeseana 47
A. roeseana forma major 48
A. banaresensis 48
A. biformis 49
A. littoralis 49
A. pulchra 49
A. thermalis 50
A. grevillei 50
Aphanothece 50
A. bullosa 51
A. microscopic 51
A. pallida 51
A. stagnina 52
A. naegelii 52
Arthrospira 56
A. jenneri 56
A. massartii 56
A. massartii var. *indica* 57
A. gomontiana 57
A. platensis 57
Aulosira 118
A. prolificica 119
A. anigmatica 119
A. pseudoramosa 120
A. implexa 120
A. fritschii 121
A. fertilissima 121
A. fertilissima var. *tenuis* 121
Azolla 151
Bacteria and blue green algae 17
Bacocytes 14
Baoyancy: Cyanophyta 11
Benecke's medium 21
Bibliography 157
BIomass and Biogas 151
Biophotolysis 151
Blue green algae as food 147
Blue green algae in Agriculture 148
Bold Basal medium 23
Calothrix 133
C. fusca 134
C. fusca var. *crassa* 134
C. parietina 134
C. ghosei 135
C. brevissima 135
C. brevissima var. *moniliforme* 135
C. braunii 136
C. scytonemicola 136
C. castellii 136
C. castellii var. *somastipurense* 136

C. elenkinii 137
C. membranacea 137
C. marchica 137
Camptylonema 145
C. indicum 145
Carotene: β carotene 7
Cell division 12
Cell sheath 4
Cell wall 4
Central body 8
Centroplasm 8
Chamaesiphon 11
Chlorogloea 54
C. fritschii 54
Chlorophyll 7
Chroococcus 42
C. minutus 42
C. pallidus 42
C. coharens 43
Chromatin 8
Chromatoplasm 7
Chu-10 21
Complementary chromatic adaptation 7
Crinallium 72
C. magnum 72
Cyanidium caldarium 19
Cyanophyta/Cyanobacteria 1
Cyanophycin granules 10
Cyanobacterial genes 16
Cyanobacterial genetical studies 153
Cylindrospermum 97
C. muscicola 98
C. muscicola var. *longispora* 98
C. stagnale 99
C. licheniforme 99
C. majus 99
C. indicum 100
C. alatosporum 100
C. michailovskoense 100
C. sphaerica 101
Distribution of cyanobacteria 1
DNA 8
Endospores 14
Filament 2
Fogg's nitrogen free medium 21
Fossil blue greens 19
Gas vacuoles 11
Gerloff's medium 22
Gloeocapsa 43
G. polydermatica 43
G. calcarea 44
G. dcorticans 44
G. rupestris 44
G. montana 44
G. luteofusca 45
G. crepidinum 45
Gloeothece 45
G. samoensis var. *major* 46
G. membranacea 46
G. rupestris 46
G. rupestris var. *maxima* 47
Gloeotrichia 138
G. natans 138
G. raciborskii 139
G. raciborskii var. *kashiense* 139
G. intrmedia 139
G. intermedia var. *kanwaensis* 140
G. pisum 140
G. pilgeri 141
G. ghosei 141
Glycogen granules 10
Habitats of cyanobactreia 1
Hapalosiphon 142
H. welwitschii 143
H. fontinalis 143
H. luteolus 144
H. intericatus 144
Heterocysts 13
Hormogones 12
Homoeothrix 132
H. juliana 132
Hydrocoleum 96
H. lyngbyaceum 96
Illustration of plates 163-172
Isolation procedure 25
Johannesbaptisia 53

J. pellucida 53
Kratz and Myer's media 23
Lyngbya 83
L. laxespiralis 84
L. spiralis 85
L. birgei 86
L. ceylanica 86
L. ceylanica var. *bankurensis* 86
L. confervoides 87
L. majuscula 87
L. martensiana 87
L. hieronymusii 88
L. allorgei 88
L. cryptovaginata 89
L. stagnina 89
L. major 89
L. perelegen 90
L. aerugineo - coerulea 90
Mass culture 27
Mastigocladus 141
M. laminosus 141
Merismopedia 52
M. convoluta 52
M. tenuissima 53
Microchacte 131
M. uberrime 131
M. tenera 132
Microcoleus 94
M. chthonoplastes 94
M. subtorulosus 95
M. vaginatus 95
M. lacustris 95
M. paludosus 96
Microcystis 40
M. longata 40
M. aeruginosa 41
M. flos-aguae 41
M. bengalensis 41
Myxosarcina 54
M. spectabilis 54
Nitrogen fixation 148
Nodularia 118
N. spumigena 118
Nostocales 55
Nostoc 102
N. humifusum 103
N. calcicola 103
N. linckia 104
N. paludosum 104
N. piscinale 105
N. punctiforme 105
N. hatei 106
N. muscorum 106
N. ellipsosporum 106
N. communutum 107
N. depressum 107
N. verrucosum 107
N. commune 107
N. spongiaeforme 108
Nucleoplasmic region 8
Oscillatoria 59
O. annae 62
O. subbrevis 62
O. subbrevis 62
O. subbrevis forma *major* 63
O. acuta 63
O. brevis var. *non-granulata* 63
O. earli 64
O. acuminata 64
O. schultzii 64
O. okeni 65
O. amphibia 65
O. obscura 65
O. princeps 66
O. curviceps 66
O. pseudogeminata 66
O. pseudogeminata 67
O. amphigranulata 67
O. proteus 67
O. quadripunctulata 68
O. anguina 68
O. proboscidea 68
O. chalybea 69
O. amoena 69
O. formosa 69
O. splendida 70

O. animalis 70
O. tenuis 70
O. raoi 71
O. limosa 71
Photosynthetic membrane 6
Phormidium 72
P. corium 74
P. corium var. *capitatum* 75
P. corium forma *major* 75
P. ampiguum 75
P. ambiguum var *major* 76
P. fragile 76
P. bohneri 77
P. jadinianum 77
P. molle 77
P. rotunda 78
P. laminosum 78
P. tenue 78
P. purpurascens 79
P. valderianum 79
P. stagnina 79
P. papyraceum 80
P. subfuscum 80
P. angustissimum 80
P. foveolarum 81
P. anomala 81
P. favosum 81
P. retzii 82
P. lucidum 82
P. uncinatum 82
Plasma membrane 6
Phycocyanin 7
Phycoerythrin 7
Phycobililisomes 8
Phylogeny 17
Plctonema 122
P. notatum 122
P. tomasinianum 123
P. nostocorum 123
Prophyrosiphon 72
P. notarisii 72
Pseudanabaena 118
P. schmidlei 118
Relationship with red algae 18
Reclamation of saline and alkaline soil 152
Reproduction 11
Reserve products 10
Ribosomes 9
Schizothrix 91
S. telephoroides 91
S. fresii 92
S. muelleri 92
S. penicillata 92
S. lacustris 92
S. arenarea 93
S. lamyi 93
Scytonema 123
S. simplex 124
S. coactile 125
S. fritschii 125
S. hofmanni 125
S. fremyii 126
S. mirabile 126
S. schmidtii 126
S. caldarium 127
S. javanicum 127
S. leptobasis 128
S. burmanicum 128
S. guyanense 128
S. stuposum 129
S. circinnatum 129
S. myochrous 129
S. tolypothrichoides 130
Sexuality 14
Spirulina 57
S. major 58
S. subsalsa 58
S. princeps 58
S. meneghiniana 59
Stichosiphon 54
S. sansibaricus 54
Stigonema 145
S. ocellatum 145
Stress compatibility of Cyanobacteria 155
Succssional pattern 34
Symploca 93

S. muralis 93
Symbiotic origin of chloroplasts 20
Synchronous culture 27
Taxonomic enumeration 29
Thallus morphology 2
Trichodesmium 71
T. thiebautii 71
Tolypothrix 130
T. nodosa 130
T. tenuis 130
Trichome fragmentation 12
Ultrastructures of cells 4
Volutin 10
Westiellopsis 144
W. prolifica 144
Wollea 101
W. bharadwajae 101
Zarrouk medium 23

ILLUSTRATION OF PLATES

Plate 1

Figs.	1	*Microcystis flos-aquae* (Wittr.) Kirchner
	2	*Chroococcus pallidus* Näg.
	3 &7	*Microcystis aeruginosa* Kütz.
	4	*Gloeocapsa crepidinum* Thuret
	5 &8	*Microcystis elongata* Desikachary
	6	*Chroococcus cohaerens* (Breb.) Näg.
	9	*Microcystis bengalensis* Banerji
	10	*Aphanothece bullosa* (Menegh.) Rabenh.
	11	*A. microscopica* Näg.
	12	*Chlorogloea fritschii* Mitra
	13	*Myxosarcina spectabilis* Geitler
	14	*Stichosiphon sansibaricus* (Hieron.) Drouet et Daily

Plate 2

Figs.	1 & 2	*Chroococcus minutus* (Kütz.) Näg.
	3	*Merismopedia tenuissima* Lemm.
	4	*Gloeocapsa polydermatica* Kütz.
	5	*G. decorticans* (A. Br.) Richter
	6	*G. montana* Kütz.
	7	*G. calcarea* Tilden
	8	*G. luteo-fusca* Martens
	9	*Gloeothece samoensis* var. *major* Wille
	10	*G. rupestris* (Lyngb.) Bornet
	11	*Merismopedia convoluta* Breb.
	12	*Gloeocapsa rupestris* Kütz.
	13	*Gloeothece rupestris* (Lyngb.) Bornet var *maxima* West
	14	*G. membranacea* (Rabenh.) Bornet

Plate 3

Figs.	1 & 2	*Aphanocapsa roeseana* de Bary
	3	*A. roeseana* f. *major* Laloraya et Mitra
	4	*A. banaresensis* Bharadwaja

	5	*A. pulchra* (Kütz) Rabenh.
	6	*A. biformis* A. Br.
	7	*A. littoralis* Hansg.
	8	*A. thermalis* (Kütz.) Brugg.
	9	*A. grevillei* (Hass.) Rabenh.
	10	*Aphanothece pallida* (Kütz.) Rabenh.
	11	*A. stagnina* (Spreng.) A. Braun
	12	*A. naegelii* Wartm.

Plate 4

Figs.	1	*Johannesbaptistia pellucida* (Dickie) Taylor et Drouet
	2	*Arthrospira jenneri* Stizenb. ex Gomont
	3	*A. platensis* (Nordst.) Gomont var. *tenuis* (Rao) Desikachary
	4	*A. massartii* var. *indica* Desikachary
	5	*A. gomontiana* Setchell
	6	*A. massartii* Kuffareth
	7	*Spirulina major* Kütz. ex Gomont
	8 & 9	*S. subsalsa* Oerst. ex Gomont
	10	*S. princeps* W. et G.S. West
	11	*Oscillatoria subbrevis* Schmidle
	12 & 13	*O. annae* van Goor
	14 & 15	*O. subbrevis* forma major West
	16	*Spirulina meneghiniana* Zanard. ex Gomont

Plate 5

Figs.	1	*Oscillatoria acuta* Bruhl et Biswas, orth. mut. Geitler
	2	*O. acuminata* Gomont
	3	*O. pseudogeminata* G. Schmid
	4	*O. earlei* Gardner
	5	*O. schultzii* Lemm.
	6	*O. brevis* (Kütz.) Gomont var. *non-granulata* Kamat
	7	*O. pseudogeminata* var. *unigranulata* Biswas
	8	*O. amphibia* Ag. ex Gomont
	9	*O. amphigranulata* van Goor
	10	*O. okeni* Ag. ex Gomont

	11	*O. princeps* Vaucher ex Gomont
	12	*O. anguina* (Bory) Gomont
	13	*O. obscura* Bruhl et Biswas
	14	*O. animalis* Ag. ex Gomont

Plate 6

Figs.	1 & 2	*Oscillatoria proboscidea* Gomont
	3 & 4	*O. chalybea* (Mertens) Gomont
	5	*O. curviceps* Ag. ex Gomont
	6	*O. raoi* De Toni
	7	*O. formosa* Bory ex Gomont
	8 & 9	*O. quadripunctulata* Bruhl et Biswas
	10	*O. limosa* Ag. ex Gom.
	11	*O. amoena* (Kütz.) Gomont
	12	*Trichodesmium thiebautii* Gomont
	13 & 14	*Phormidium ambiguum* Gomont
	15	*Oscillatoria anguina* (Bory) Gomont forma
	16-18	*Phormidium ambiguum* var. *major* Lemm.

Plate 7

Figs.	1	*Phormidium corium* (Ag.) Gomont
	2	*P. angustissimum* W. et G.S. West
	3	*P. molle* (Kütz.) Gomont
	4	*Oscillatoria tenuis* Ag. ex Gom.
	5	*Phormidium jadinianum* Gomont
	6	*Porphyrosiphon notarisii* (Menegh.) Kütz ex Gomont
	7	*Phormidium corium* forma *major* Santra et Pal
	8	*Oscillatoria proteus* Skuja
	9	*Phormidium fragile* (Meneghini) Gomont
	10	*Oscillatoria splendida* Grev. ex Gomont
	11	*Phormidium bohneri* Schmidle

Plate 8

Figs.	1 & 2	*Phormidium subfuscum* Kütz. ex Gomont
	3	*P. stagnina* Rao, C.B.
	4 & 5	*P. tenue* (Menegh.) Gomont

	6	*P. purpurascens* (Kütz.) Gomont
	7 & 8	*P. papyraceum* (Ag.) Gomont
	9 & 11	*Lyngbya laxespiralis* Skuja
	10	*L. lagerheimii* forma *lacerata* Prasad et al.
	12 & 13	*Phormidium laminosum* Gomont
	14-16	*P. rotunda* Santra et Pal
	17	*P. valderianum* (Delp.) Gomont

Plate 9

Figs.	1	*Crinalium magnum* Fritsch et John
	2	*Phormidium corium* var. *capitatum* Gardner
	3 & 4	*P. foveolarum* (Mont.) Gom.
	5	*P. anomala* Rao, C.B.
	6 & 7	*P. favosum* (Bory) Gom.
	8 & 9	*P. retzii* (Ag.) Gom.
	10	*P. lucidum* Kütz. ex Gom.
	11 & 12	*P. uncinatum* (Ag.) Gom.
	13	*Lyngbya ceylanica* Wille var. *bankurensis* Sinha & Mukherjee
	14	*L. aerugineo-coerulea* (Kütz.) Gom.
	15	*Schizothrix arenaria* (Berk.) Gomont

Plate 10

Figs.	1	*Lyngbya spiralis* Geitler
	2, 3, 5 &6	*L. birgei* Smith, G.M.
	4	*L. majuscula* Harvey ex Gomont
	7	*L. ceylanica* Wille

Plate 11

Figs.	1	*Lyngbya confervoides* C. Ag. ex Gomont
	2-4	*L. martensiana* Menegh. ex Gomont
	5 & 6	*L. hieronymusii* Lemm.
	7 & 8	*L. allorgei* Fremy
	9 & 10	*L. cryptovaginata* Schkorbatow
	11	*L. stagnina* Kutz.
	12 & 13	*L. perelegans* Lemm.
	14	*L. lagerheimii* (Mob.) Gomont

	15	*Oscillatoria probascidea* Gomont

Plate 12

Figs.	1-3	*Lyngbya major* Menegh. ex Gomont
	4 & 5	*Schizothrix penicillata* (Kütz.) Gom.
	6	*S. lamyi* Gomont
	7	*S. telephoroides* (Mont.) Gomont
	8 & 10	*S. muelleri* Näg. ex Gomont
	9	*S. friesii* (Ag.) Gomont

Plate 13

Figs.	1 & 2	*Symploca muralis* Kütz. ex Gomont
	3	*Microcoleus chthonoplastes* Thuret ex Gomont
	4 & 5	*M. subtorulosus* (Breb.) Gomont
	6	*M. lacustris* (Rabenh.) Farlow
	7 & 9	*M. paludosus* (Kütz.) Gomont
	8	*M. vaginatus* (Vaucher) Gomont

Plate 14

Figs.	1 & 2	*Cylindrospermum michailoskoense* Elenkin
	3, 6 & 7	*Hydrocoleum lyngbyaceum* Kütz. ex Gom.
	4 & 5	*Cylindrospermum licheniforme* Kütz. ex Born. et Flah.
	8-10	*C. muscicola* Kütz. ex Born. et Flah.
	11-13	*C. indicum* Rao, orth. mut. De Toni
	14 & 15	*C. sphaerica* Prasad
	16 & 17	*C. majus* Kütz. ex Born. et Flah.

Plate 15

Figs.	1	*Schizothrix lacustris* A. Br. ex Gomont
	2 & 3	*Cylindrospermum stagnale* (Kütz.) Born. et Flah.
	4-6	*C. alatosporum* Fritsch
	7 & 8	*Wollea bharadwajae* Singh
	9-11	*Cylindrospermum muscicola* var. *longispora* Dixit
	12	*Nostoc humifusum* Carm. ex Born. et Flah.
	13	*N. punctiforme* (Kütz.) Hariot
	14	*N. piscinale* Kutz ex Born. et Flah.

Plate 16

Figs.	1	*Nostoc calcicola* Breb. ex Born. et Flah.
	2	*N. hatei* Dixit
	3	*N. verrucosum* Vaucher ex Born. et Flah
	4	*N. paludosum* Kütz. ex Born. et Flah.
	5	*N. linckia* var. *arvense* Rao, C.B.

Plate 17

Figs.	1	*Nostoc muscorum* Ag. ex Born. et Flah.
	2	*N. ellipsosporum* (Desm.). Rabenh. ex Born. et Flah.
	3	*N. comminutum* Kütz.
	4	*N. spongiaeforme* Agardh ex Born. et Flah.
	5	*N. commune* Vaucher ex Born. et Flah.
	6	*N. depressum* Wood

Plate 18

Figs.	1	*Anabaena oryzae* Fritsch
	2	*A. variabilis* var. *ellipsospora* Fritsch
	3	*A. khannae* Skuja
	4	*A. naviculoides* Fritsch
	5	*A. vaginicola* f. *fertilissima* Prasad
	6	*A. fertilissima* Rao, C.B.

Plate 19

Figs.	1 & 2	*Anabaena oscillarioides* Bory ex Born. et Flah.
	3 & 4	*A. iyengarii* Bharadwaja
	5 & 6	*A. torulosa* (Carm.) Lagerh. ex Born. et Flah.
	7 & 8	*A. fuellebornii* Schmidle
	9-11	*A. orientalis* Dixit

Plate 20

Figs.	1-4	*Anabaena volzii* Lemm.
	5 & 6	*A. gelatinicola* Ghose
	7	*A. vaginicola* Fritsch et Rich
	8 & 9	*A. iyengarii* var. *tenuis* Rao
	10 & 11	*A. variabilis* Kütz. ex Born. et Flah.

Plate 21

Figs.	1	*Anabaena ambigua* Rao
	2	*A. utermohlii* Geitler
	3 & 4	*A. spiroides* Klebahn
	5 & 6	*Aulosira fertilissima* Ghose
	7-10	*A. fertilissima* var. *tenuis* Rao
	11	*Nodularia spumigena* Mertens ex Born. et Flah.

Plate 22

Figs.	1	*Aulosira prolifica* Bharadwaja
	2 & 3	*A. aenigmatica* Fremy
	4 & 5	*A. implexa* Bornet et Flahault
	6 & 9	*A. pseudoramosa* Bharadwaja
	7 & 8	*A. fritschii* Bharadwaja
	10	*Anabaena iyengarii* var. *attenuata* Rao

Plate 23

Figs.	1-4	*Scytonema simplex* Bharadwaja
	5-7	*S. fritschii* Ghose

Plate 24

Figs.	1 & 2	*Scytonema fremyii* Desikachary
	3-5	*S. coactile* Mont. ex Born. et Flah.

Plate 25

Figs.	1	*Scytonema schmidtii* Gom.
	2	*S. caldarium* Setchell
	3	*S. hofmanni* Ag. ex Born. et Flah.
	4-6	*S. leptobasis* Ghose

Plate 26

Figs.	1	*Scytonema mirabile* (Dillw.) Born.
	2 & 3	*S. tolypothrichoides* Kütz. ex Born. et Flah.
	4	*Plectonema tomasinianum* (Kütz.) Born. ex Gom.

Plate 27

Figs.	1	*Scytonema guyanense* (Mont.) Born. et Flah.

	2	*Plectonema nostocorum* Born. ex Gom.
	3	*P. notatum* Schmidle
	4	*Tolypothrix nodosa* Bharadwaja

Plate 28

Figs.	1-3	*Scytonema burmanicum* Skuja
	4 & 5	*S. javanicum* (Kütz.) Bornet ex Born. et Flah.

Plate 29

Figs.	1	*Scytonema cincinnatum* Thuret ex Born. et Flah.
	2 & 3	*S. myochrous* (Dillw.) Ag. ex Born. et Flah.

Plate 30

Figs.	1	*Anabaena circinalis* var. *crassa* Ghose
	2	*Pseudanabaena schmidlei* Jaag. O.
	3	*Calothrix parietina* Thuret ex Born. et Flah.
	4	*C. fusca* var. *crassa* Rao
	5	*Scytonema stuposum* (Kütz.) Bornet ex Born. et. Flah.
	6	*Calothrix brevissima* West

Plate 31

Figs.	1-3	*Microchaete uberrima* Carter
	4	*Homoeothrix juliana* (Menegh.) Krichn.
	5	*Calothrix brevissima* var. *moniliforme* Ghose Orth. mut. Geitler
	6	*Microchaete tenera* Thuret ex Born. et Flah.
	7	*Calothrix ghosei* Bharadwaja
	8 & 9	*C. marchica* Lemm.
	10	*C. castellii* (Massal.) Born. et Flah.
	11	*C. membranacea* Schmidle

Plate 32

Figs.	1	*Tolypothrix tenuis* (Kütz.) Johs. Schmidt. em.
	2	*Calothrix fusca* (Kütz.) Born. et Flah.
	3	*C. brauni* (A. Br.) Born. et Flah.
	4	*C. castellii* var. *somastipurense* Rao
	5	*Gloeotrichia pilgeri* Schmidle
	6	*G. pisum* Thuret ex Born. et Flah.

Plate 33

Figs.	1	*Gloeotrichia raciborskii* Woloszynska
	2	*G. raciborskii* var. *kashiense* Rao
	3	*G. natans* Raben. ex Born. et Flah.
	4 & 5	*G. intermedia* var. *kanwaensis* Rao
	6	*G. intermedia* (Lemm.) Geitler
	7 & 8	*G. ghosei* Singh

Plate 34

Figs.	1	*Mastigocladus laminosus* Cohn
	2	*Calothrix Scytonemicola* Tilden
	3-5	*Westiellopsis prolifica* Janet
	6-8	*C. elenkinii* Kossinskaja

Plate 35

Figs.	1	*Hapalosiphon welwitschii* W. et G.S. West
	2 & 3	*H. luteolus* W. & G.S. West

Plate 36

Figs.	1 & 2	*Camptylonema indicum* Schmidle
	3	*Hapalosiphon intricatus* W. et G.S. West
	4	*H. fontinalis* (Ag.) Bornet

Plate 37

Figs.	1-4	*Stigonema ocellatum* (Dillw.) Thuret ex Born. et Flah.

Plate 38

A	—	*Microcystis aeruginosa* Kütz.
B	—	*Aphanocapsa banaresensis* Bharadwaja
C	—	*A. biformis* A. Br.
D	—	*A. grevillei* (Hass.) Rabenh.

Plate 39

A	—	*Merismopedia tenuissima* Lemm.
B	—	*Spirulina subsalsa* Oerst. ex Gom.
C	—	*Oscillatoria annae* van Goor
D	—	*O. princeps* Vaucher ex Gomont

Plate 40

A — *Porphyrosiphon notarisii* (Menegh.) Kütz.
B — *Phormidium corium* forma *major* Santra et Pal
C — *P. ambiguum* Gomont
D — *P. ambiguum* var. *major* Lemm.

Plate 41

A — *Phormidium rotunda* Santra et Pal
B — *Lyngbya confervoides* C. Ag. ex Gomont
C — *L. majuscula* Harvey ex Gom.
D — *Schizothrix telephoroides* (Mont.) Gomont.

Plate 42

A — *Schizothrix friesii* (Ag.) Gomont
B — *S. muelleri* Näg. ex Gomont
C — *Hydrocoleum lyngbyaceum* Kütz. ex Gomont.
D — *Cylindrospermum majus* Kütz. ex Born. et Flah.

Plate 43

A — *Anabaena naviculoides* Fritsch
B — *A. orientalis* Dixit
C — *A. fuellebornii* Schmidle
D — *A. volzii* Lemm.

Plate 44

A — *Aulosira prolifica* Bharadwaja
B — *A. fertilissima* var. *tenuis* Rao, C.B.
C — *Scytonema simplex* Bharadwaja
D — *S. coactile* Montagne ex Born. et Flah.

Plate 45

A — *Scytonema javanicum* (Kütz.) Born. ex Born. et Flah.
B — *S. leptobasis* Ghose
C — *Microchaete uberrima* Carter
D — *Gloetorichia natans* Rabenhorst ex Born. et Flah.

PLATE I

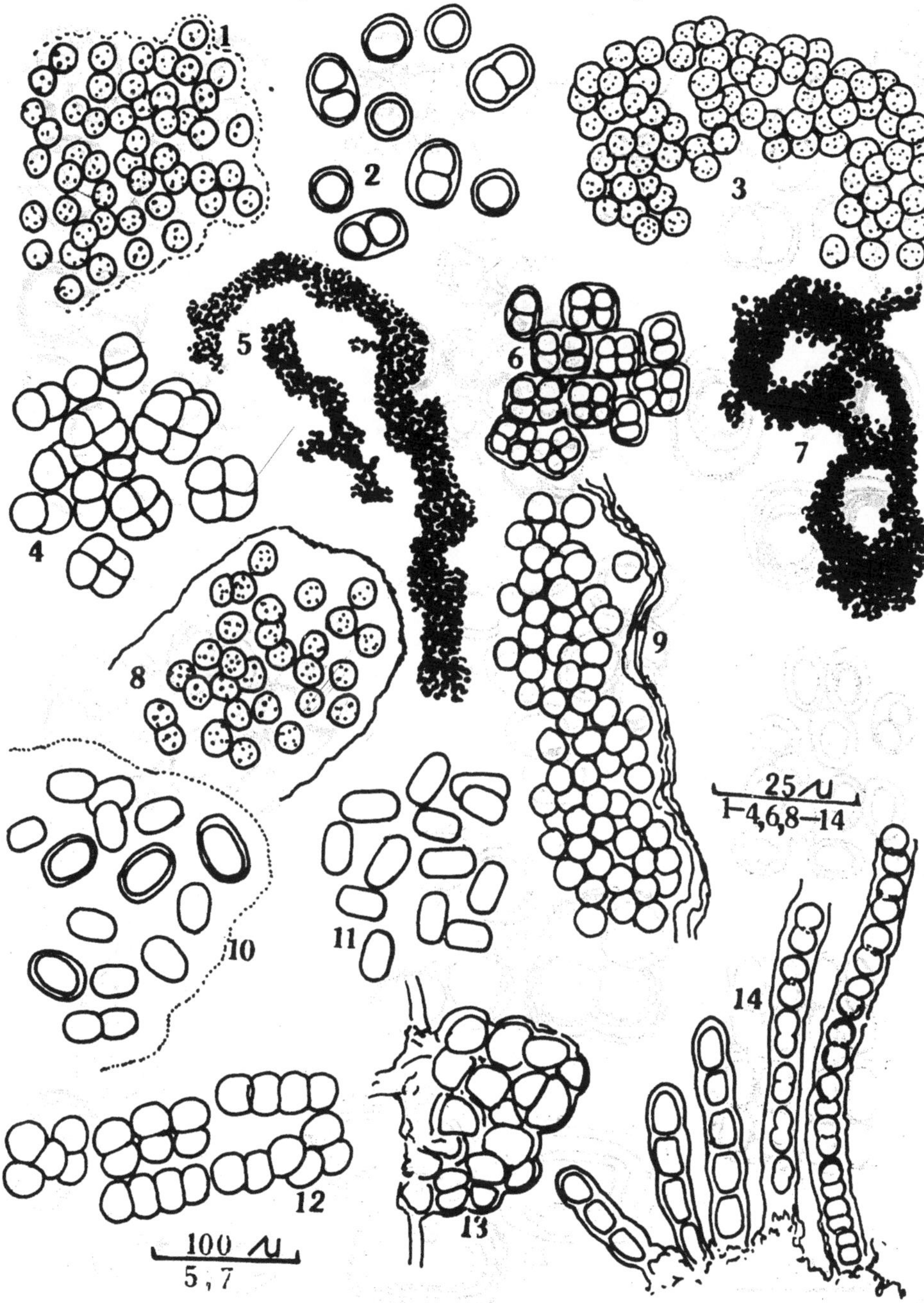

PLATE 2

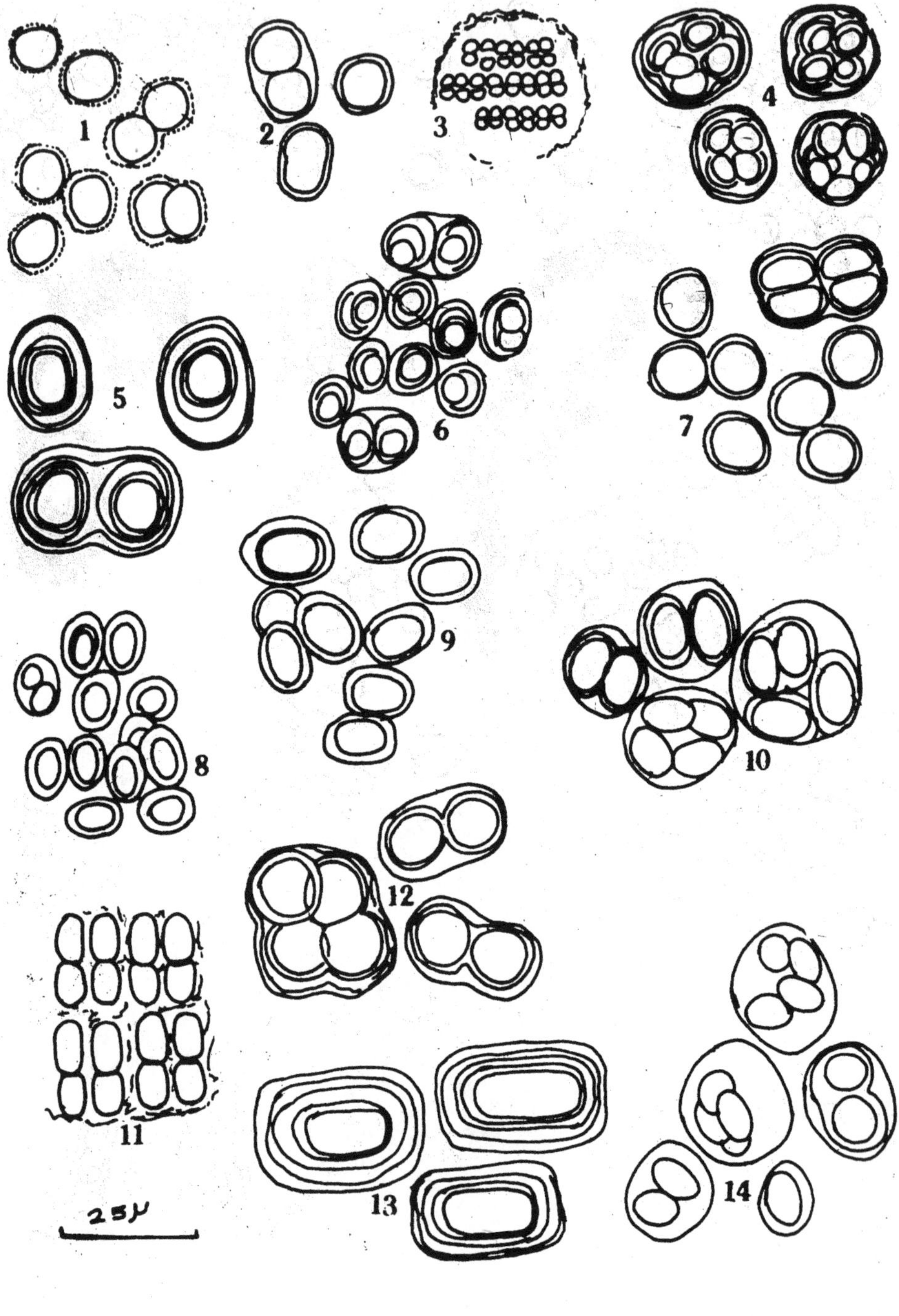

PLATE 3

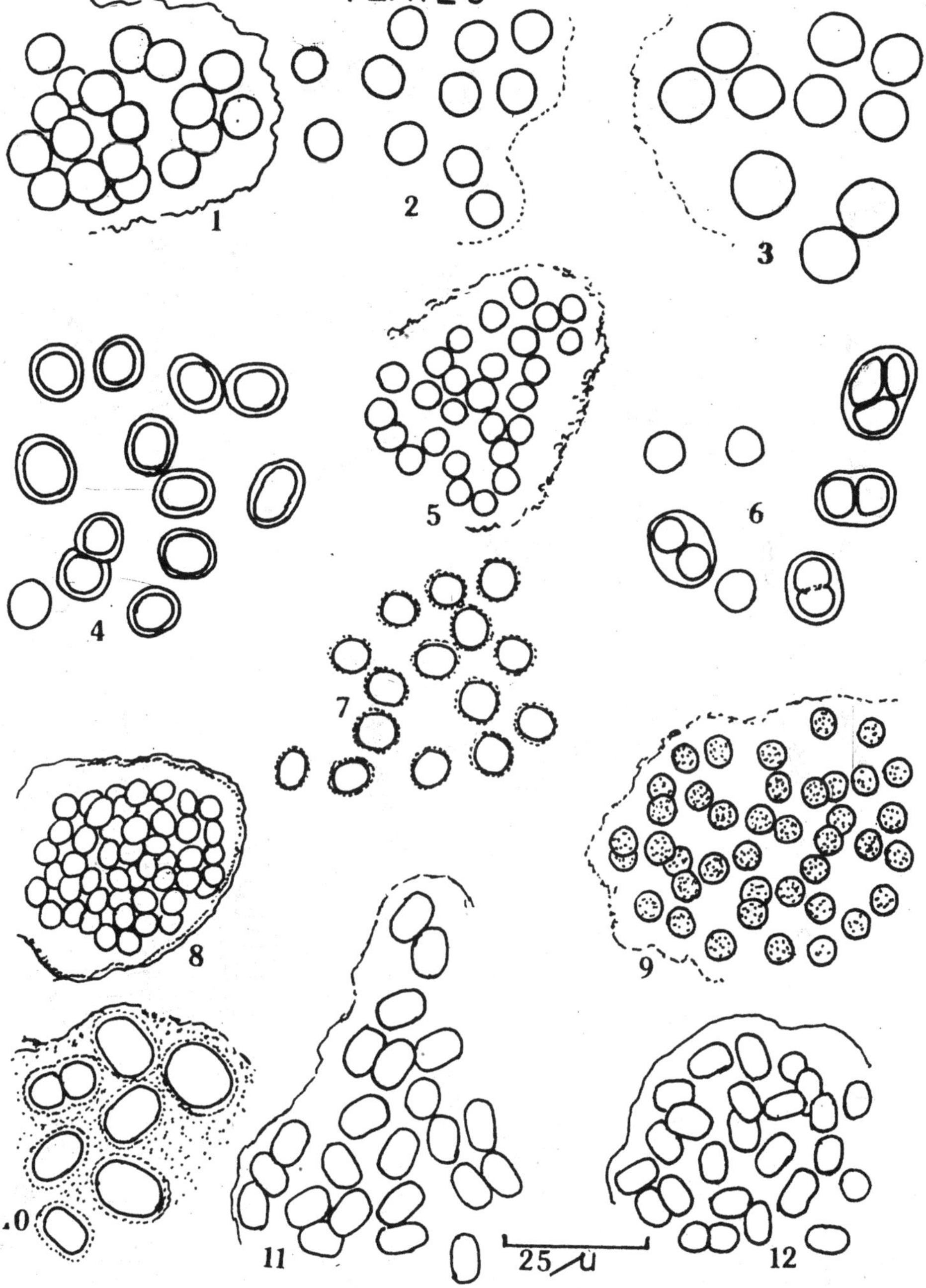

PLATE 4

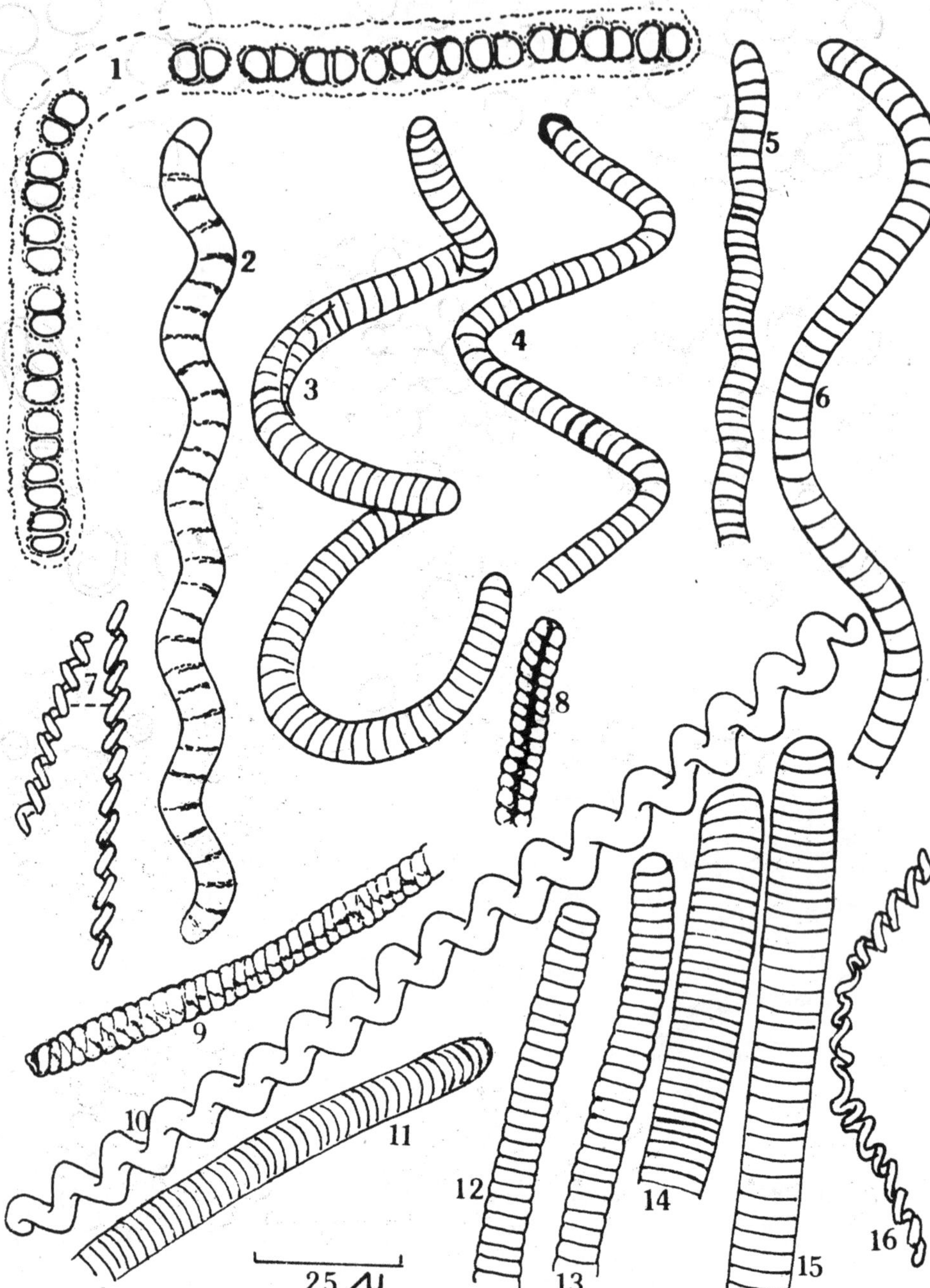

PLATE 5

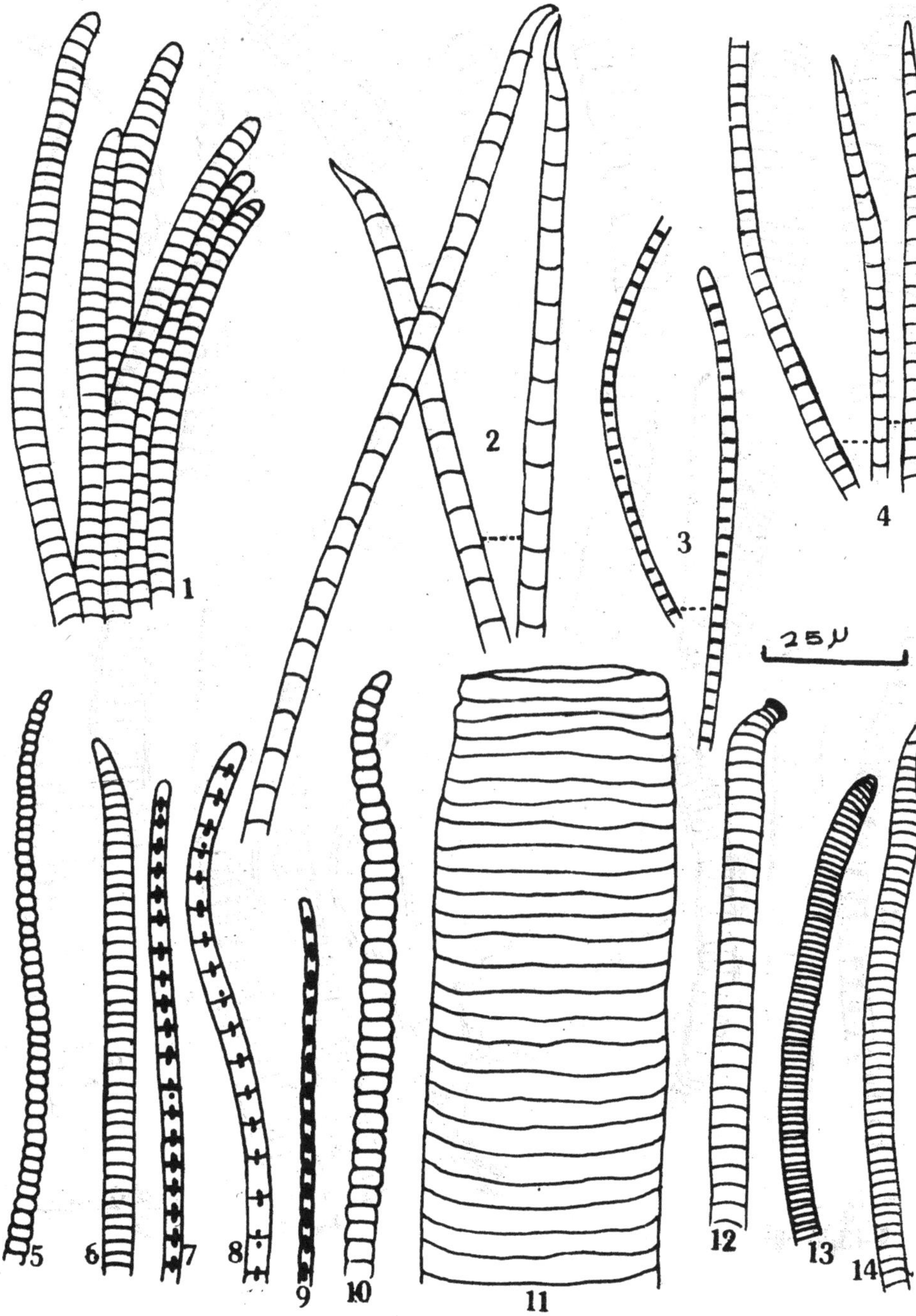

PLATE 6

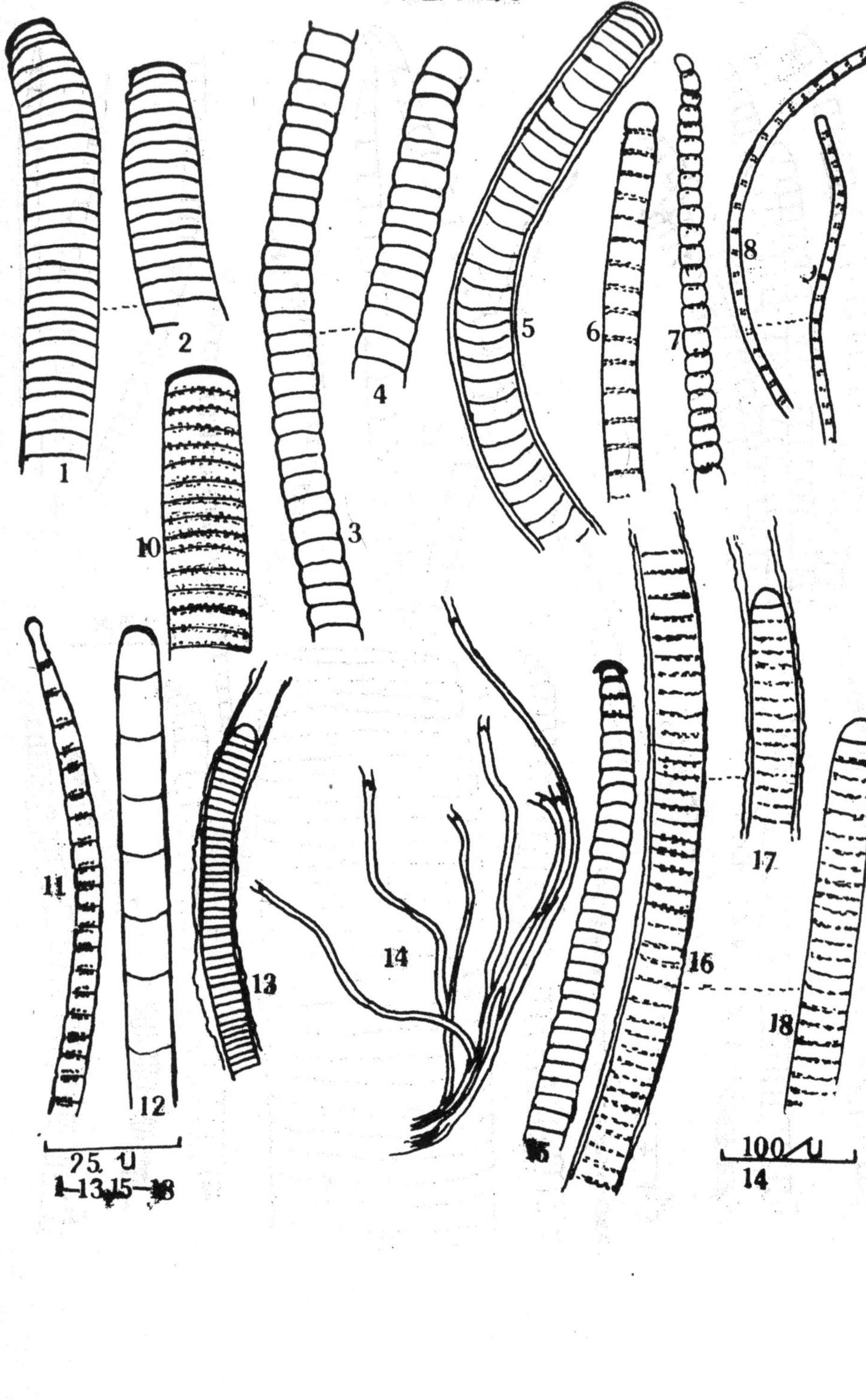

PLATE 7

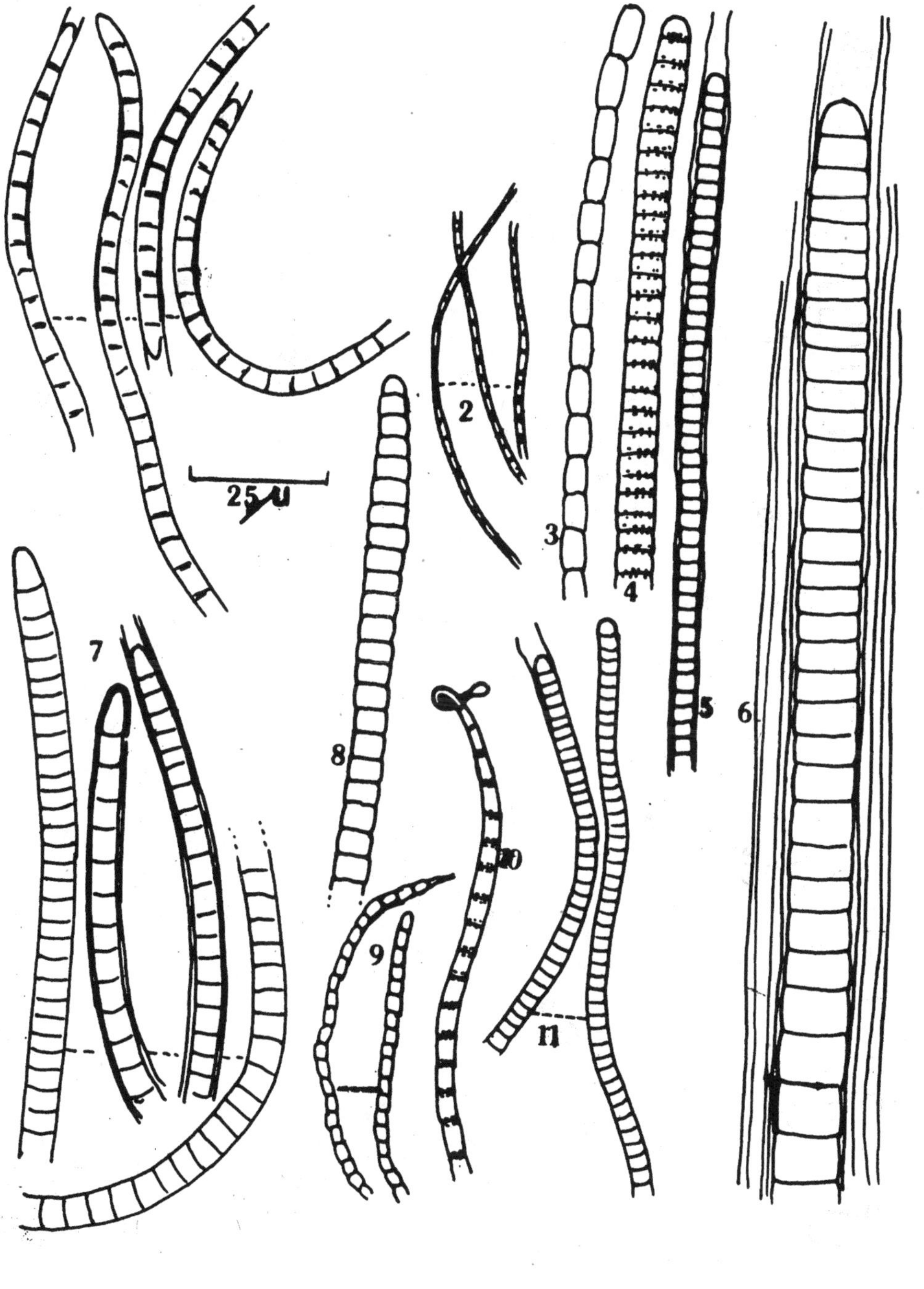

PLATE 8

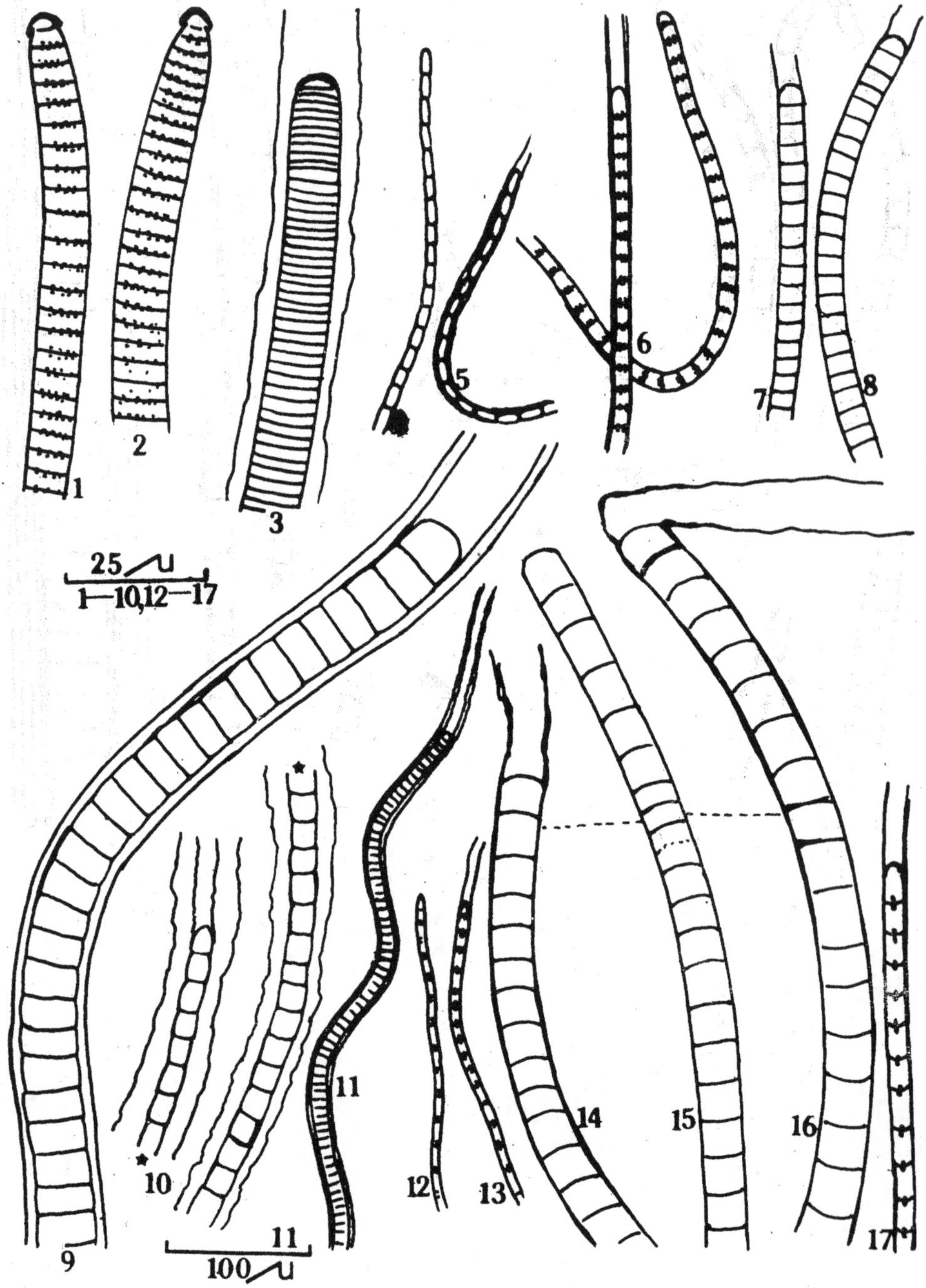

PLATE 9

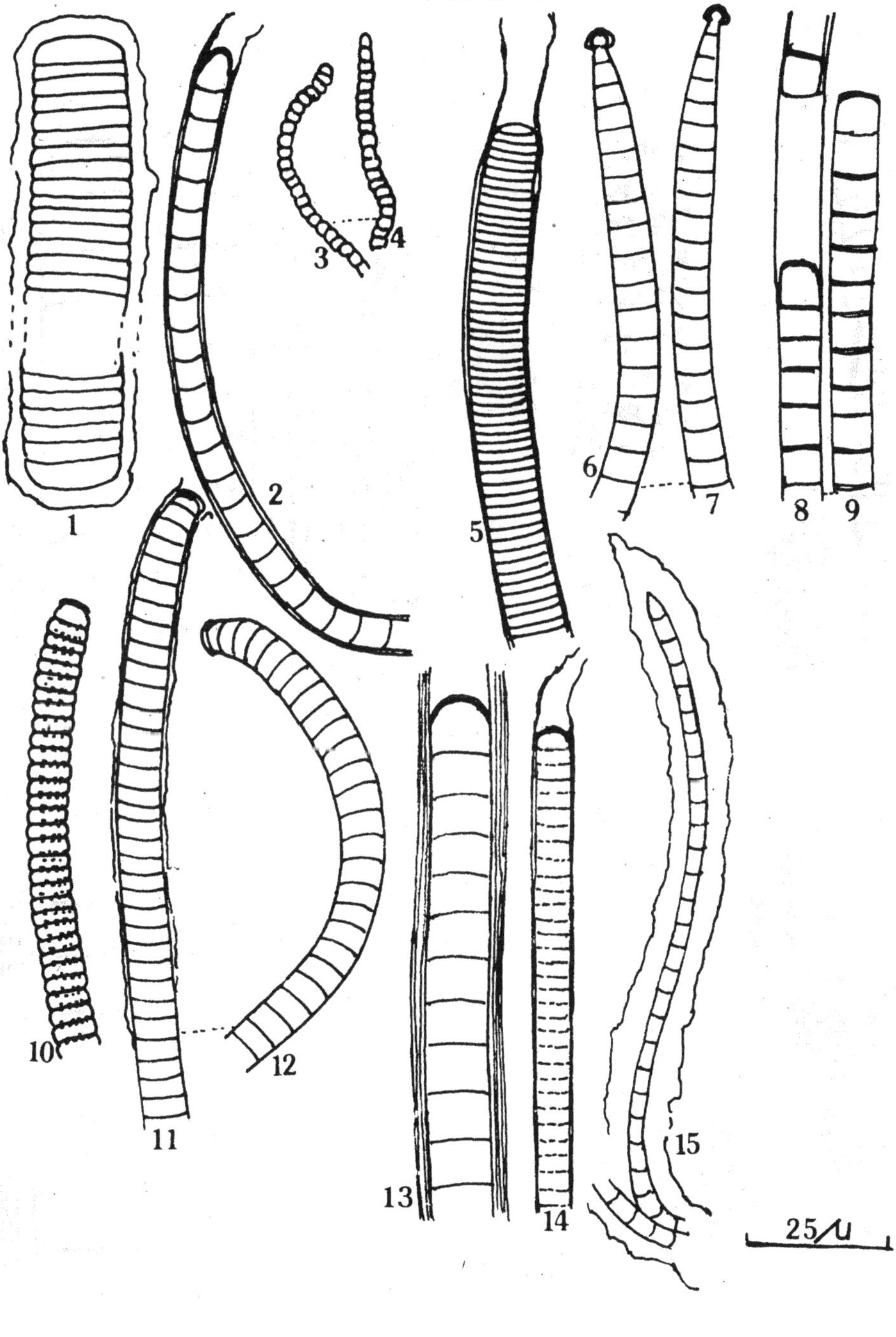

PLATE 10

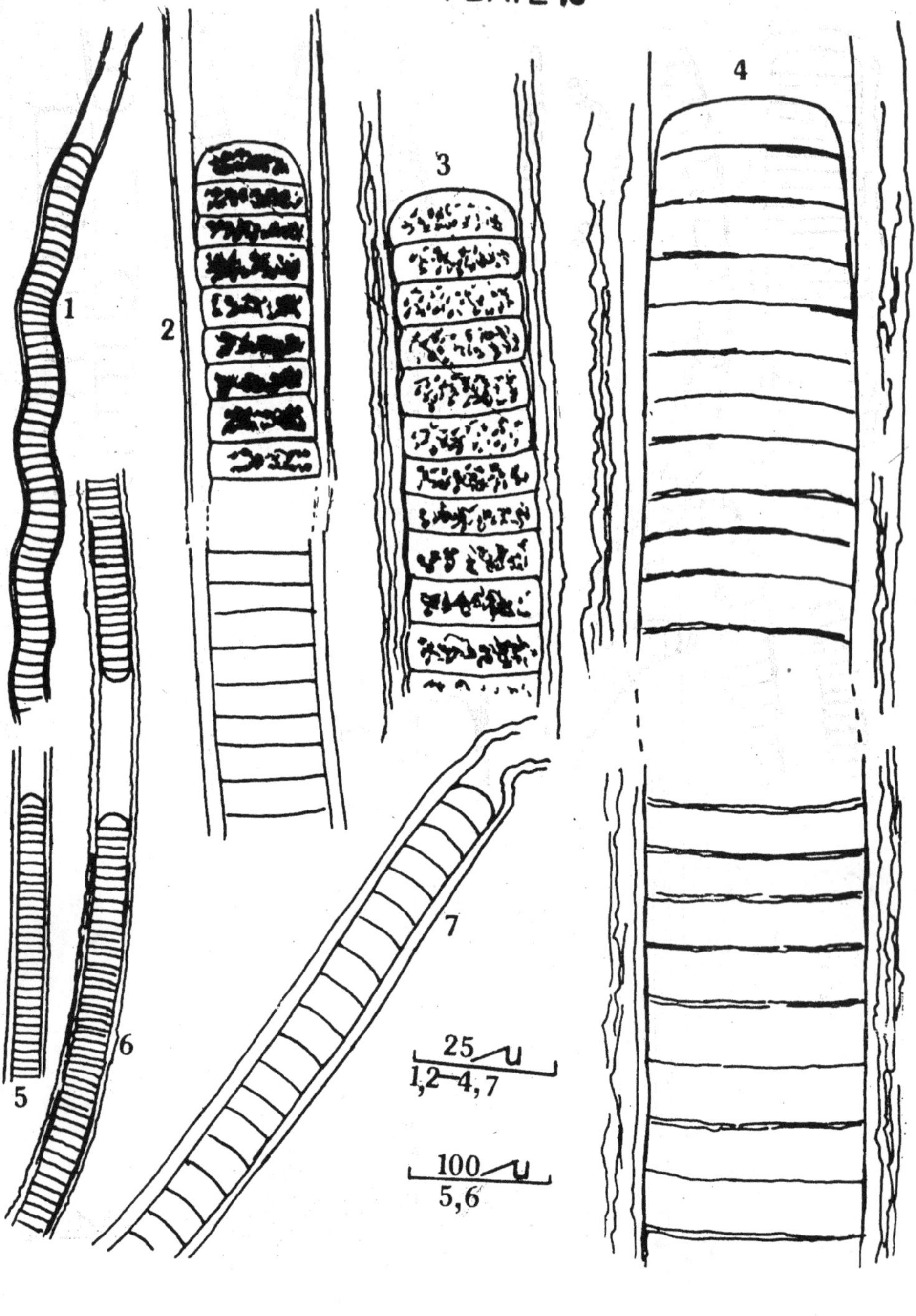

LATE11

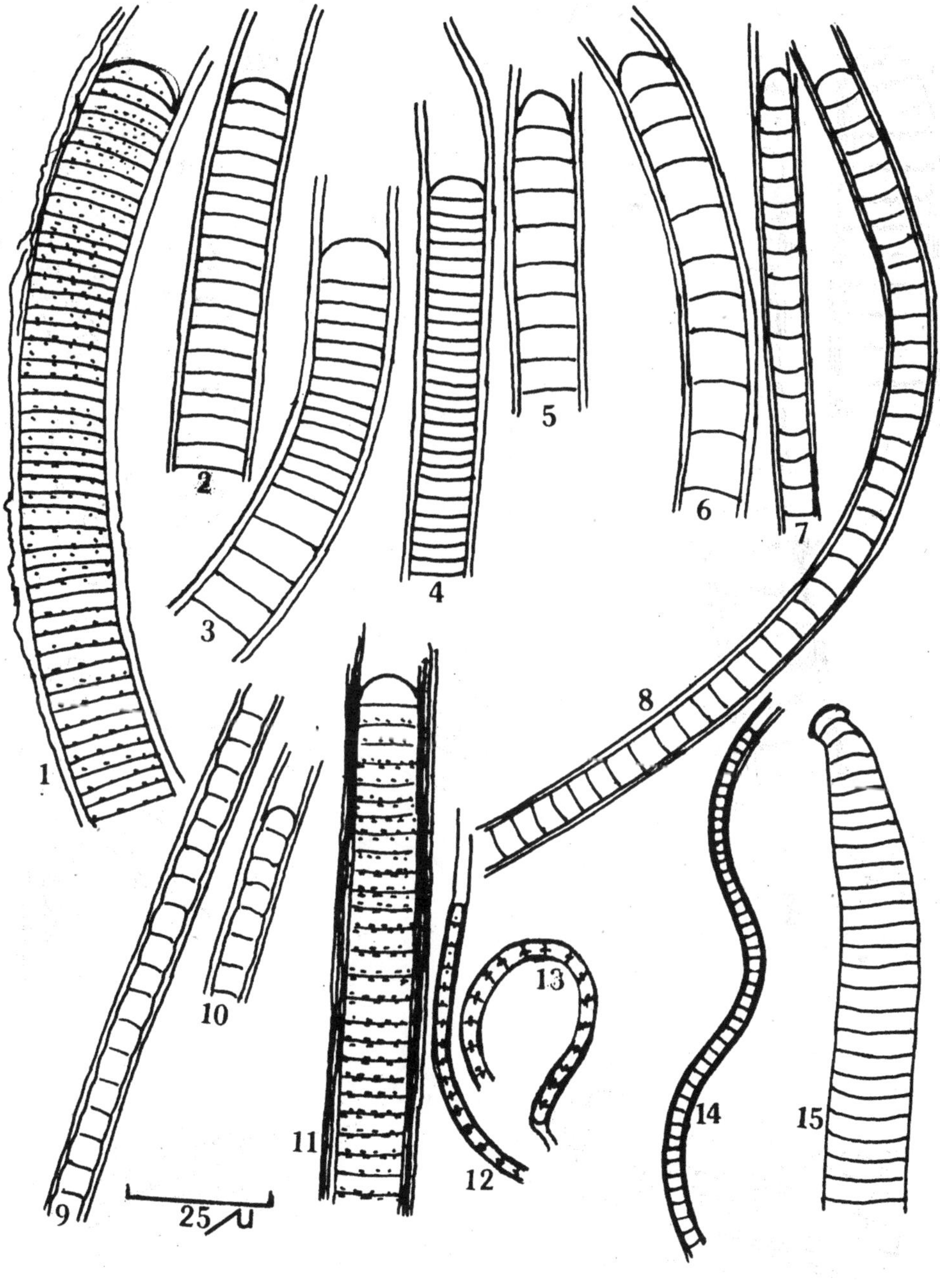

PLATE 12

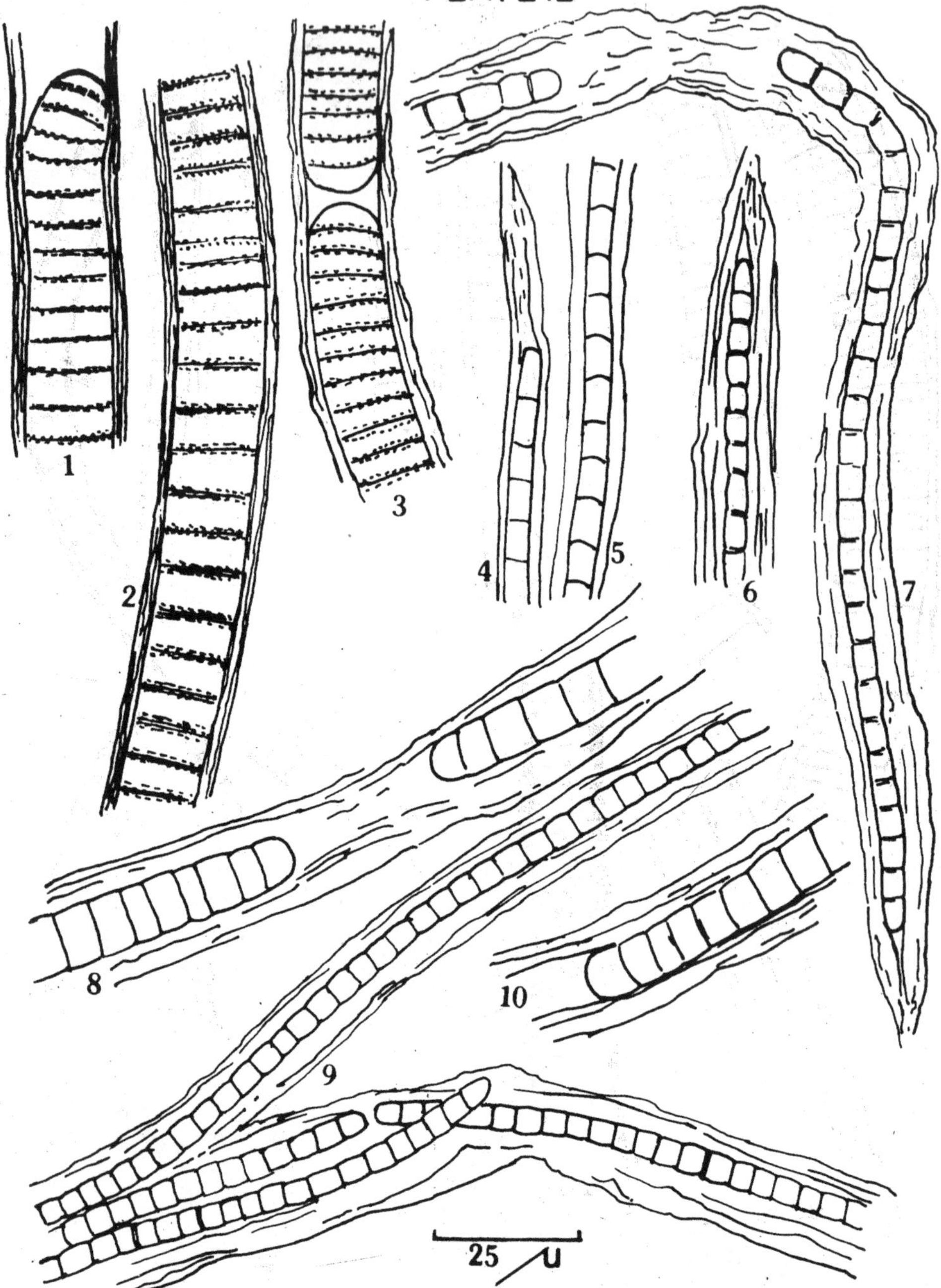

PLATE13

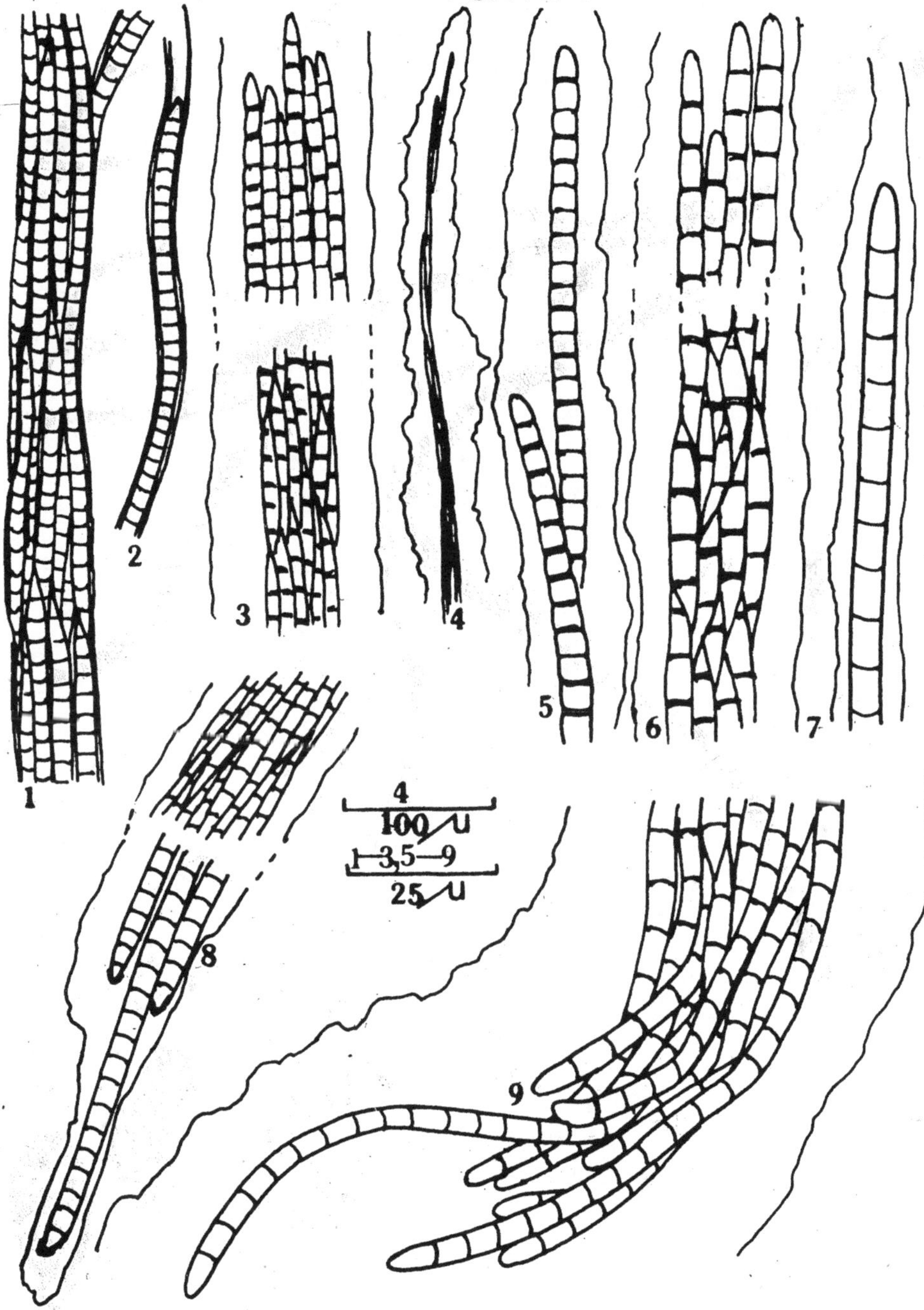

PLATE 14

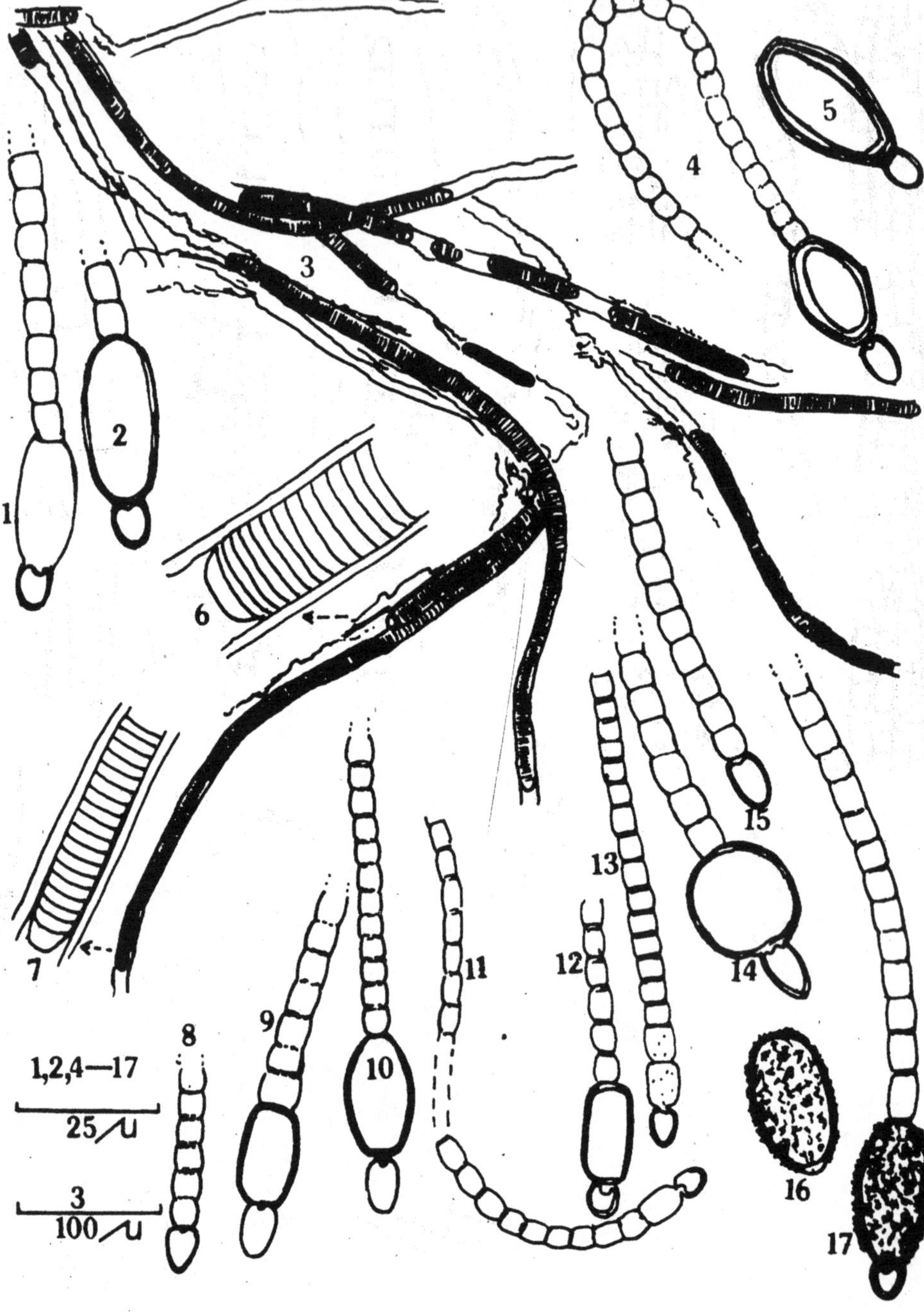

PLATE15

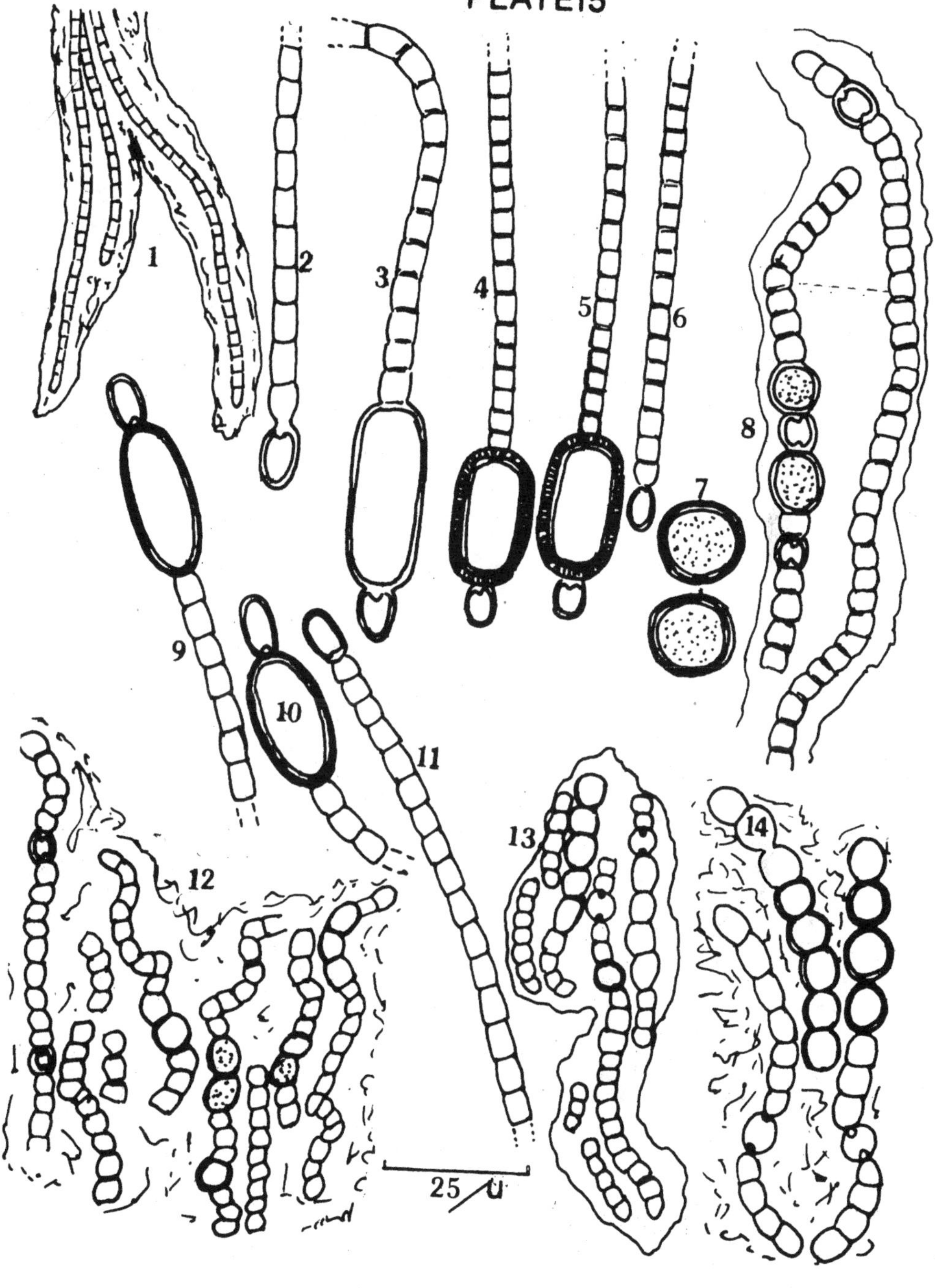

PLATE16

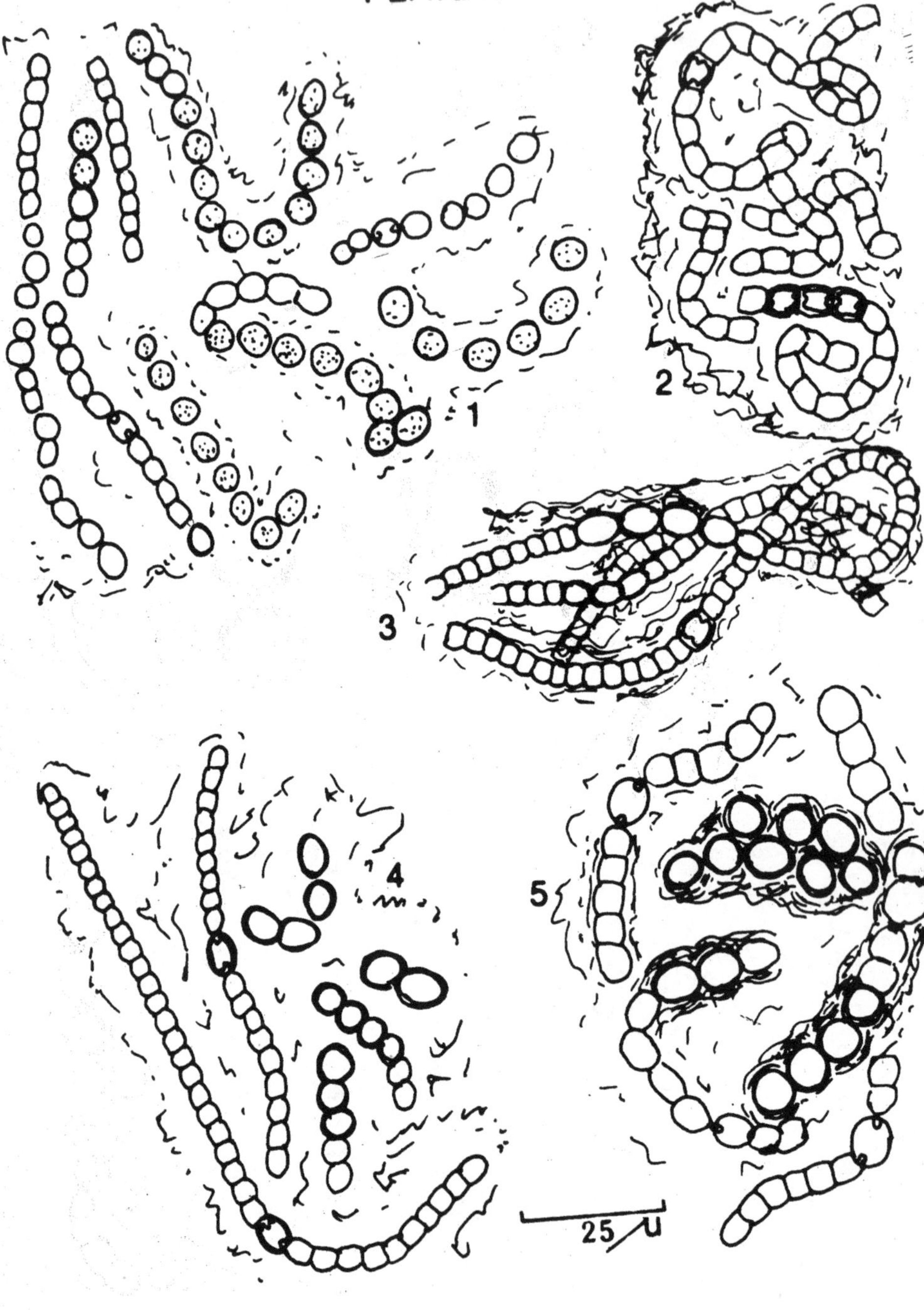

PLATE17

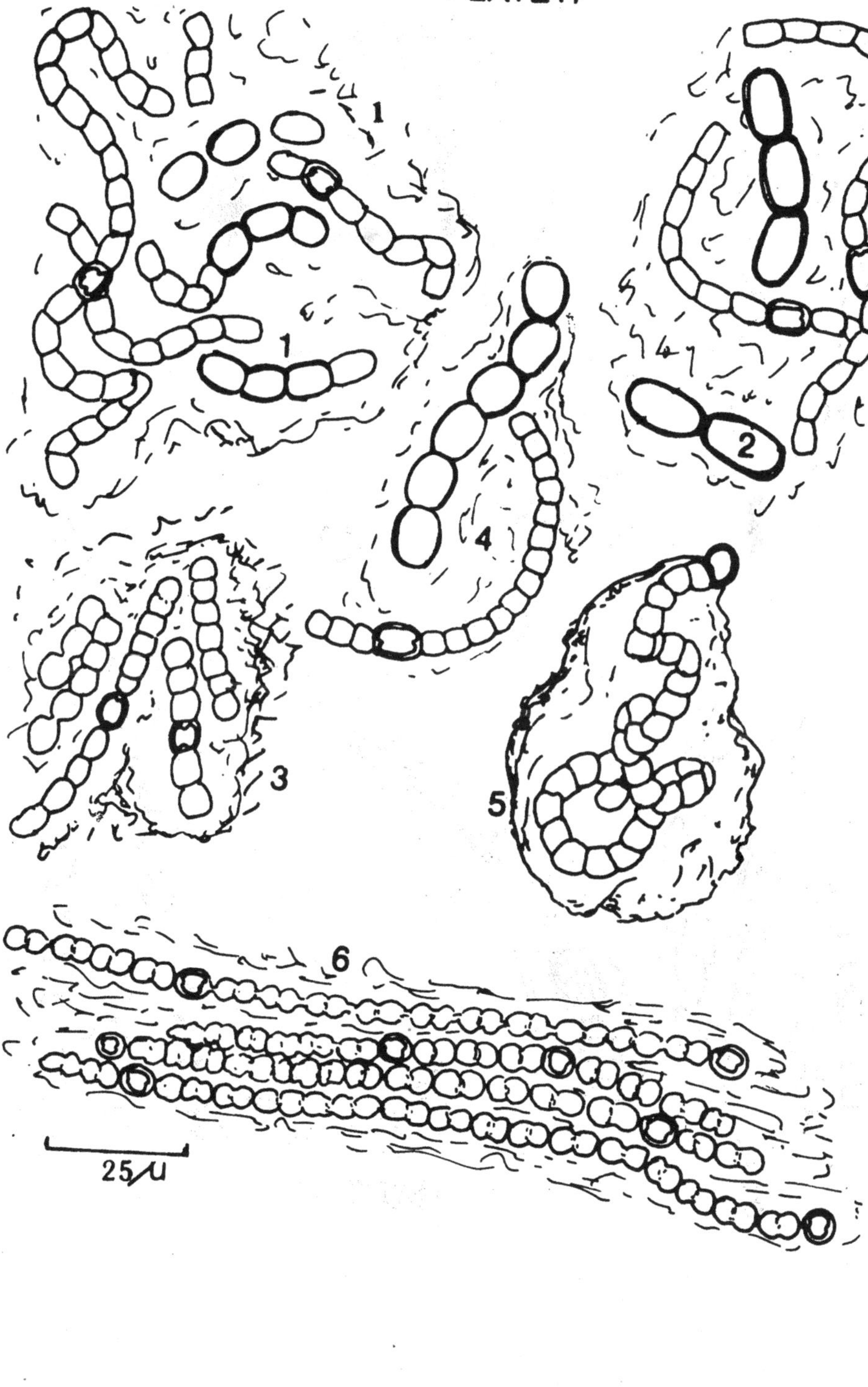

PLATE 18

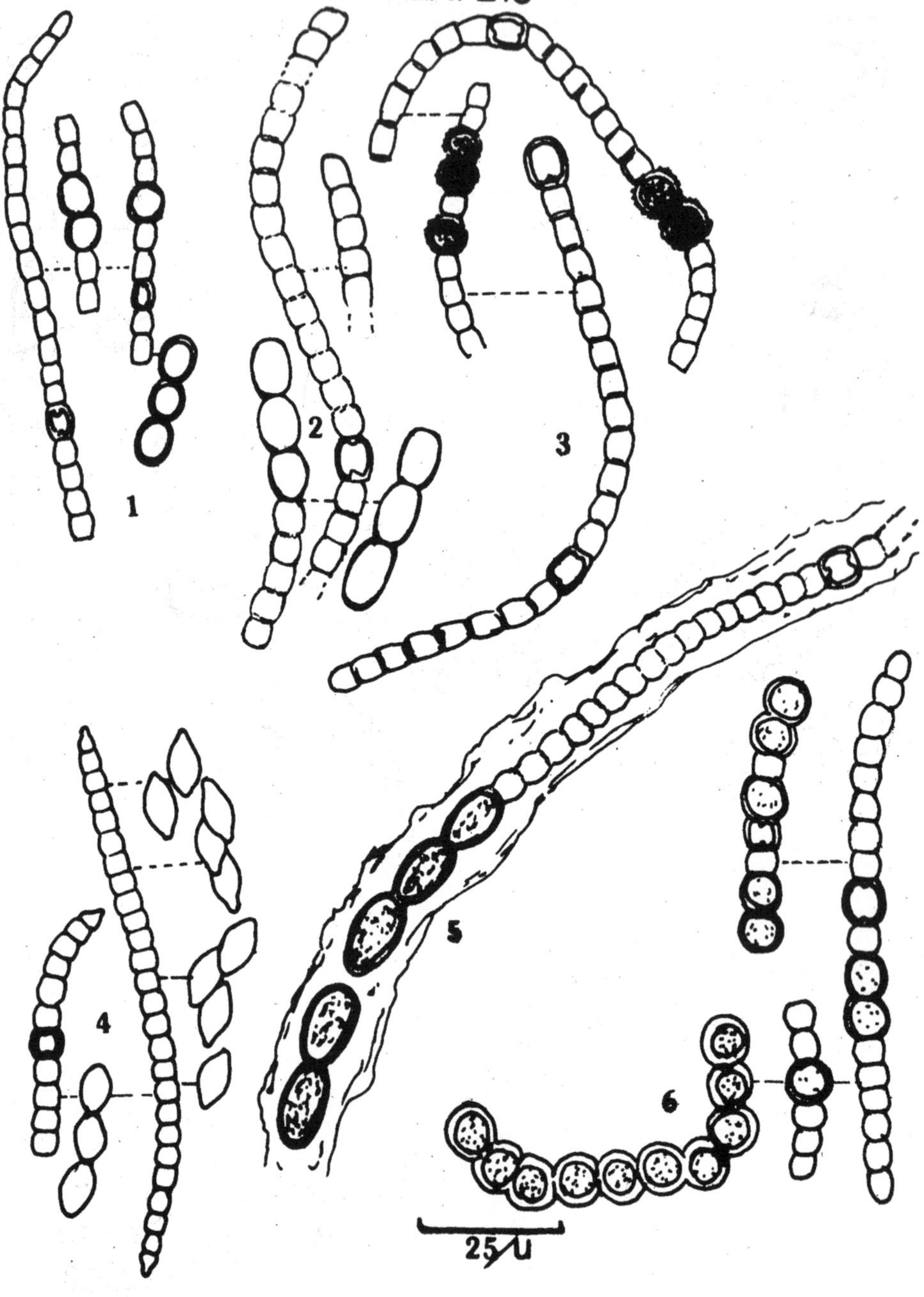

PLATE 19

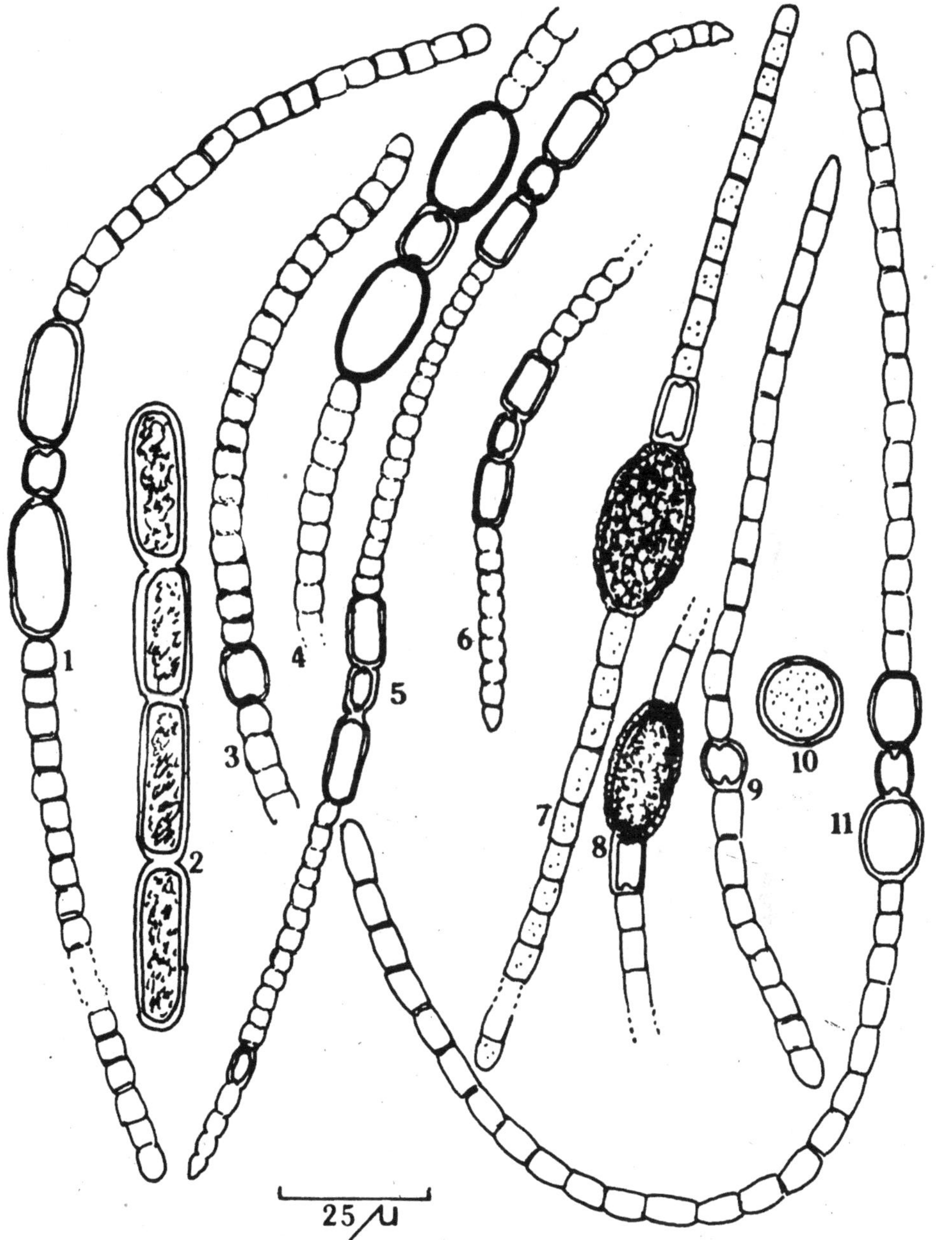

PLATE 20

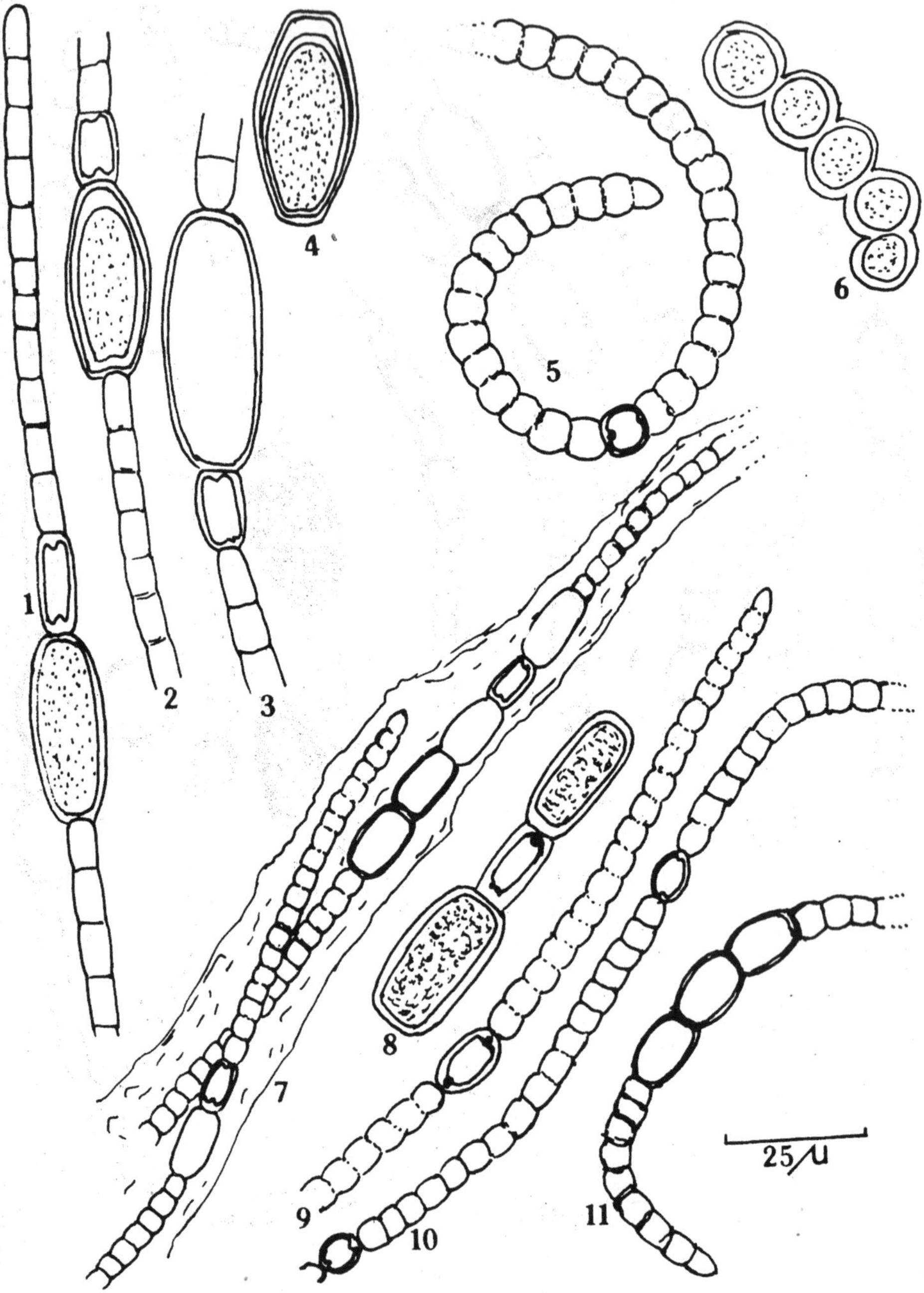

PLATE 21

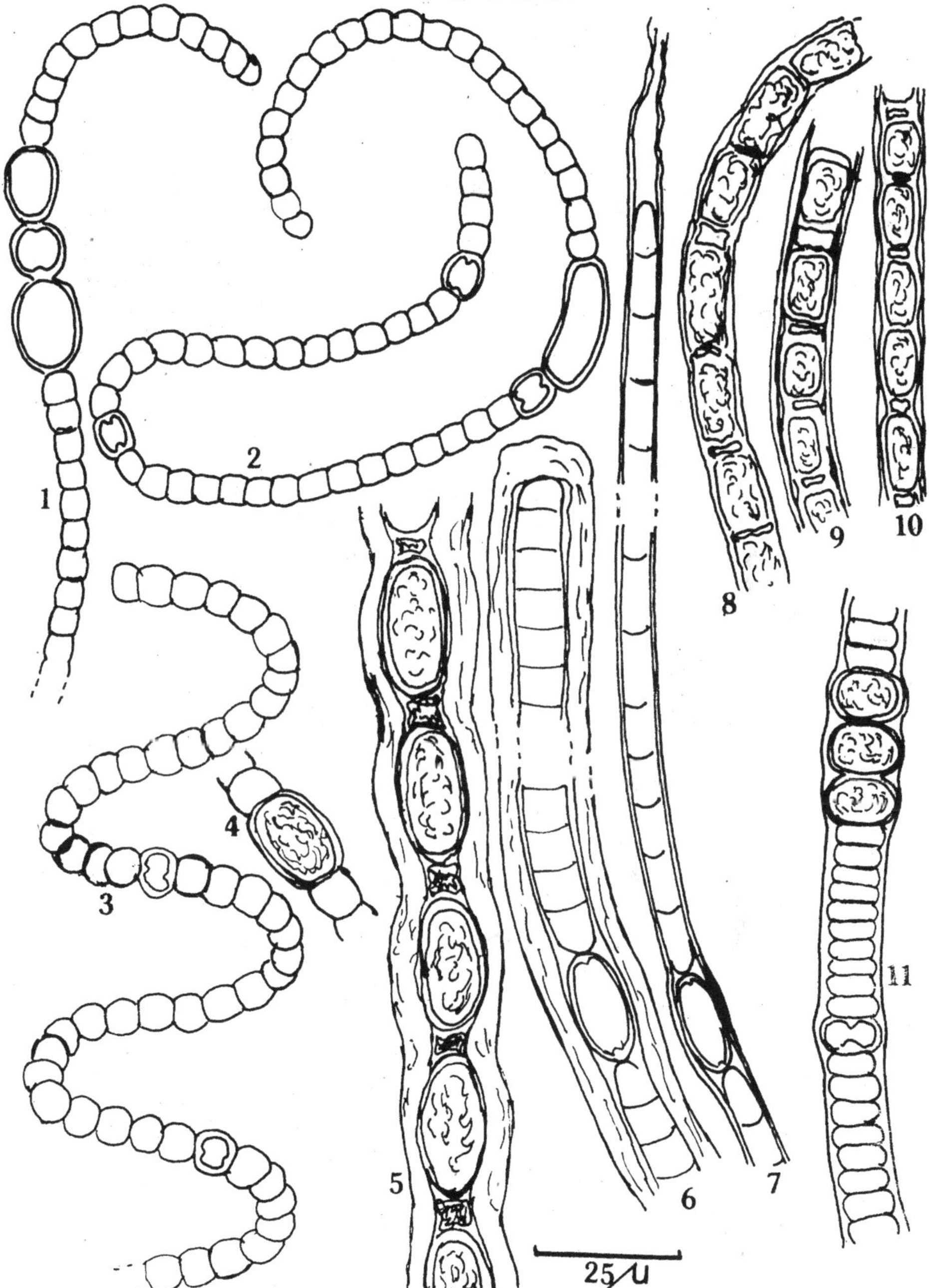

PLATE 22

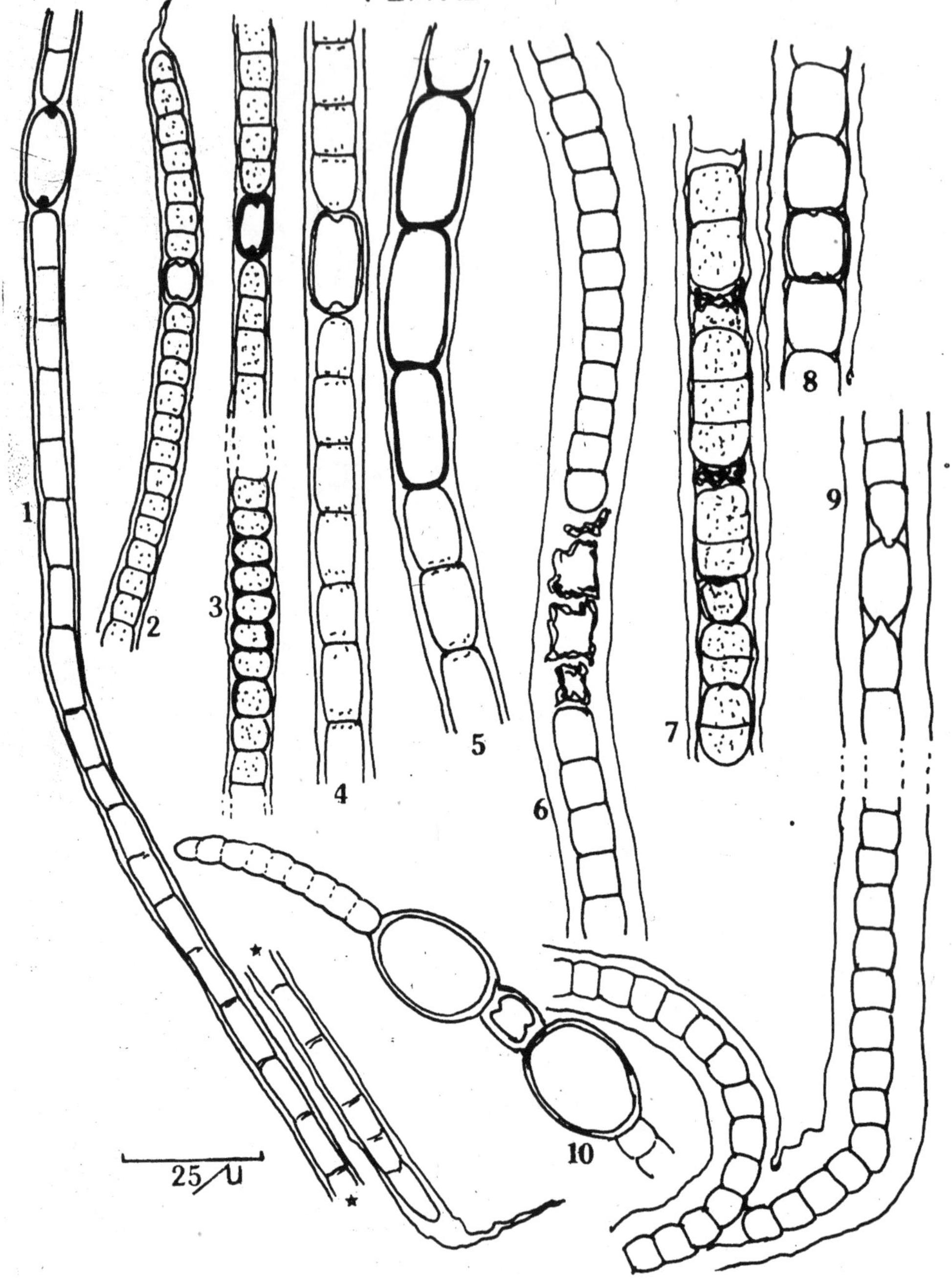

PLATE 23

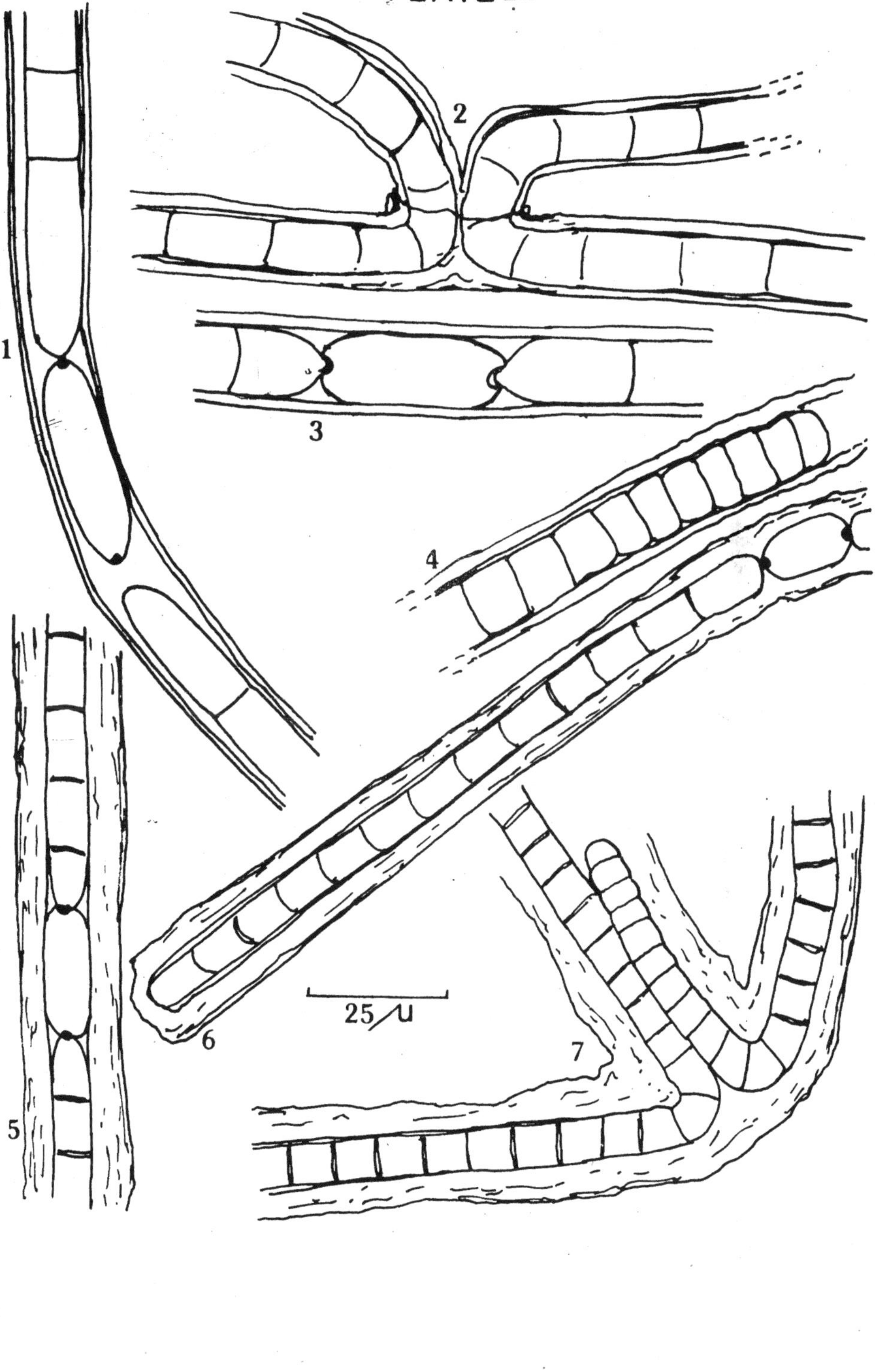

PLATE24

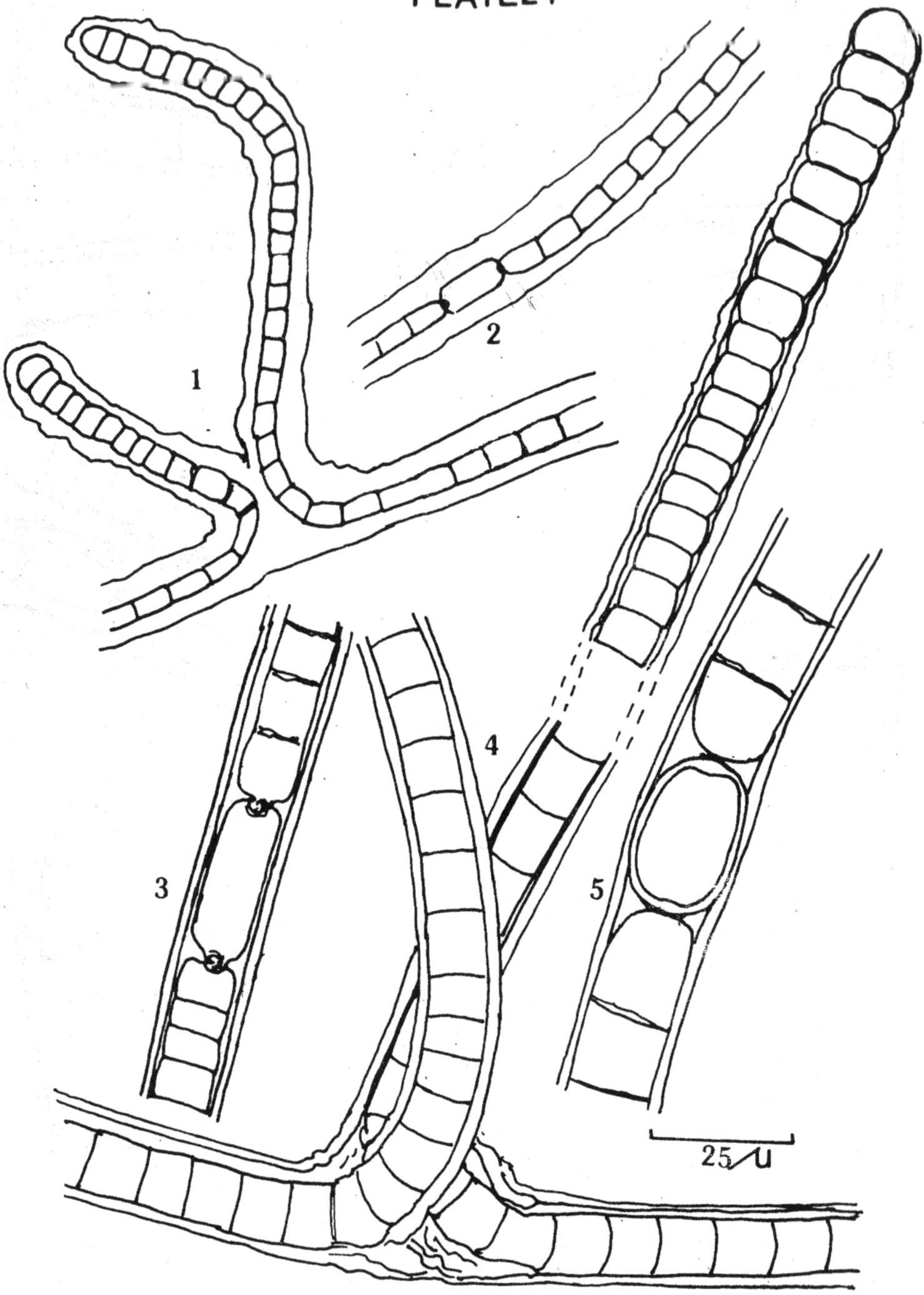

PLATE 25

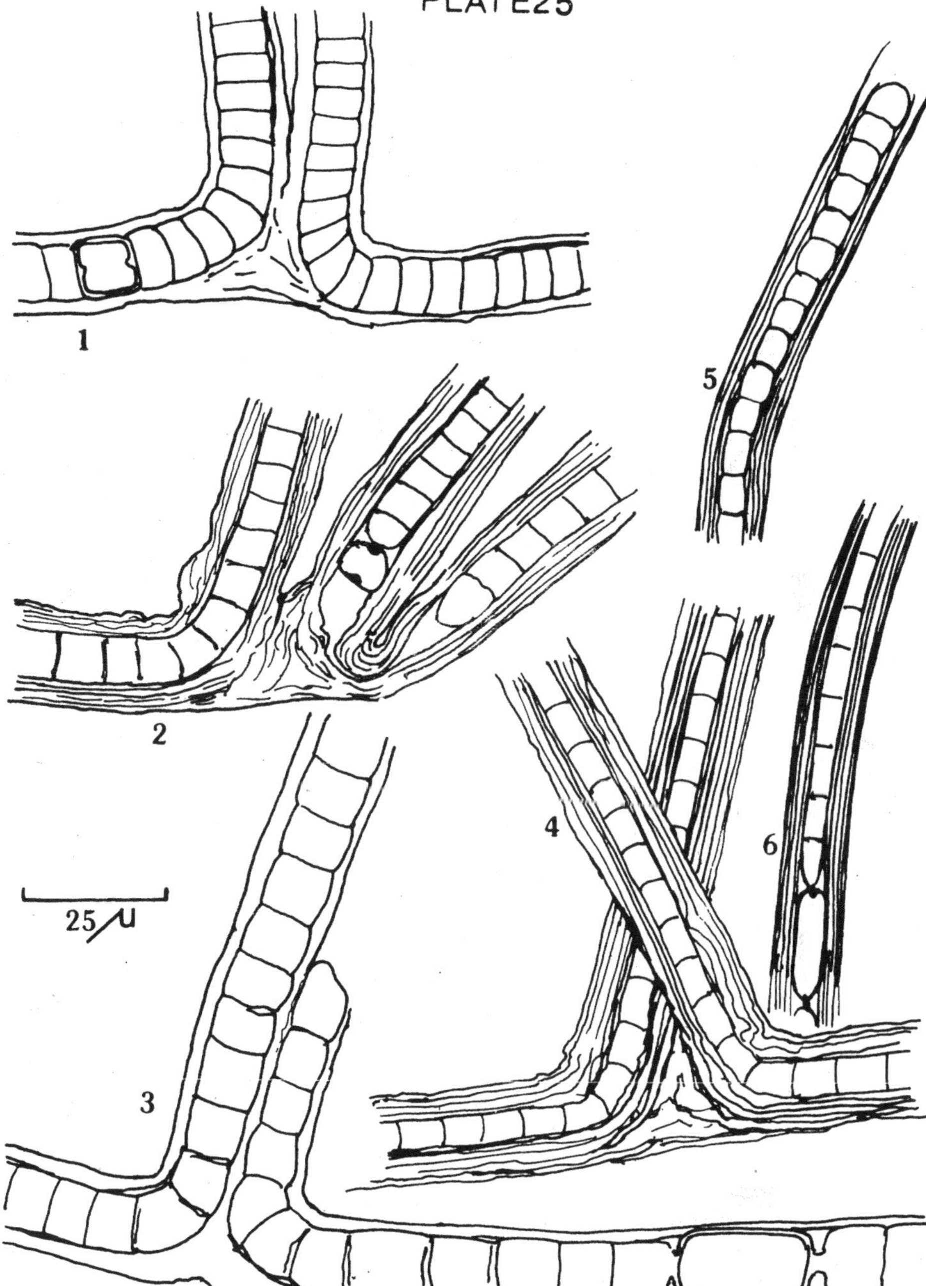

PLATE 26

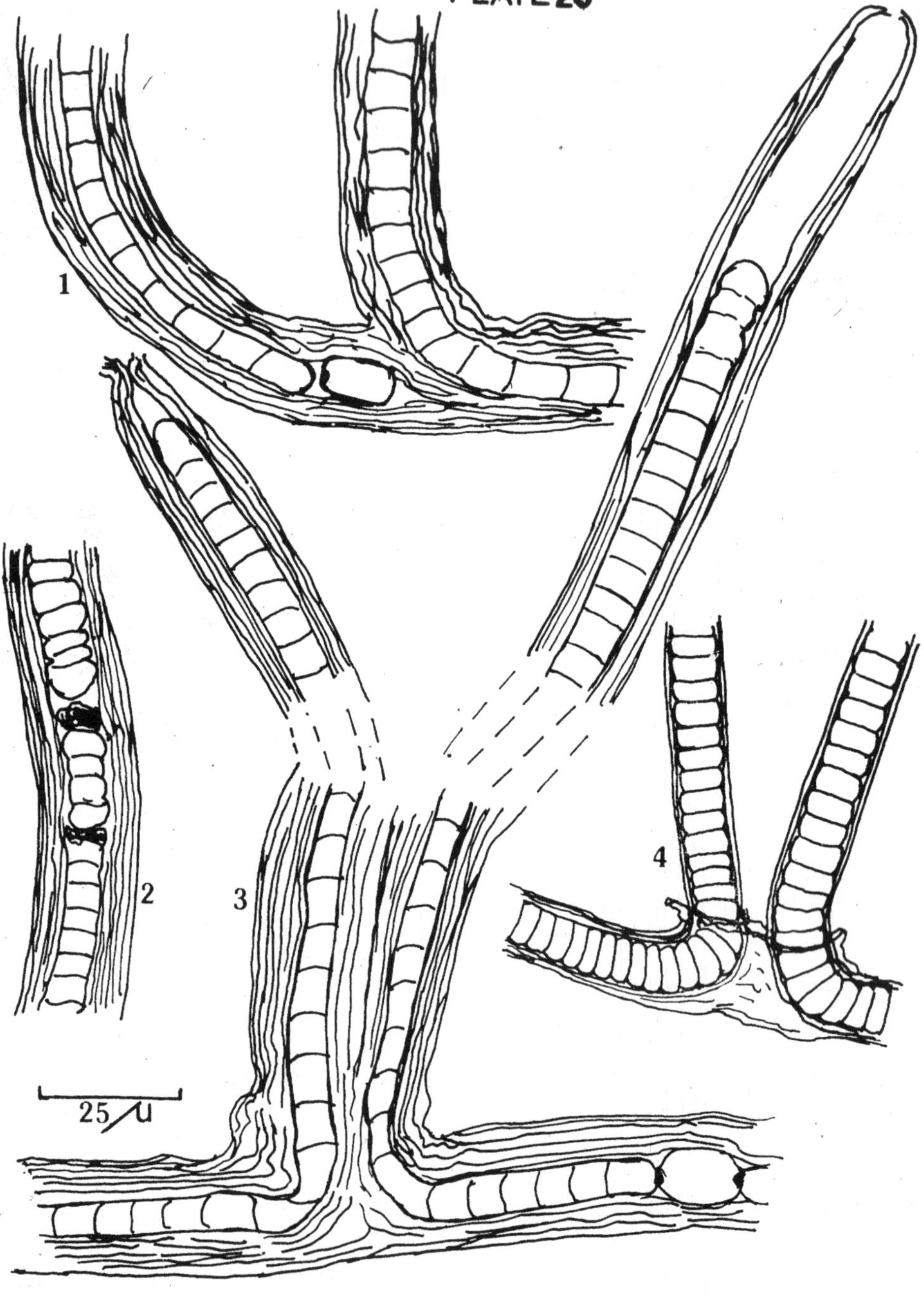

PLATE 27

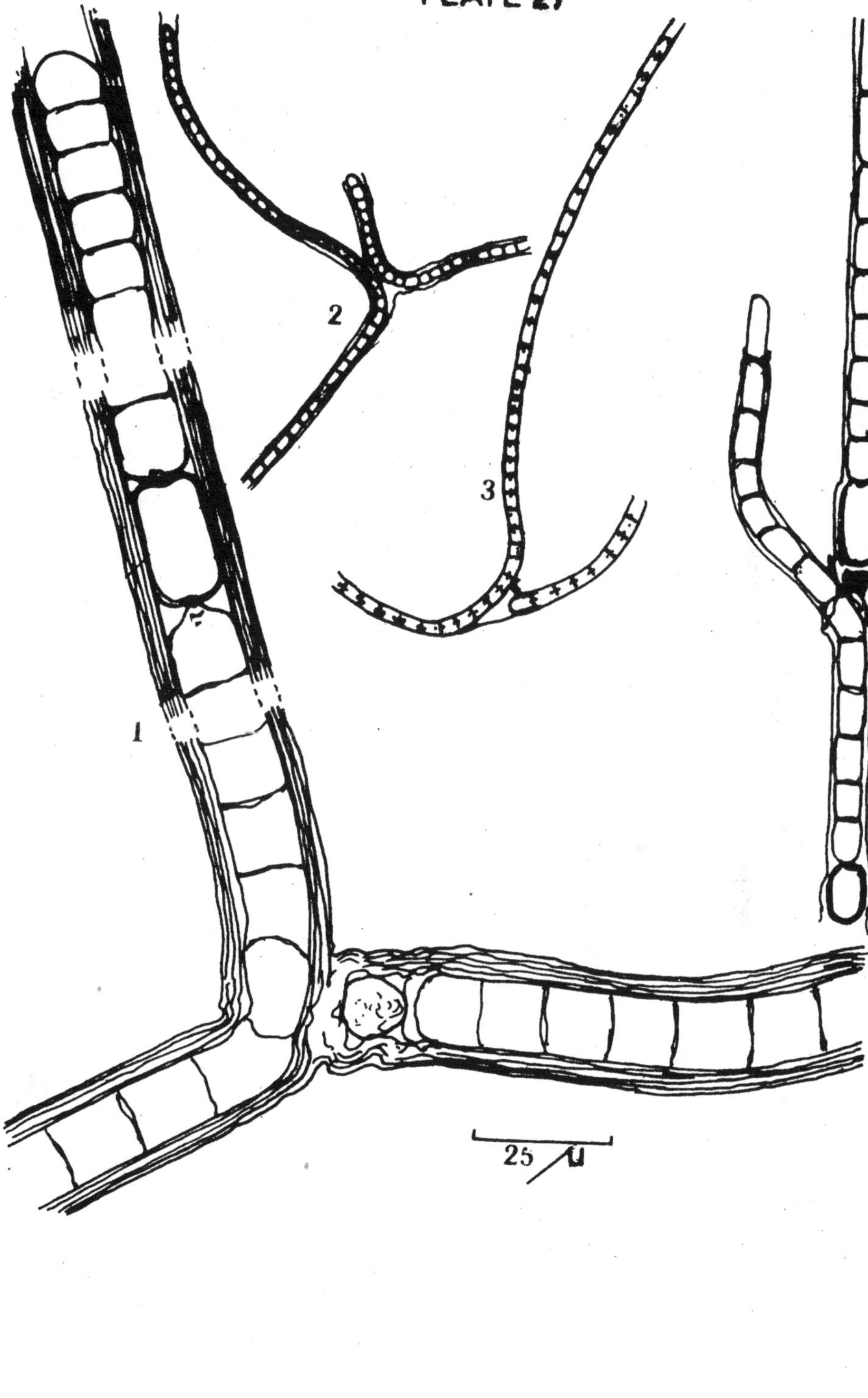

PLATE 28

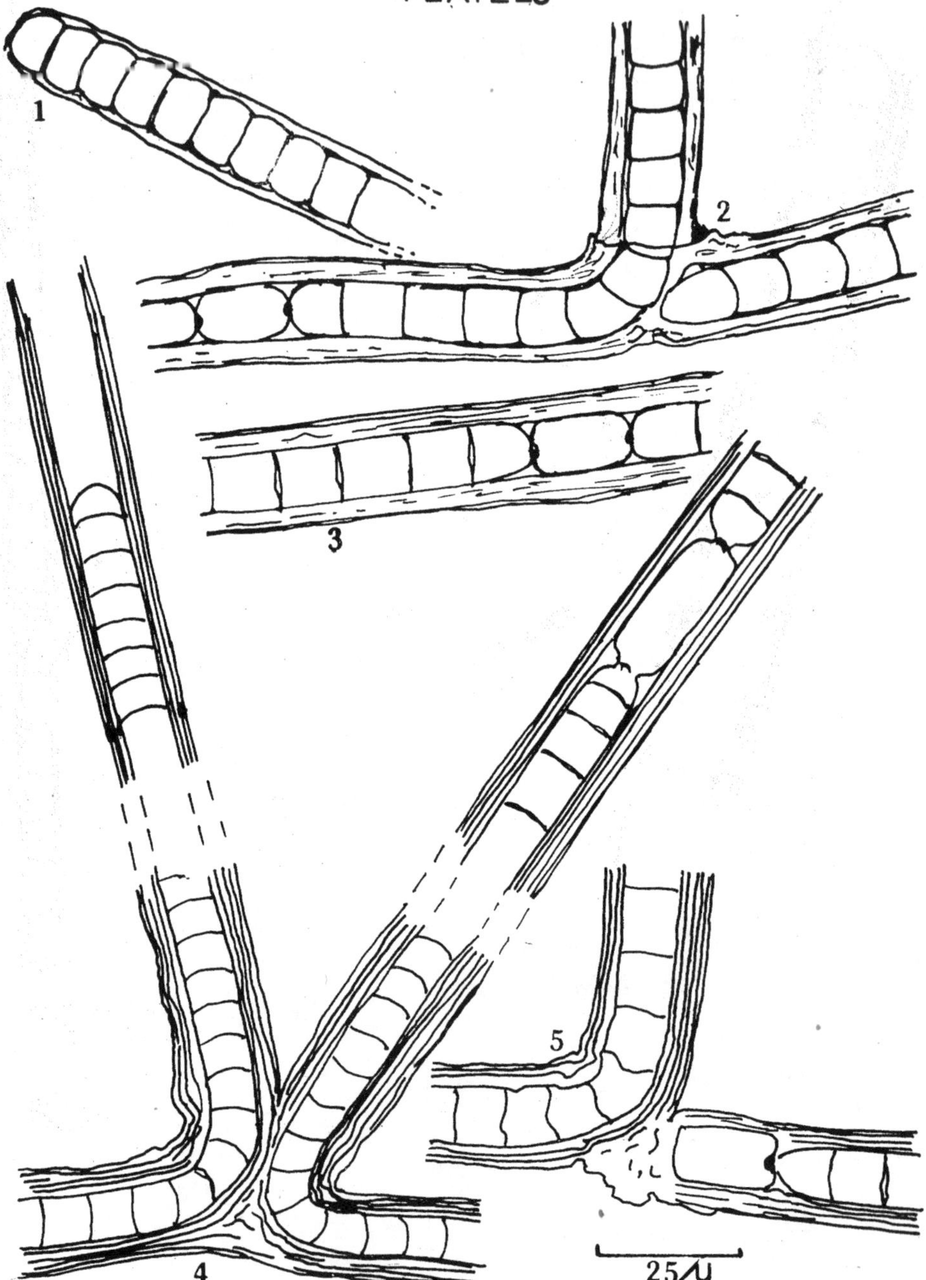

PLATE 29

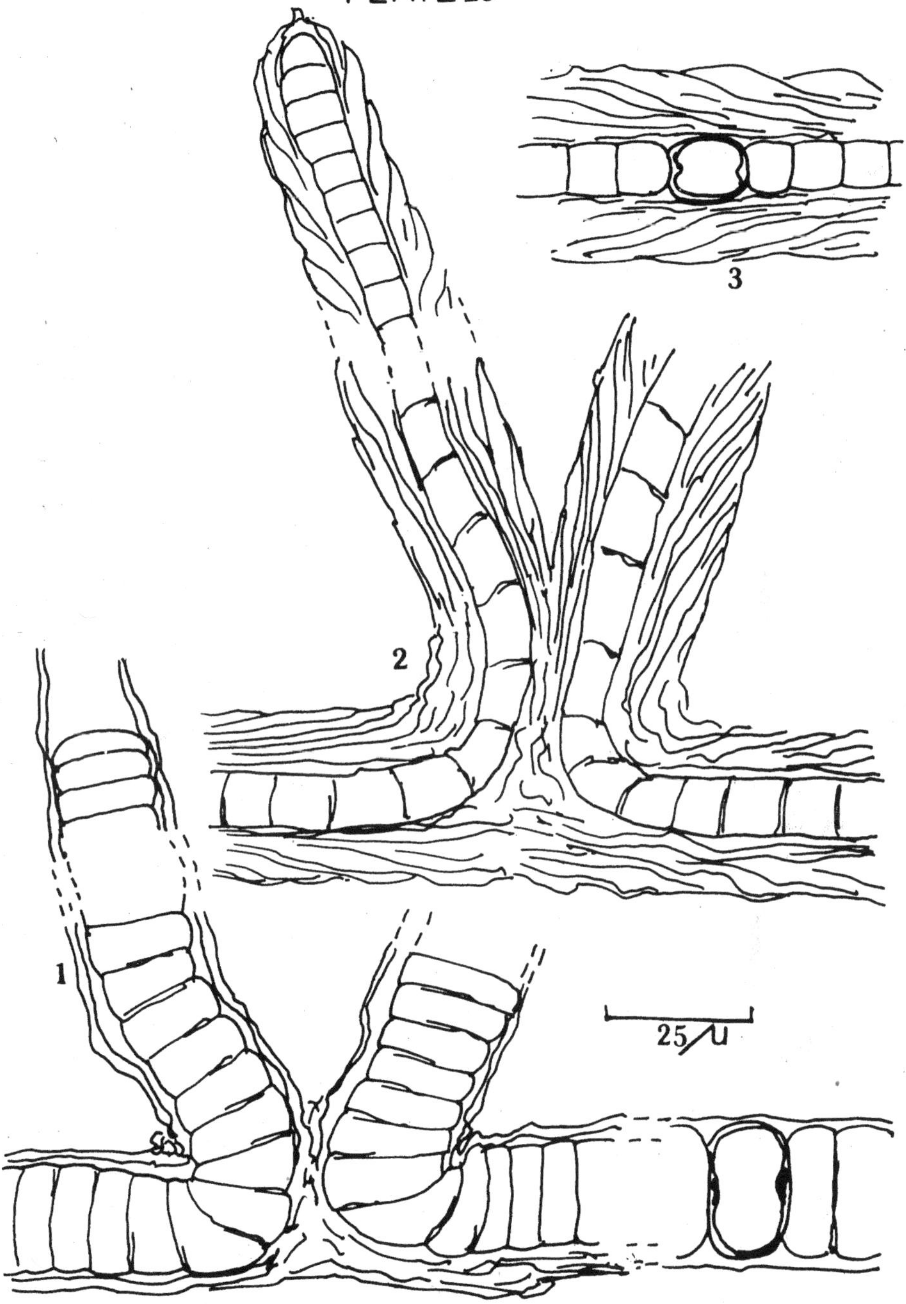

PLATE30

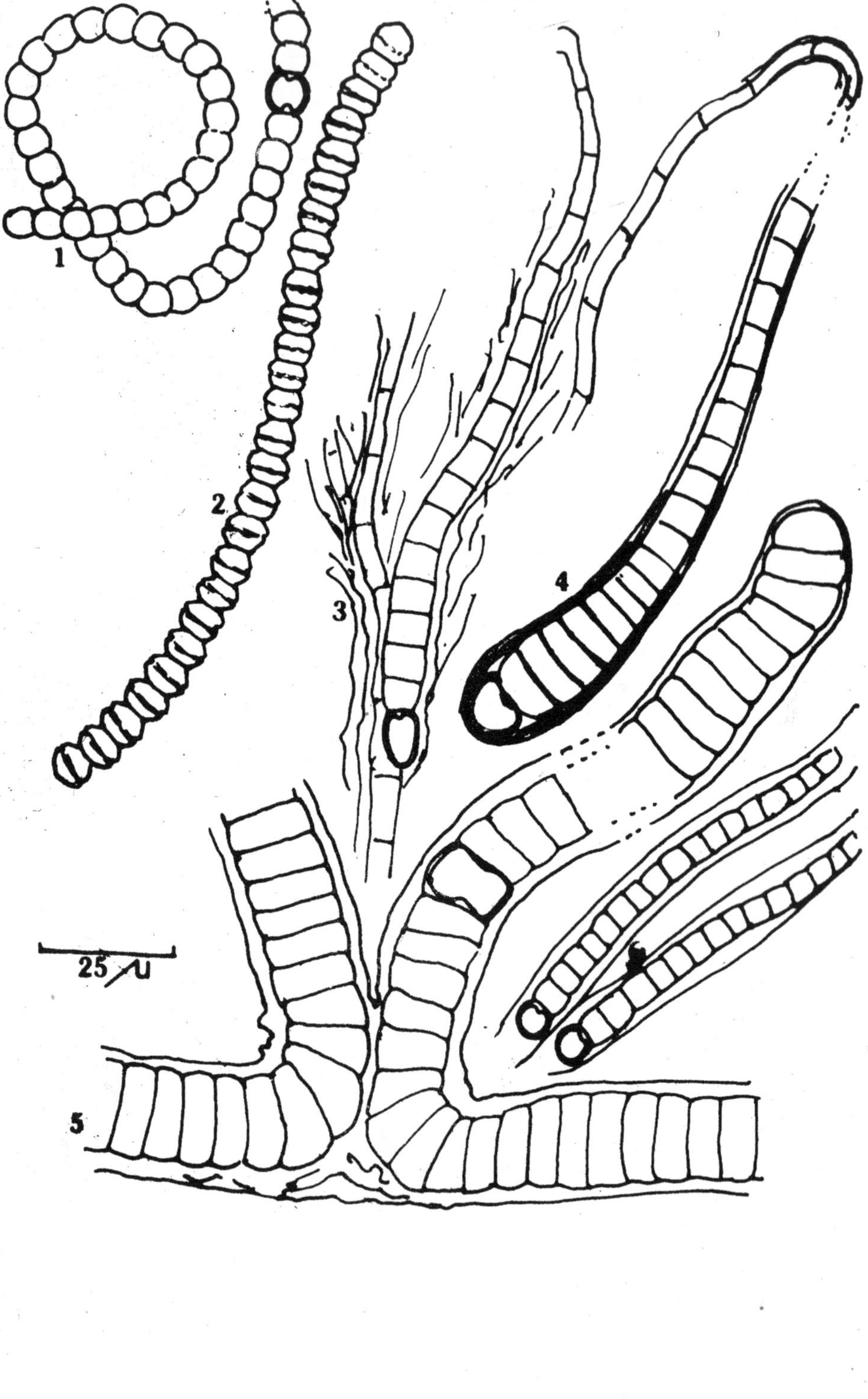

PLATE 31

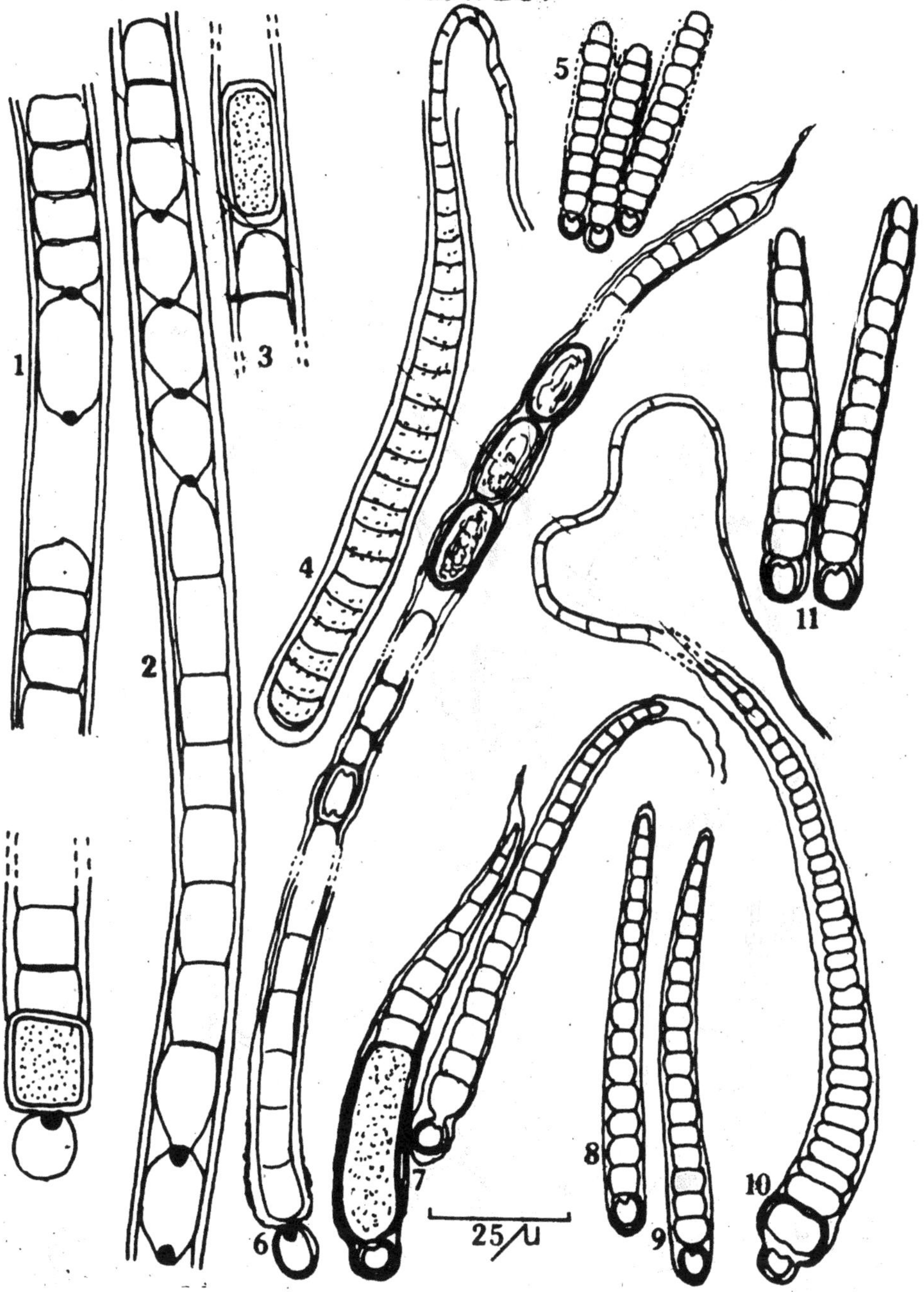

PLATE 32

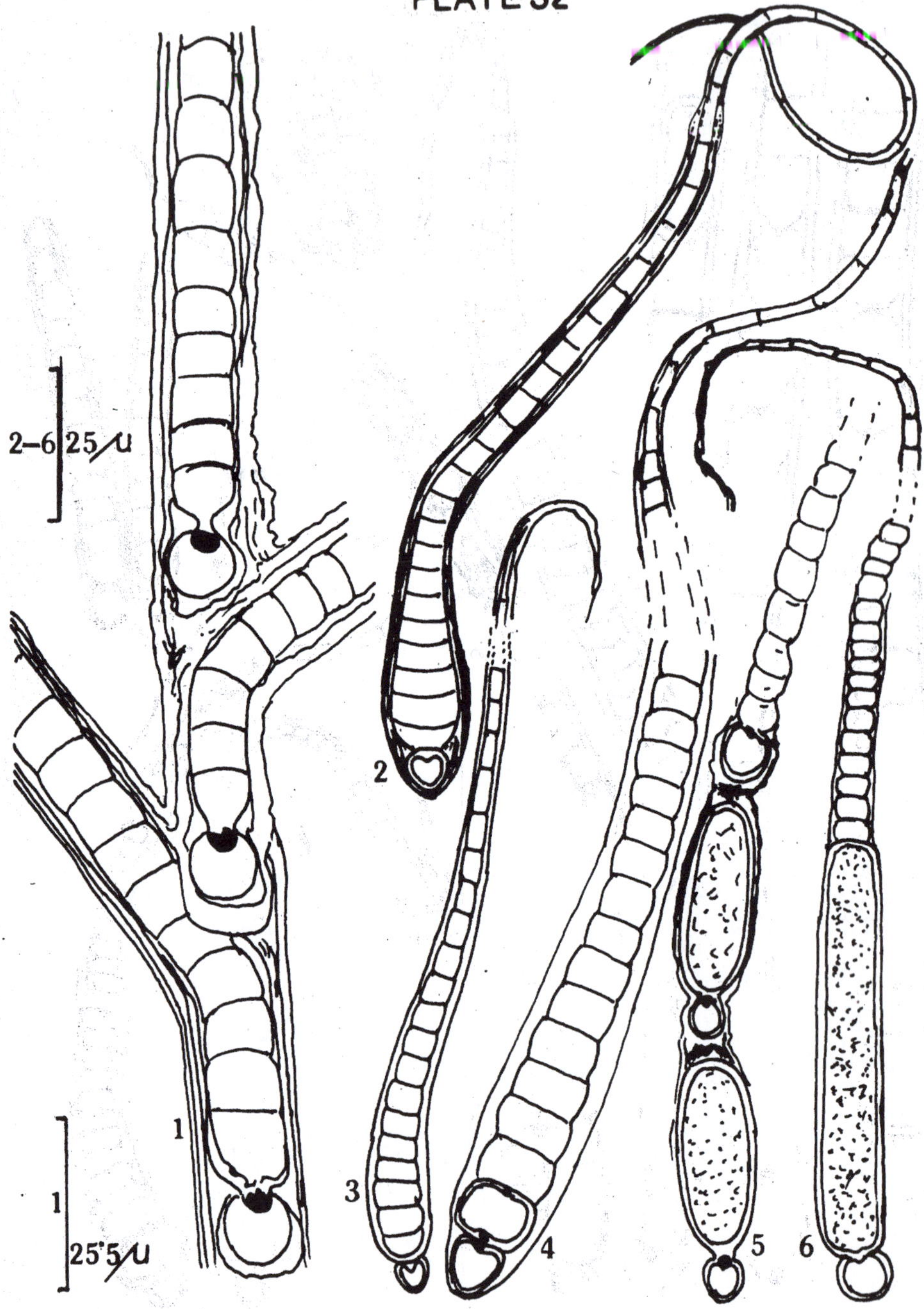

PLATE33

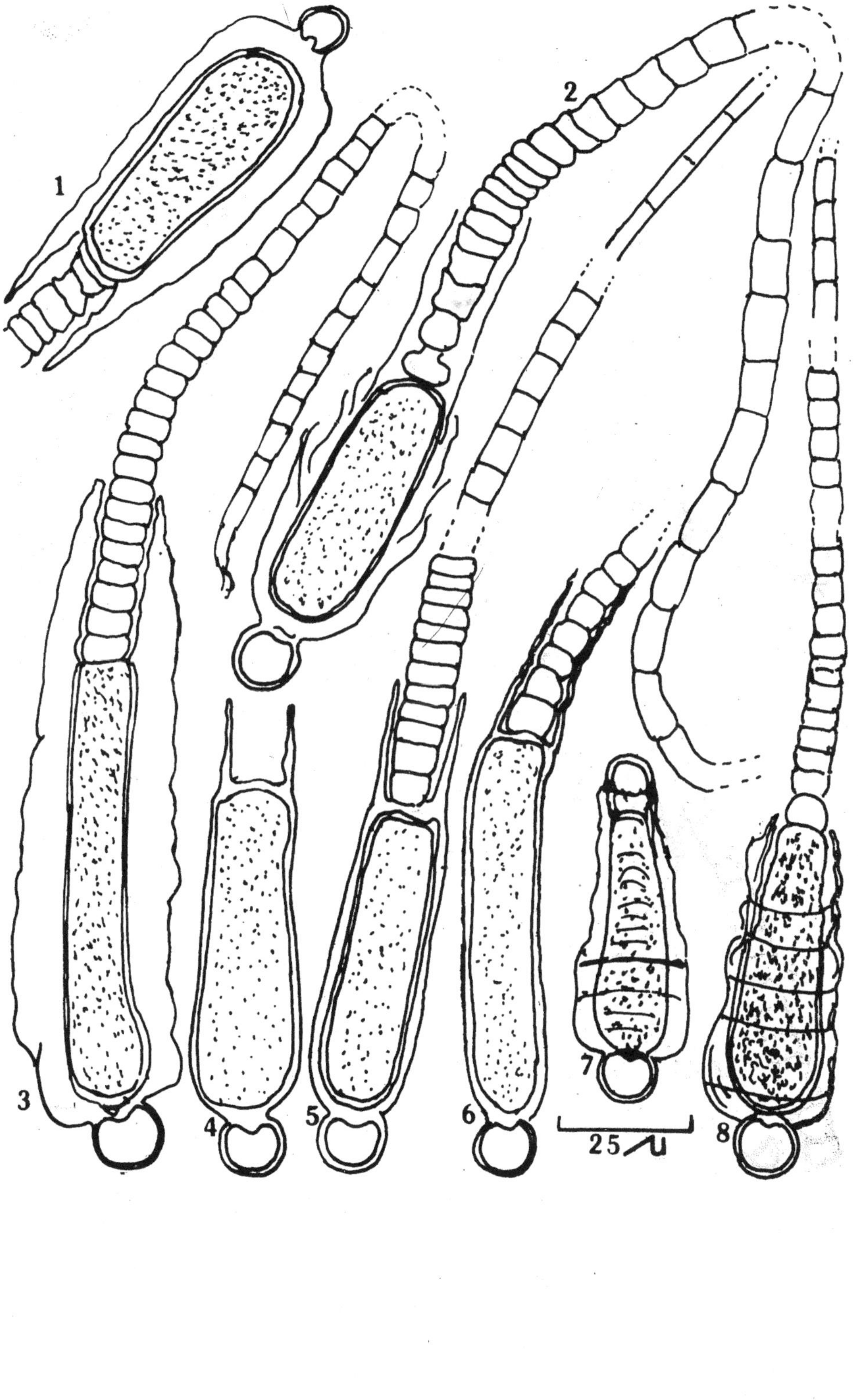

PLATE 34

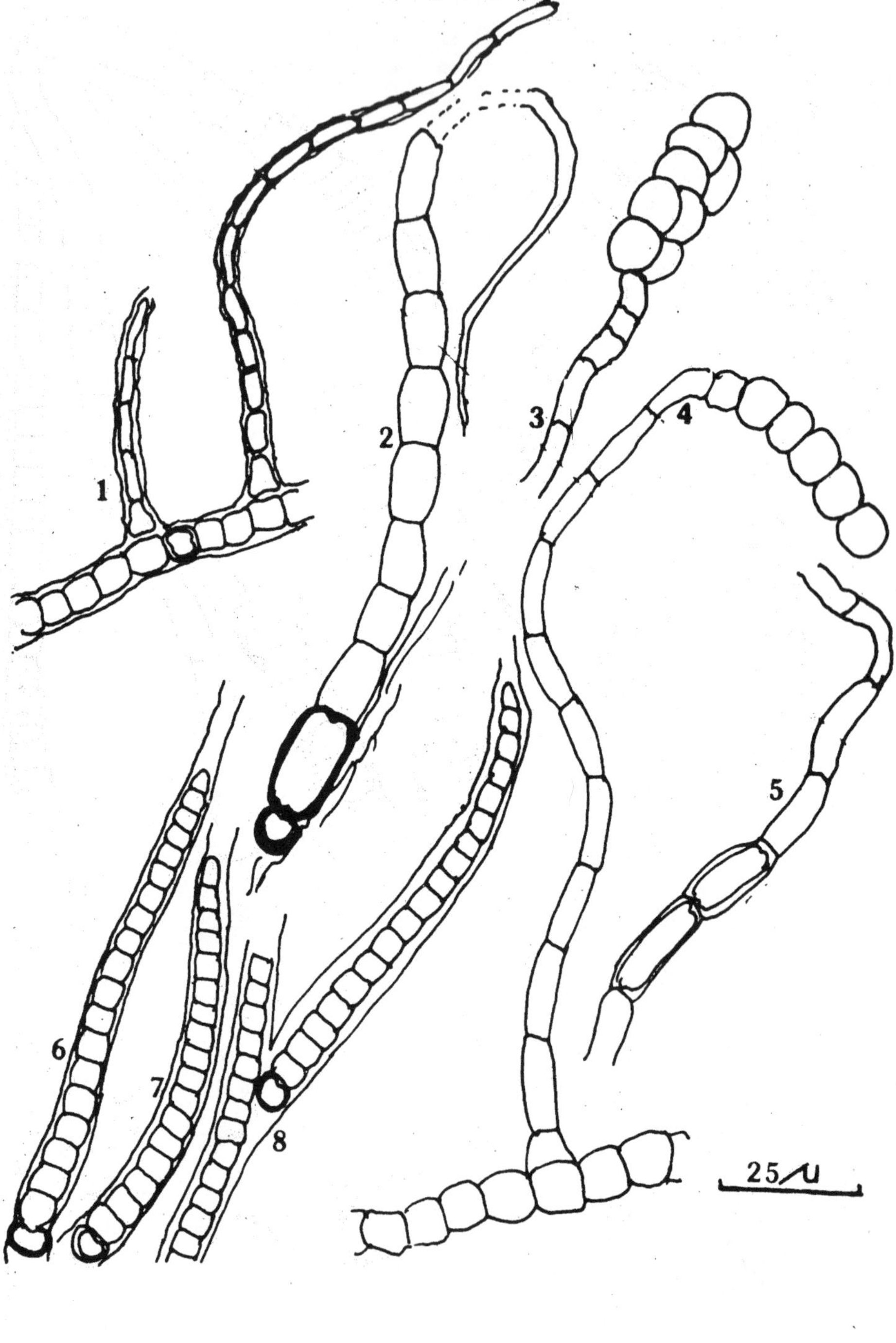

PLATE 35

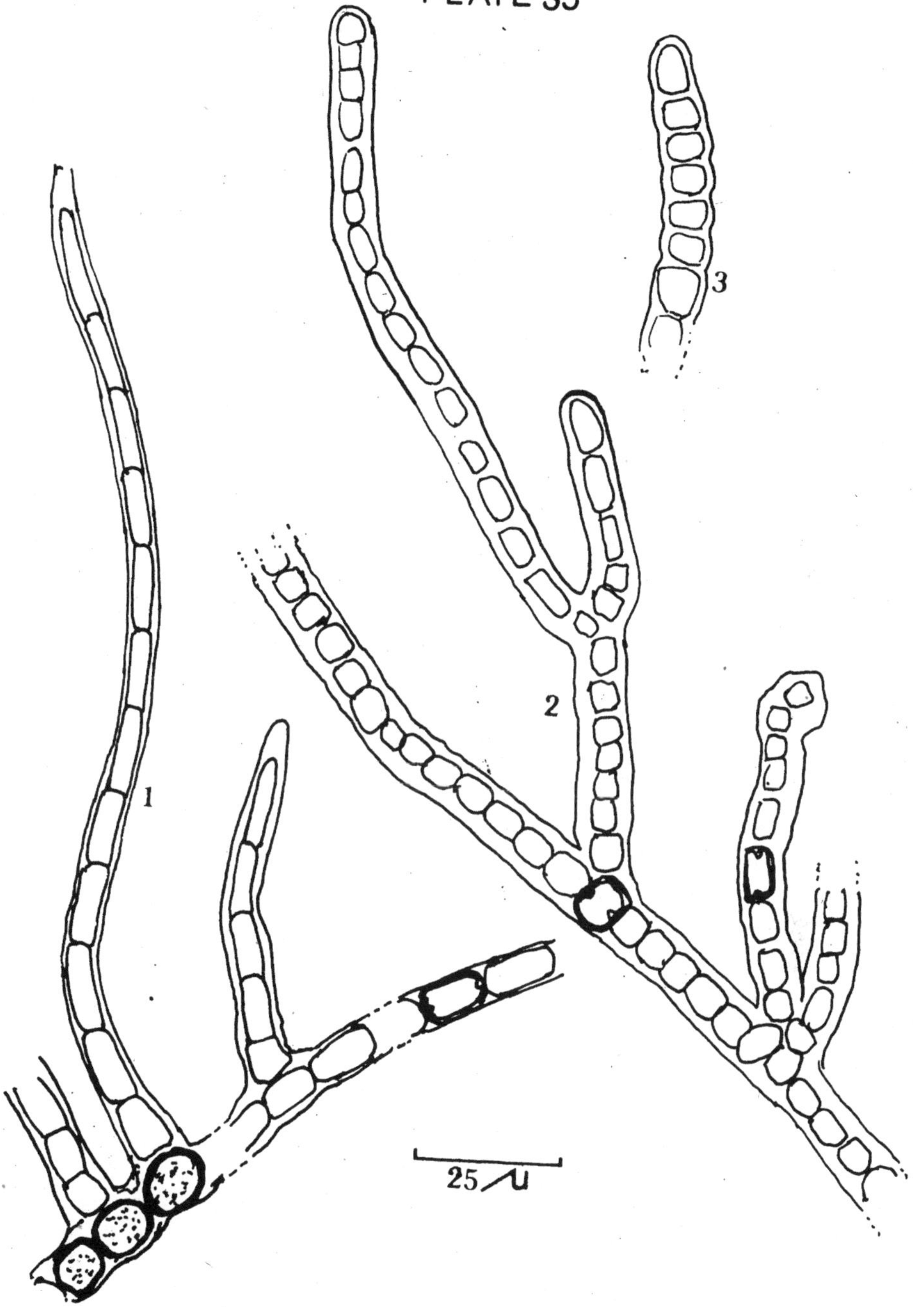

PLATE 6

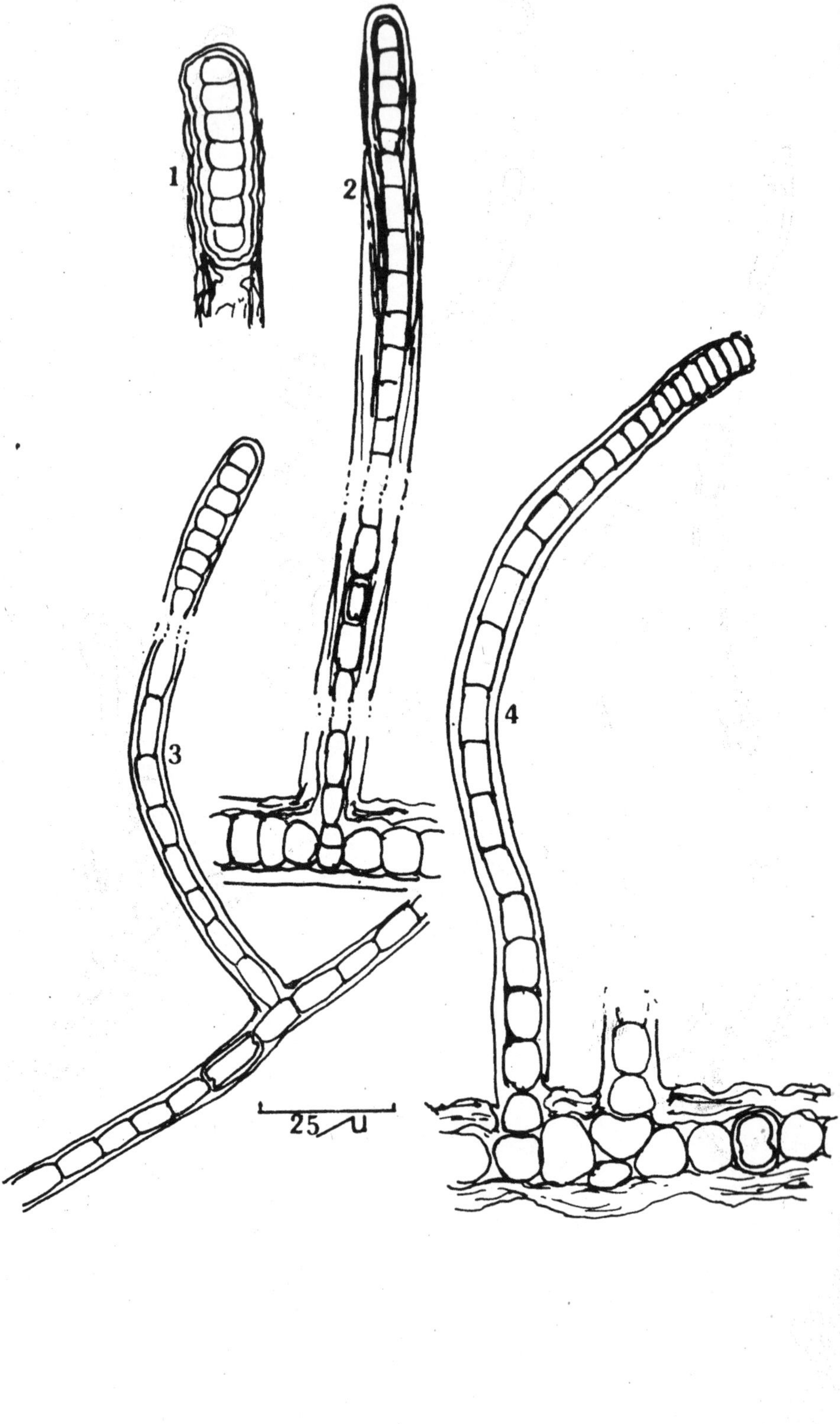

PLATE 37

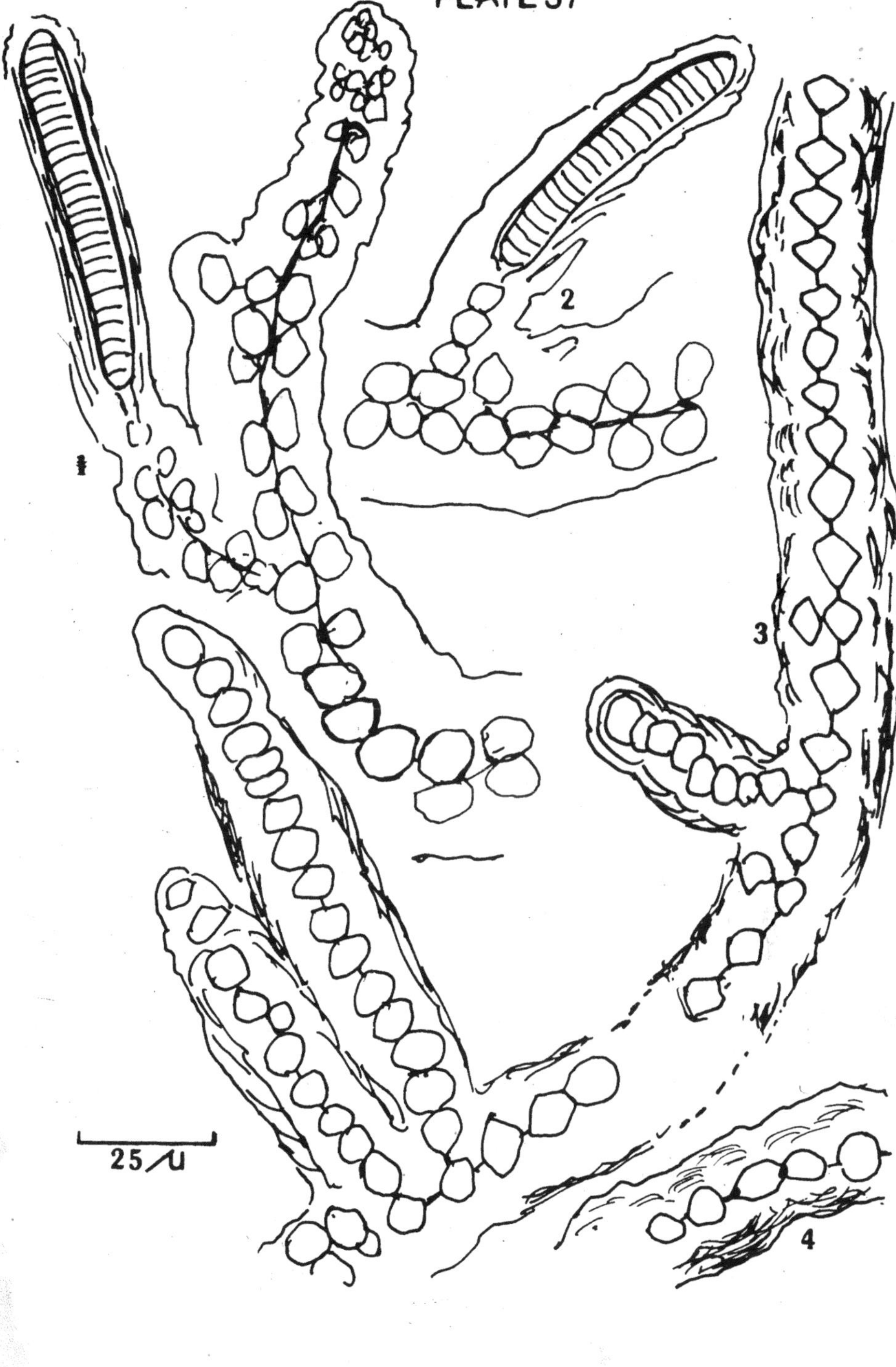

PLATE 38

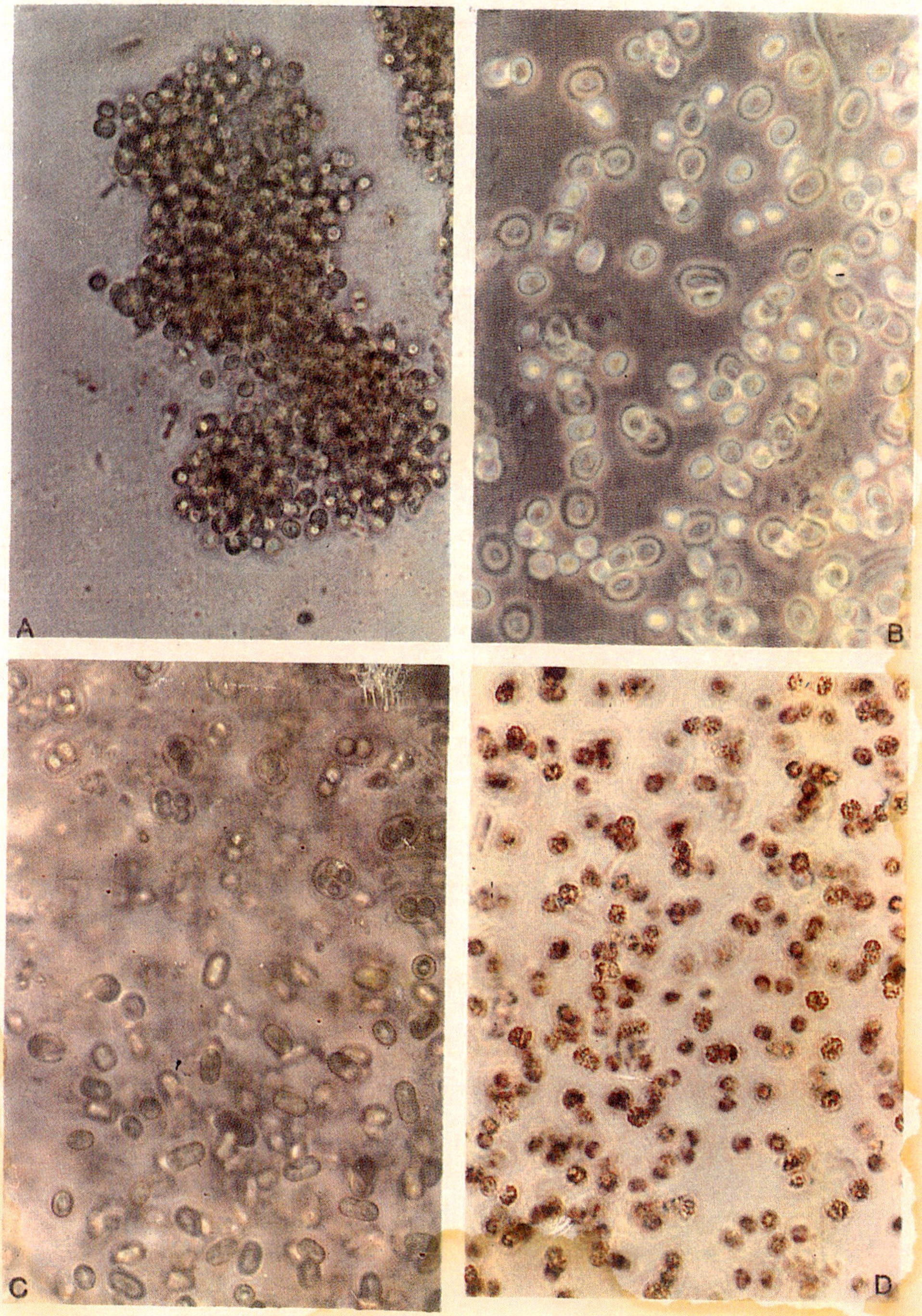

PLATE 39

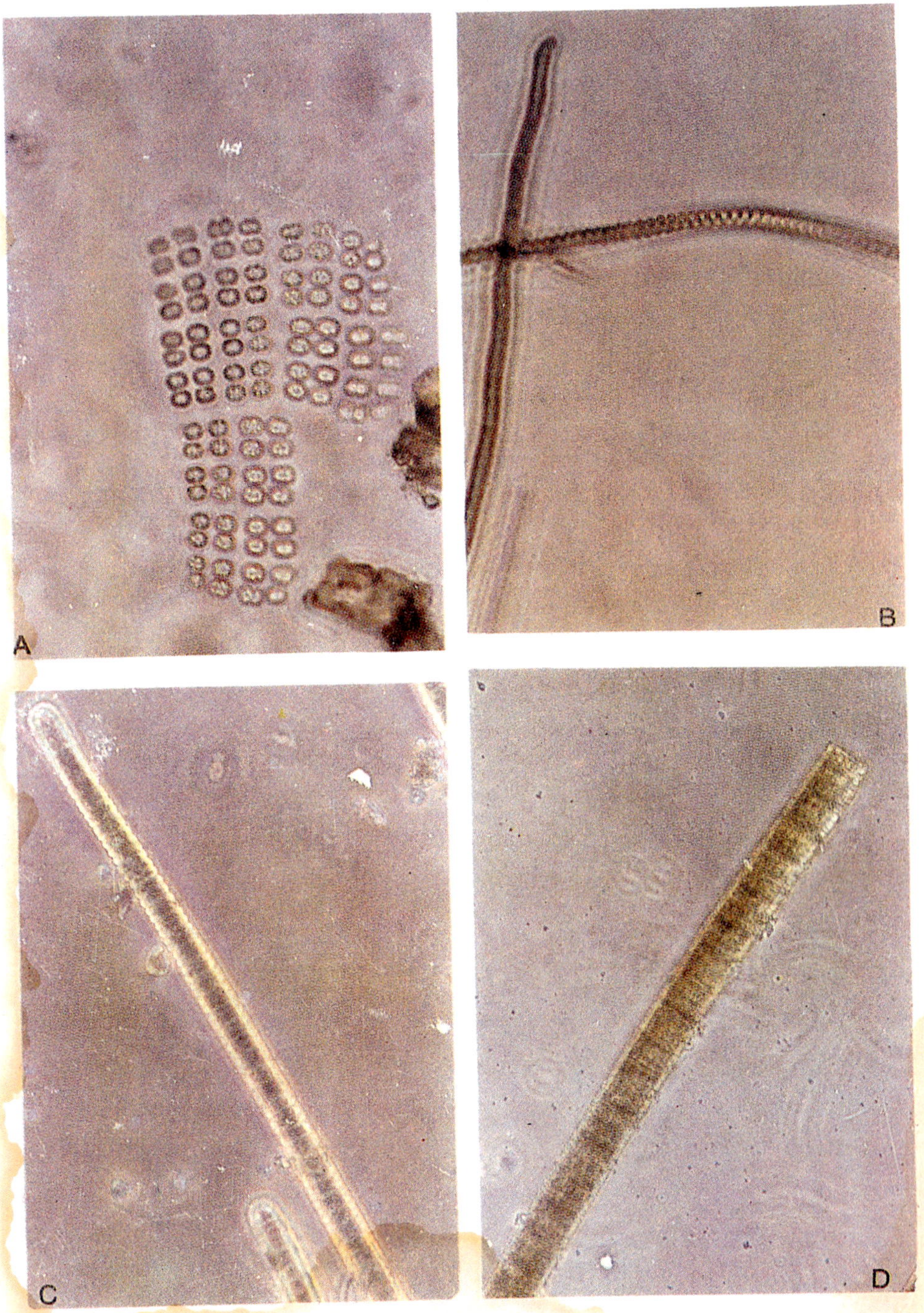

PLATE 40

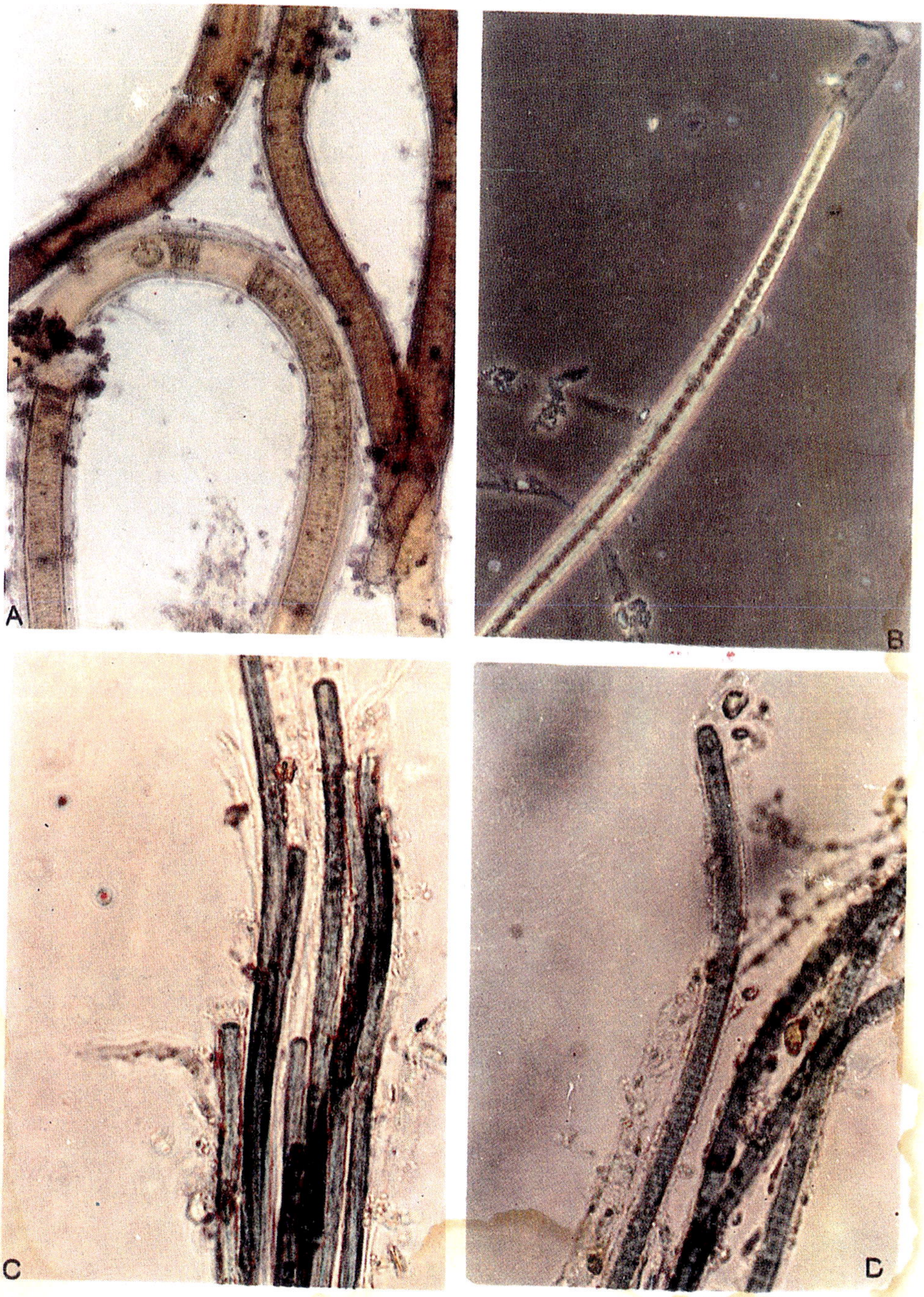

PLATE 41

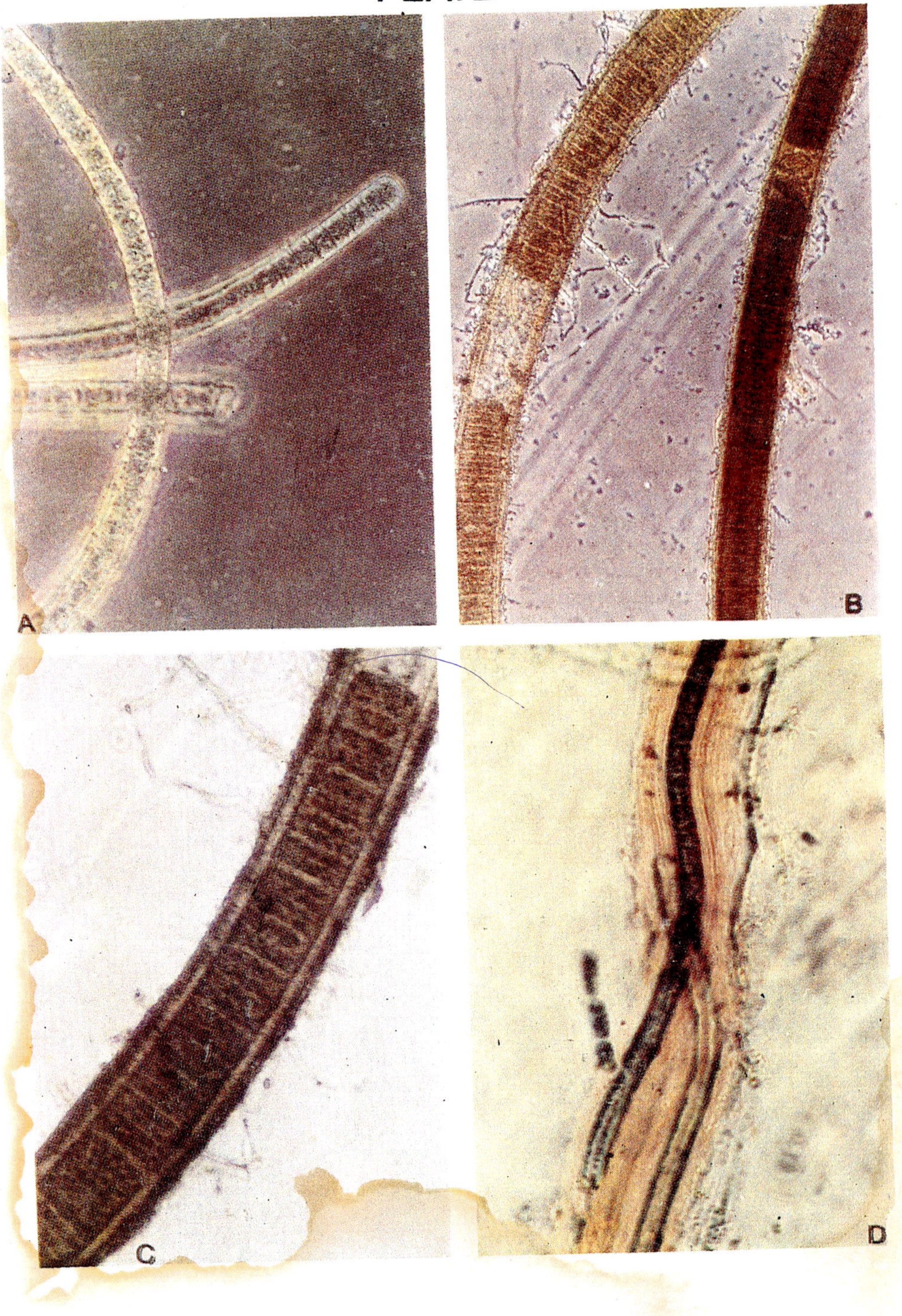

PLATE 42

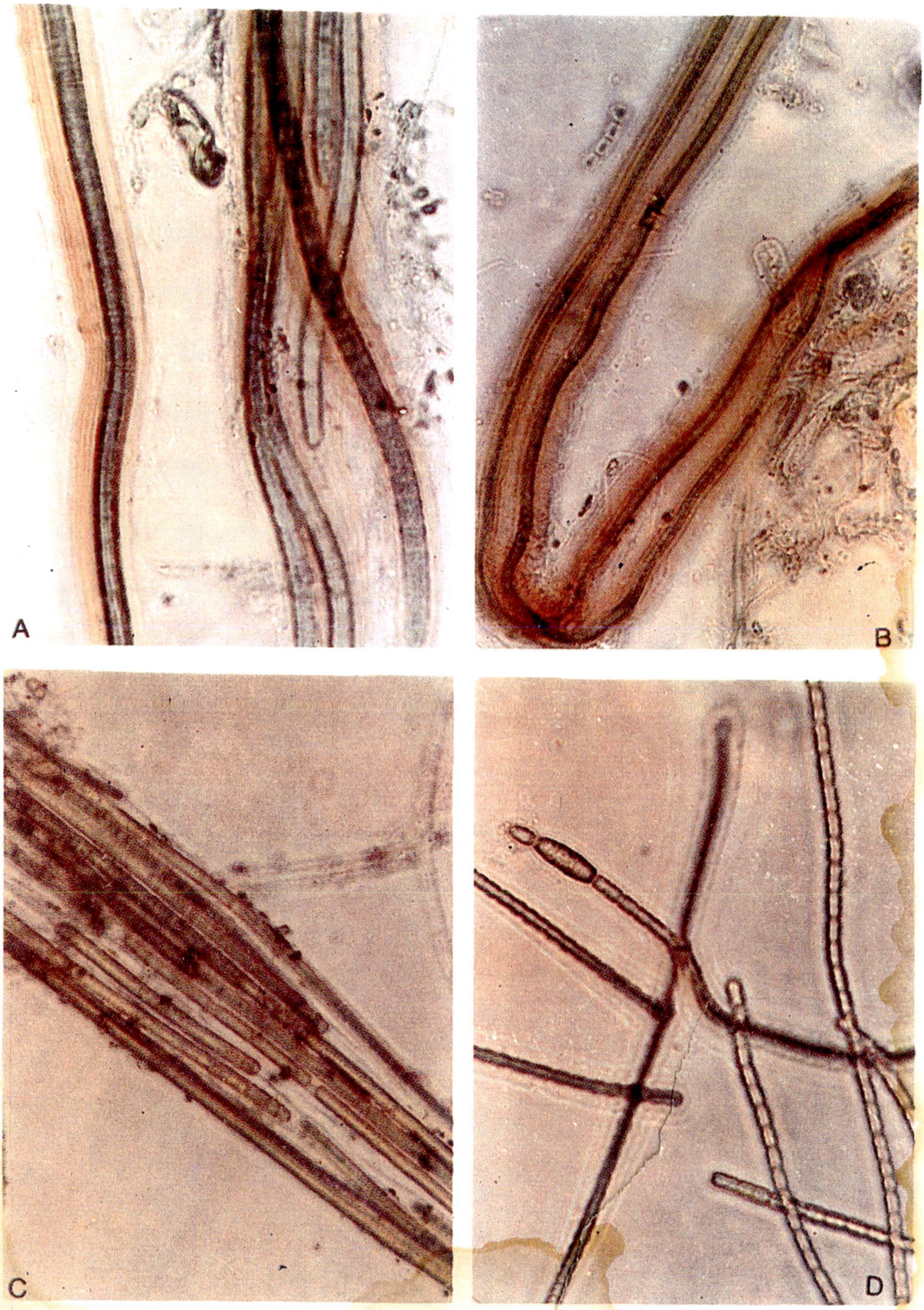

PLATE 43

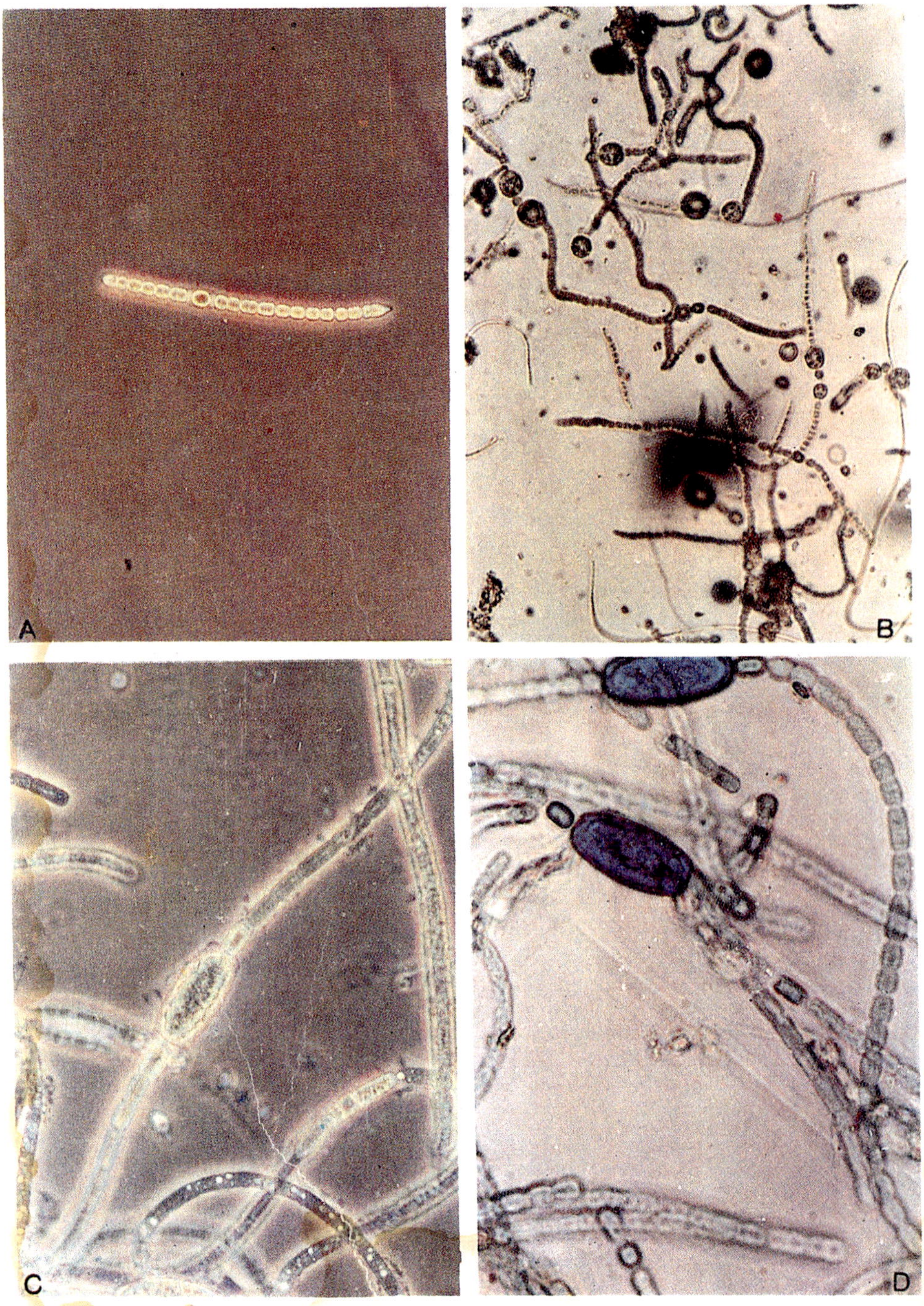

PLATE 44

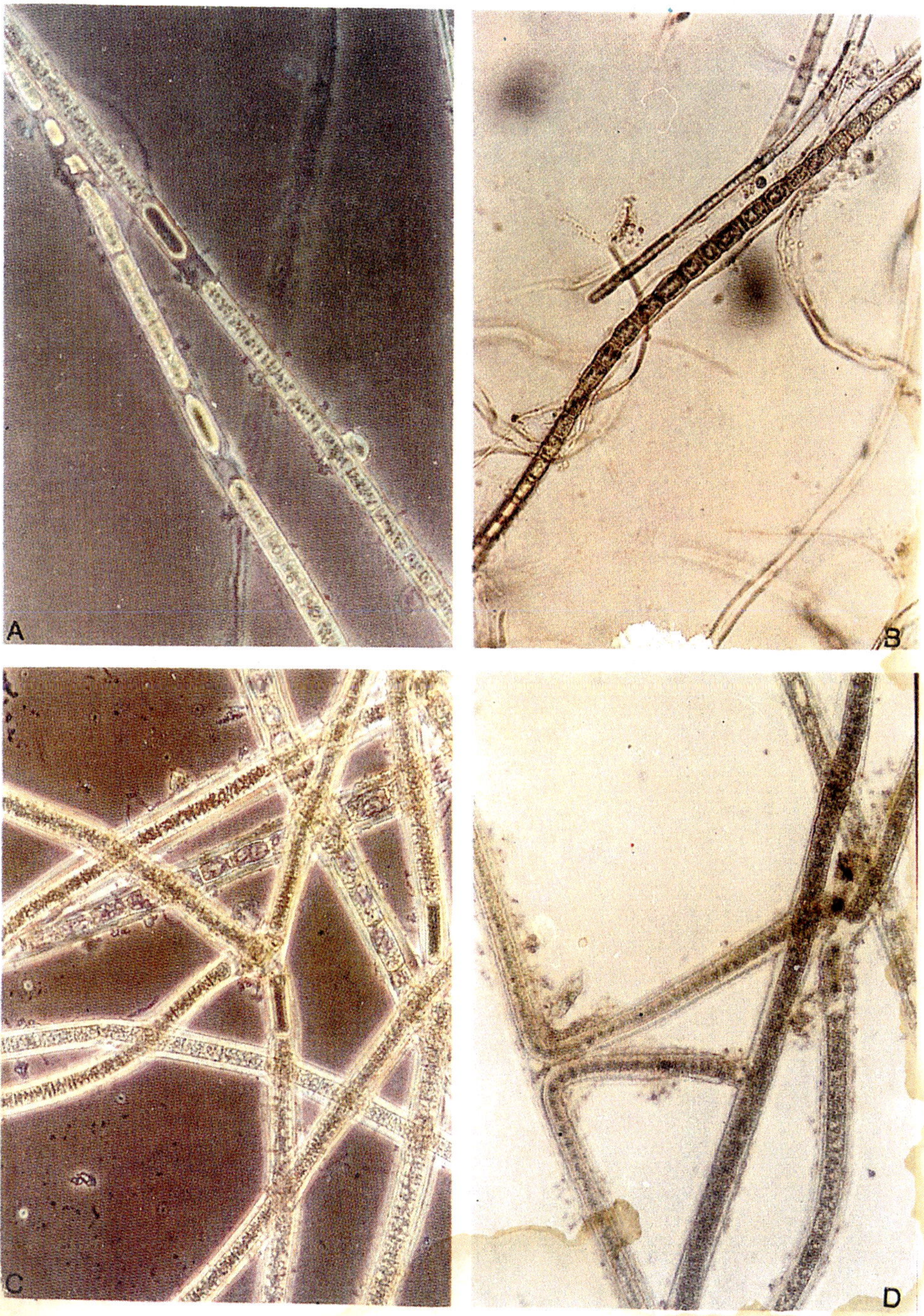

PLATE 45

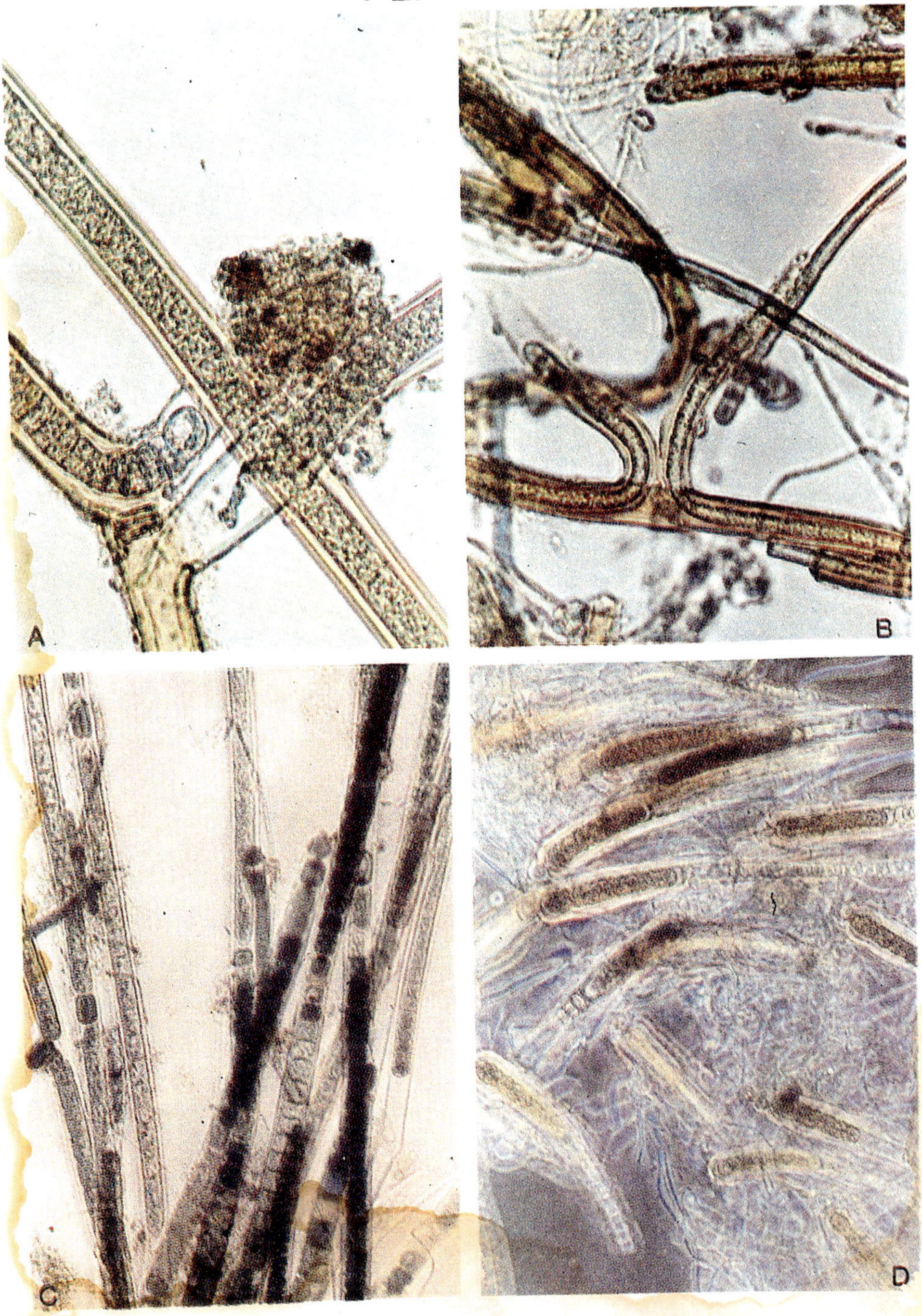